FREE Test Taking Tips DVD Offer

To help us better serve you, we have developed a Test Taking Tips DVD that we would like to give you for FREE. **This DVD covers world-class test taking tips that you can use to be even more successful when you are taking your test.**

All that we ask is that you email us your feedback about your study guide. Please let us know what you thought about it – whether that is good, bad or indifferent.

To get your **FREE Test Taking Tips DVD**, email freedvd@studyguideteam.com with "FREE DVD" in the subject line and the following information in the body of the email:

    a. The title of your study guide.

    b. Your product rating on a scale of 1-5, with 5 being the highest rating.

    c. Your feedback about the study guide. What did you think of it?

    d. Your full name and shipping address to send your free DVD.

If you have any questions or concerns, please don't hesitate to contact us at freedvd@studyguideteam.com.

Thanks again!

HiSET 2021 and 2022 Preparation Book

HiSET Exam Prep with Practice Questions
for the High School Equivalency Test
[6th Edition Study Guide]

TPB Publishing

Written and edited by TPB Publishing.

TPB Publishing is not associated with or endorsed by any official testing organization. TPB Publishing is a publisher of unofficial educational products. All test and organization names are trademarks of their respective owners. Content in this book is included for utilitarian purposes only and does not constitute an endorsement by TPB Publishing of any particular point of view.

Interested in buying more than 10 copies of our product? Contact us about bulk discounts: bulkorders@studyguideteam.com

ISBN 13: 9781628456882
ISBN 10: 1628456884

Table of Contents

Quick Overview

As you draw closer to taking your exam, effective preparation becomes more and more important. Thankfully, you have this study guide to help you get ready. Use this guide to help keep your studying on track and refer to it often.

This study guide contains several key sections that will help you be successful on your exam. The guide contains tips for what you should do the night before and the day of the test. Also included are test-taking tips. Knowing the right information is not always enough. Many well-prepared test takers struggle with exams. These tips will help equip you to accurately read, assess, and answer test questions.

A large part of the guide is devoted to showing you what content to expect on the exam and to helping you better understand that content. In this guide are practice test questions so that you can see how well you have grasped the content. Then, answer explanations are provided so that you can understand why you missed certain questions.

Don't try to cram the night before you take your exam. This is not a wise strategy for a few reasons. First, your retention of the information will be low. Your time would be better used by reviewing information you already know rather than trying to learn a lot of new information. Second, you will likely become stressed as you try to gain a large amount of knowledge in a short amount of time. Third, you will be depriving yourself of sleep. So be sure to go to bed at a reasonable time the night before. Being well-rested helps you focus and remain calm.

Be sure to eat a substantial breakfast the morning of the exam. If you are taking the exam in the afternoon, be sure to have a good lunch as well. Being hungry is distracting and can make it difficult to focus. You have hopefully spent lots of time preparing for the exam. Don't let an empty stomach get in the way of success!

When travelling to the testing center, leave earlier than needed. That way, you have a buffer in case you experience any delays. This will help you remain calm and will keep you from missing your appointment time at the testing center.

Be sure to pace yourself during the exam. Don't try to rush through the exam. There is no need to risk performing poorly on the exam just so you can leave the testing center early. Allow yourself to use all of the allotted time if needed.

Remain positive while taking the exam even if you feel like you are performing poorly. Thinking about the content you should have mastered will not help you perform better on the exam.

Once the exam is complete, take some time to relax. Even if you feel that you need to take the exam again, you will be well served by some down time before you begin studying again. It's often easier to convince yourself to study if you know that it will come with a reward!

Test-Taking Strategies

1. Predicting the Answer

When you feel confident in your preparation for a multiple-choice test, try predicting the answer before reading the answer choices. This is especially useful on questions that test objective factual knowledge. By predicting the answer before reading the available choices, you eliminate the possibility that you will be distracted or led astray by an incorrect answer choice. You will feel more confident in your selection if you read the question, predict the answer, and then find your prediction among the answer choices. After using this strategy, be sure to still read all of the answer choices carefully and completely. If you feel unprepared, you should not attempt to predict the answers. This would be a waste of time and an opportunity for your mind to wander in the wrong direction.

2. Reading the Whole Question

Too often, test takers scan a multiple-choice question, recognize a few familiar words, and immediately jump to the answer choices. Test authors are aware of this common impatience, and they will sometimes prey upon it. For instance, a test author might subtly turn the question into a negative, or he or she might redirect the focus of the question right at the end. The only way to avoid falling into these traps is to read the entirety of the question carefully before reading the answer choices.

3. Looking for Wrong Answers

Long and complicated multiple-choice questions can be intimidating. One way to simplify a difficult multiple-choice question is to eliminate all of the answer choices that are clearly wrong. In most sets of answers, there will be at least one selection that can be dismissed right away. If the test is administered on paper, the test taker could draw a line through it to indicate that it may be ignored; otherwise, the test taker will have to perform this operation mentally or on scratch paper. In either case, once the obviously incorrect answers have been eliminated, the remaining choices may be considered. Sometimes identifying the clearly wrong answers will give the test taker some information about the correct answer. For instance, if one of the remaining answer choices is a direct opposite of one of the eliminated answer choices, it may well be the correct answer. The opposite of obviously wrong is obviously right! Of course, this is not always the case. Some answers are obviously incorrect simply because they are irrelevant to the question being asked. Still, identifying and eliminating some incorrect answer choices is a good way to simplify a multiple-choice question.

4. Don't Overanalyze

Anxious test takers often overanalyze questions. When you are nervous, your brain will often run wild, causing you to make associations and discover clues that don't actually exist. If you feel that this may be a problem for you, do whatever you can to slow down during the test. Try taking a deep breath or counting to ten. As you read and consider the question, restrict yourself to the particular words used by the author. Avoid thought tangents about what the author *really* meant, or what he or she was *trying* to say. The only things that matter on a multiple-choice test are the words that are actually in the question. You must avoid reading too much into a multiple-choice question, or supposing that the writer meant something other than what he or she wrote.

5. No Need for Panic

It is wise to learn as many strategies as possible before taking a multiple-choice test, but it is likely that you will come across a few questions for which you simply don't know the answer. In this situation, avoid panicking. Because most multiple-choice tests include dozens of questions, the relative value of a single wrong answer is small. As much as possible, you should compartmentalize each question on a multiple-choice test. In other words, you should not allow your feelings about one question to affect your success on the others. When you find a question that you either don't understand or don't know how to answer, just take a deep breath and do your best. Read the entire question slowly and carefully. Try rephrasing the question a couple of different ways. Then, read all of the answer choices carefully. After eliminating obviously wrong answers, make a selection and move on to the next question.

6. Confusing Answer Choices

When working on a difficult multiple-choice question, there may be a tendency to focus on the answer choices that are the easiest to understand. Many people, whether consciously or not, gravitate to the answer choices that require the least concentration, knowledge, and memory. This is a mistake. When you come across an answer choice that is confusing, you should give it extra attention. A question might be confusing because you do not know the subject matter to which it refers. If this is the case, don't eliminate the answer before you have affirmatively settled on another. When you come across an answer choice of this type, set it aside as you look at the remaining choices. If you can confidently assert that one of the other choices is correct, you can leave the confusing answer aside. Otherwise, you will need to take a moment to try to better understand the confusing answer choice. Rephrasing is one way to tease out the sense of a confusing answer choice.

7. Your First Instinct

Many people struggle with multiple-choice tests because they overthink the questions. If you have studied sufficiently for the test, you should be prepared to trust your first instinct once you have carefully and completely read the question and all of the answer choices. There is a great deal of research suggesting that the mind can come to the correct conclusion very quickly once it has obtained all of the relevant information. At times, it may seem to you as if your intuition is working faster even than your reasoning mind. This may in fact be true. The knowledge you obtain while studying may be retrieved from your subconscious before you have a chance to work out the associations that support it. Verify your instinct by working out the reasons that it should be trusted.

8. Key Words

Many test takers struggle with multiple-choice questions because they have poor reading comprehension skills. Quickly reading and understanding a multiple-choice question requires a mixture of skill and experience. To help with this, try jotting down a few key words and phrases on a piece of scrap paper. Doing this concentrates the process of reading and forces the mind to weigh the relative importance of the question's parts. In selecting words and phrases to write down, the test taker thinks about the question more deeply and carefully. This is especially true for multiple-choice questions that are preceded by a long prompt.

9. Subtle Negatives

One of the oldest tricks in the multiple-choice test writer's book is to subtly reverse the meaning of a question with a word like *not* or *except*. If you are not paying attention to each word in the question, you can easily be led astray by this trick. For instance, a common question format is, "Which of the following is...?" Obviously, if the question instead is, "Which of the following is not...?," then the answer will be quite different. Even worse, the test makers are aware of the potential for this mistake and will include one answer choice that would be correct if the question were not negated or reversed. A test taker who misses the reversal will find what he or she believes to be a correct answer and will be so confident that he or she will fail to reread the question and discover the original error. The only way to avoid this is to practice a wide variety of multiple-choice questions and to pay close attention to each and every word.

10. Reading Every Answer Choice

It may seem obvious, but you should always read every one of the answer choices! Too many test takers fall into the habit of scanning the question and assuming that they understand the question because they recognize a few key words. From there, they pick the first answer choice that answers the question they believe they have read. Test takers who read all of the answer choices might discover that one of the latter answer choices is actually *more* correct. Moreover, reading all of the answer choices can remind you of facts related to the question that can help you arrive at the correct answer. Sometimes, a misstatement or incorrect detail in one of the latter answer choices will trigger your memory of the subject and will enable you to find the right answer. Failing to read all of the answer choices is like not reading all of the items on a restaurant menu: you might miss out on the perfect choice.

11. Spot the Hedges

One of the keys to success on multiple-choice tests is paying close attention to every word. This is never truer than with words like almost, most, some, and sometimes. These words are called "hedges" because they indicate that a statement is not totally true or not true in every place and time. An absolute statement will contain no hedges, but in many subjects, the answers are not always straightforward or absolute. There are always exceptions to the rules in these subjects. For this reason, you should favor those multiple-choice questions that contain hedging language. The presence of qualifying words indicates that the author is taking special care with his or her words, which is certainly important when composing the right answer. After all, there are many ways to be wrong, but there is only one way to be right! For this reason, it is wise to avoid answers that are absolute when taking a multiple-choice test. An absolute answer is one that says things are either all one way or all another. They often include words like *every*, *always*, *best*, and *never*. If you are taking a multiple-choice test in a subject that doesn't lend itself to absolute answers, be on your guard if you see any of these words.

12. Long Answers

In many subject areas, the answers are not simple. As already mentioned, the right answer often requires hedges. Another common feature of the answers to a complex or subjective question are qualifying clauses, which are groups of words that subtly modify the meaning of the sentence. If the question or answer choice describes a rule to which there are exceptions or the subject matter is complicated, ambiguous, or confusing, the correct answer will require many words in order to be expressed clearly and accurately. In essence, you should not be deterred by answer choices that seem excessively long. Oftentimes, the author of the text will not be able to write the correct answer without

offering some qualifications and modifications. Your job is to read the answer choices thoroughly and completely and to select the one that most accurately and precisely answers the question.

13. Restating to Understand

Sometimes, a question on a multiple-choice test is difficult not because of what it asks but because of how it is written. If this is the case, restate the question or answer choice in different words. This process serves a couple of important purposes. First, it forces you to concentrate on the core of the question. In order to rephrase the question accurately, you have to understand it well. Rephrasing the question will concentrate your mind on the key words and ideas. Second, it will present the information to your mind in a fresh way. This process may trigger your memory and render some useful scrap of information picked up while studying.

14. True Statements

Sometimes an answer choice will be true in itself, but it does not answer the question. This is one of the main reasons why it is essential to read the question carefully and completely before proceeding to the answer choices. Too often, test takers skip ahead to the answer choices and look for true statements. Having found one of these, they are content to select it without reference to the question above. Obviously, this provides an easy way for test makers to play tricks. The savvy test taker will always read the entire question before turning to the answer choices. Then, having settled on a correct answer choice, he or she will refer to the original question and ensure that the selected answer is relevant. The mistake of choosing a correct-but-irrelevant answer choice is especially common on questions related to specific pieces of objective knowledge. A prepared test taker will have a wealth of factual knowledge at his or her disposal, and should not be careless in its application.

15. No Patterns

One of the more dangerous ideas that circulates about multiple-choice tests is that the correct answers tend to fall into patterns. These erroneous ideas range from a belief that B and C are the most common right answers, to the idea that an unprepared test-taker should answer "A-B-A-C-A-D-A-B-A." It cannot be emphasized enough that pattern-seeking of this type is exactly the WRONG way to approach a multiple-choice test. To begin with, it is highly unlikely that the test maker will plot the correct answers according to some predetermined pattern. The questions are scrambled and delivered in a random order. Furthermore, even if the test maker was following a pattern in the assignation of correct answers, there is no reason why the test taker would know which pattern he or she was using. Any attempt to discern a pattern in the answer choices is a waste of time and a distraction from the real work of taking the test. A test taker would be much better served by extra preparation before the test than by reliance on a pattern in the answers.

FREE DVD OFFER

Don't forget that doing well on your exam includes both understanding the test content and understanding how to use what you know to do well on the test. We offer a completely FREE Test Taking Tips DVD that covers world class test taking tips that you can use to be even more successful when you are taking your test.

All that we ask is that you email us your feedback about your study guide. To get your **FREE Test Taking Tips DVD**, email freedvd@studyguideteam.com with "FREE DVD" in the subject line and the following information in the body of the email:

- The title of your study guide.
- Your product rating on a scale of 1-5, with 5 being the highest rating.
- Your feedback about the study guide. What did you think of it?
- Your full name and shipping address to send your free DVD.

Introduction to the HiSET

Function of the Test

The High School Equivalency Test (HiSET) was introduced by the Educational Testing Service (ETS) in 2014 as an affordable alternative to the GED exam. It is intended for individuals who have not received a high school diploma as a way to demonstrate that they have knowledge and skills equivalent to someone who has successfully graduated from high school. While the GED exam has long been offered nationwide, the HiSET is thus far available only in certain states: California, Colorado, Hawaii, Illinois, Iowa, Louisiana, Maine, Massachusetts, Mississippi, Missouri, Montana, Nevada, New Hampshire, New Jersey, New Mexico, North Carolina, Oklahoma, Pennsylvania, Tennessee, and Wyoming. Additionally, each state's rules relating to the HiSET may vary.

The test is available in both English and Spanish, but the large majority of test takers take it in English.

Test Administration

The HiSET is administered at various community colleges and testing centers in the jurisdictions in which it is currently accepted. The test is typically available on any day that a given test center is open for business. Students who pass the test receive a high school equivalency certification. Students who do not pass may take it up to two more times within one year to attempt to pass.

The cost of the HiSET is determined by the individual states in which it is offered. The states in turn often allow the individual testing centers to set the price of the test. Some states charge for each individual subsection of the test, while others charge one price to take the whole thing. In the end, the typical total cost for taking the entire test is usually in the vicinity of $30 to $50.

ETS will provide reasonable accommodations for documented disabilities including but not limited to attention deficit/hyperactivity disorder, psychological or psychiatric disorders, learning and other cognitive disabilities, physical disorders/chronic health disabilities, intellectual disabilities, and hearing and visual impairment.

Test Format

The content of the HiSET exam is intended to cover the fundamental material that a student would gain mastery over during a typical high school education. It is broken down into five sections that are summarized below. A test taker can take one section at a time, or schedule several back-to-back. The

test can be taken either by computer or in a pencil-and-paper form, and in either English or Spanish. Test takers can answer questions in any order they choose.

Section	Questions	Time
Language Arts- Reading	50	65 minutes
Language Arts- Writing	61	120 minutes
Mathematics	55	90 minutes
Science	60	80 minutes
Social Studies	60	70 minutes

Scoring

In order to pass the HiSET, test takers must get at least the minimum passing score on each of the five sections, the minimum passing score on the essay portion of the Writing section, and also get at least the minimum passing overall score. The minimum score on each section is an 8, the minimum score on the essay portion of the Writing section is a 2, and the overall minimum score is a 45.

Note that not all questions will be scored. Some questions are experimental and are being tested out. However, you have no way of knowing which are real and which are not, so be sure to do your best on every question.

The score is based on the total number of correct answers on the scored questions, with no deductions for incorrect responses, so there is no guessing penalty. Make sure you answer every question! If you begin to run out of time, just mark something down to at least give yourself a chance.

Recent/Future Developments

When the HiSET was created in 2014, it was intended as an alternative or replacement for the GED test in states that chose to adopt it. In 2016, ETS changed the structure of the Writing portion of the exam. As a result, students may now switch back and forth between the essay and multiple-choice portions of the exam as they like. This permits test takers who get through the multiple-choice questions quickly to spend more time on the essay, or vice-versa. ETS also adjusted the essay prompt such that it now includes two opposing arguments, one of which the test taker must adopt and defend.

Language Arts: Reading

Comprehension

Understanding Explicit Details

Readers want to draw a conclusion about what the author has presented. Drawing a conclusion will help the reader to understand what the writer intended as well as whether he or she agrees with what the author has said. There are a few ways to determine the logical conclusion, but careful reading is the most important. The passage should be read a few times, and readers should highlight or take notes on the details that they deem important to the meaning of the piece. Readers may draw a conclusion that is different than what the writer intended, or they may draw more than one conclusion. Readers should look carefully at the details to see if their conclusion matches up with what the writer has presented and intended for readers to understand.

Textual evidence can help readers to draw a conclusion about a passage. Textual evidence refers to information such as facts and examples that support the main point. Textual evidence will likely come from outside sources and can be in the form of quoted or paraphrased material. Details should be precise, descriptive, and factual. Readers should look to this evidence and its credibility and validity in relation to the main idea to draw a conclusion about the writing.

The author may state the conclusion directly in the passage. Inferring the author's conclusion is useful, especially when it is not overtly stated, but inferences should not outweigh the information that is directly stated. Alternatively, when readers are trying to draw a conclusion about a text, it may not always be directly stated.

As mentioned before, summary is another effective way to draw a conclusion from a passage. Summary is a shortened version of the original text, written in one's own words. It should focus on the main points of the original text, including only the relevant details. It's important to be brief but thorough in a summary. While the summary should always be shorter than the original passage, it should still retain the meaning of the original source.

Like summary, paraphrasing can also help a reader to fully understand a part of a reading. Paraphrase calls for the reader to take a small part of the passage and to say it in their own words. Paraphrase is more than rewording the original passage, though. It should be written in one's own way, while still retaining the meaning of the original source. When a reader's goal is to write something in their own words, deeper understanding of the original source is required. Again, applying summary and paraphrase to the passages during the test may not be the most efficient use of the test taker's time. However, these tools should be considered when one is practicing comprehending passages. Test takers who are familiar with carefully selecting important aspects of the passage will benefit from this experience on test day.

Meaning of Words and Phrases

Another useful vocabulary skill is being able to understand meaning in context. A word's *context* refers to other words and information surrounding it, which can have a big impact on how readers interpret that word's meaning. Of course, many words have more than one definition. For example, consider the meaning of the word "engaged." The first definition that comes to mind might be "promised to be married," but consider the following sentences:

a. The two armies engaged in a conflict that lasted all night.

b. The three-hour lecture flew by because students were so engaged in the material.

c. The busy executive engaged a new assistant to help with his workload.

Were any of those sentences related to marriage? In fact, "engaged" has a variety of other meanings. In these sentences, respectively, it can mean: "battled," "interested or involved," and "appointed or employed." Readers may wonder how to decide which definition to apply. The appropriate meaning is prioritized based on context. For example, sentence *C* mentions "executive," "assistant," and "workload," so readers can assume that "engaged" has something to do with work—in which case, "appointed or employed" is the best definition for this context. Context clues can also be found in sentence *A*. Words like "armies" and "conflicts" show that this sentence is about a military situation, so in this context, "engaged" is closest in meaning to "battled." By using context clues—the surrounding words in the sentence—readers can easily select the most appropriate definition.

Context clues can also help readers when they don't know *any* meanings for a certain word. Test writers will deliberately ask about unfamiliar vocabulary to measure your ability to use context to make an educated guess about a word's meaning.

Which of the following is the closest in meaning to the word "loquacious" in the following sentence?

The *loquacious* professor was notorious for always taking too long to finish his lectures.
a. knowledgeable
b. enthusiastic
c. approachable
d. talkative

Even if the word "loquacious" seems completely new, it's possible to utilize context to make a good guess about the word's meaning. Grammatically, it's apparent that "loquacious" is an adjective that modifies the noun "professor"—so "loquacious" must be some kind of quality or characteristic. A clue in this sentence is "taking too long to finish his lectures." Readers should then consider qualities that might cause a professor's lectures to run long. Perhaps he's "disorganized," "slow," or "talkative"—all words that might still make sense in this sentence. Choice *D*, therefore, is a logical choice for this sentence—the professor talks too much, so his lectures run late. In fact, "loquacious" means "talkative or wordy."

One way to use context clues is to think of potential replacement words before considering the answer choices. You can also turn to the answer choices first and try to replace each of them in the sentence to see if the sentence is logical and retains the same meaning.

Another way to use context clues is to consider clues in the word itself. Most students are familiar with prefixes, suffixes, and root words—the building blocks of many English words. A little knowledge goes a long way when it comes to these components of English vocabulary, and these words can point readers in the right direction when they need help finding an appropriate definition.

Word Choices

Just as one word may have different meanings, the same meaning can be conveyed by different words or synonyms. However, there are very few synonyms that have *exactly* the same definition. Rather, there are slight nuances in usage and meaning. In this case, a writer's *diction*, or word choice, is important to the meaning meant to be conveyed.

Many words have a surface *denotation* and a deeper *connotation*. A word's *denotation* is the literal definition of a word that can be found in any dictionary (an easy way to remember this is that "denotation" and "dictionary definition" all begin with the letter "D"). For example, if someone looked up the word "snake" in the dictionary, they'd learn that a snake is a common reptile with scales, a long body, and no limbs.

A word's *connotation* refers to its emotional and cultural associations, beyond its literal definition. Some connotations are universal, some are common within a particular cultural group, and some are more personal. Let's go back to the word "snake." A reader probably already knows its denotation—a slithering animal—but readers should also take a moment to consider its possible connotations. For readers from a Judeo-Christian culture, they might associate a snake with the serpent from the Garden of Eden who tempts Adam and Eve into eating the forbidden fruit. In this case, a snake's connotations might include deceit, danger, and sneakiness.

Consider the following character description:

He slithered into the room like a snake.

Does this sound like a character who can be trusted? It's the connotation of the word "snake" that implies untrustworthiness. Connotative language, then, helps writers to communicate a deeper, more emotional meaning.

Read the following excerpt from "The Lamb," a poem by William Blake.

Little lamb, who made thee?
Dost thou know who made thee,
Gave thee life, and bid thee feed
By the stream and o'er the mead;
Gave thee clothing of delight,
Softest clothing, woolly, bright;
Gave thee such a tender voice,
Making all the vales rejoice?
Little lamb, who made thee?
Dost thou know who made thee?

Think about the connotations of a "lamb." Whereas a snake might make readers think of something dangerous and dishonest, a lamb tends to carry a different connotation: innocence and purity. Blake's poem contains other emotional language—"delight," "softest," "tender," "rejoice"—to support this impression.

Some words have similar denotations but very different connotations. "Weird" and "unique" can both describe something distinctive and unlike the norm. But they convey different emotions:

You have such a weird fashion sense!

You have such a unique fashion sense!

Which sentence is a compliment? Which sentence is an insult? "Weird" generally has more negative connotations, whereas "unique" is more positive. In this way, connotative language is a powerful way for writers to evoke emotion.

A writer's diction also informs their tone. *Tone* refers to the author's attitude toward their subject. A writer's tone might be critical, curious, respectful, dismissive, or any other possible attitude. The key to understanding tone is focusing not just on *what* is said, but on *how* it's said.

a. Although the latest drug trial did not produce a successful vaccine, medical researchers are one step further on the path to eradicating this deadly virus.

b. Doctors faced yet another disappointing setback in their losing battle against the killer virus; their most recent drug trial has proved as unsuccessful as the last.

Both sentences report the same information: the latest drug trial was a failure. However, each sentence presents this information in a different way, revealing the writer's tone. The first sentence has a more hopeful and confident tone, downplaying the doctors' failure ("although" it failed) and emphasizing their progress ("one step further"). The second sentence has a decidedly more pessimistic and defeatist tone, using phrases like "disappointing setback" and "losing battle." The details a writer chooses to include can also help readers to identify their attitude towards their subject matter.

Identifying emotional or connotative language can be useful in determining the tone of a text. Readers can also consider questions such as, "Who is the speaker?" or "Who is their audience?" (Remember, particularly in fiction, that the speaker or narrator may not be the same person as the author.) For example, in an article about military conflict written by a notable anti-war activist, readers might expect their tone to be critical, harsh, or cynical. If they are presented with a poem written between newlyweds, readers might expect the tone to be loving, sensitive, or infatuated. If the tone seems wildly different from what's expected, consider if the writer is using *irony*. When a writer uses irony, they say one thing but imply the opposite meaning.

Inference and Interpretation

Inferences from the Text

Readers should be able to make *inferences*. Making an inference requires the reader to read between the lines and look for what's *implied* rather than what's directly stated. Using information that is known from the text, the reader is able to make a logical assumption about information that isn't directly stated but is probably true. Read the following passage:

"Hey, do you wanna meet my new puppy?" Jonathan asked.

"Oh, I'm sorry but please don't—" Jacinta began to protest, but before she could finish Jonathan had already open the passenger side door of his car and a perfectly white ball of fur came bouncing towards Jacinta.

"Isn't he the cutest?" beamed Jonathan.

"Yes—achoo!—he's pretty—aaaachooo!!—adora—aaa—aaaachoo!" Jacinta managed to say in between sneezes. "But if you don't mind, I—I—achoo!—need to go inside."

Which of the following can be inferred from Jacinta's reaction to the puppy?
    a. She hates animals.
    b. She is allergic to dogs.
    c. She prefers cats to dogs.
    d. She is angry at Jonathan.

In order to make an inference, the reader must first consider the information presented and then form an idea about what is probably true. Based on the details in the passage, what's the best answer? Important details include the tone of Jacinta's dialogue, which is polite and apologetic, as well as her reaction itself, which is a long string of sneezes. Choices *A* and *D* both express strong emotions ("hates" and "angry") that aren't evident in Jacinta's speech or actions. Choice *C* mentions cats, but there isn't anything in the passage to indicate Jacinta's feelings about those animals. Choice *B* is the most logical choice. As she began sneezing as soon as the dog approached her, it makes sense to guess that Jacinta might be allergic to dogs. So even though Jacinta never directly states, "Sorry, I'm allergic to dogs!" using the clues in the passage, it's still reasonable to guess this is true.

Making inferences is crucial for readers because literary texts often avoid presenting complete and direct information about characters' thoughts or feelings, leaving the reader to interpret clues in the text. In order to make inferences, readers should ask:

- What details are presented in the text?
- Is there any important information that seems to be missing?
- Based on the information that the author does include, what else is probably true?
- Is this inference reasonable based on what is already known?

Drawing Conclusions Not Explicitly Present in the Text

It's also useful to infer meaning from informative texts. Scientists and researchers make inferences every day in order to develop new theories based on facts and observations. Readers of informative texts should also understand how inferences are applied in academic research. Generally speaking, there are two main types of reasoning—*deductive* and *inductive*. An inference based on deductive reasoning considers a principle that is generally believed to be true and then applies it to a specific situation ("All English majors love reading. Annabelle is an English major. Therefore, I can infer that Annabelle loves reading."). Inductive reasoning makes an inference by using specific evidence to make a general inference ("Trina, Arnold, and Uchenna are all from Florida. Trina, Arnold, and Uchenna all love to swim. Therefore, I can infer that people from Florida usually love swimming."). Both deductive and inductive reasoning use what is *known* to be true to make a logical guess about what is *probably* true.

Inferring the Traits, Feelings, and Motives of Characters

Inferences are useful in gaining a deeper understanding of characters in a narrative. Readers can use the same strategies outlined above—paying attention to details and using them to make reasonable guesses about the text—to read between the lines and get a more complete picture of how (and why) characters are thinking, feeling, and acting. Read the following passage from O. Henry's story "The Gift of the Magi":

> One dollar and eighty-seven cents. That was all. And sixty cents of it was in pennies. Pennies saved one and two at a time by bulldozing the grocer and the vegetable man and the butcher until one's cheeks burned with the silent imputation of parsimony that such close dealing implied. Three times Della counted it. One dollar and eighty-seven cents. And the next day would be Christmas.

> There was clearly nothing to do but flop down on the shabby little couch and howl. So Della did it.

These paragraphs introduce the reader to the character Della. Even though the author doesn't include a direct description of Della, the reader can already form a general impression of her personality and emotions. One detail that should stick out to the reader is repetition: "one dollar and eighty-seven cents." This amount is repeated twice in the first paragraph, along with other descriptions of money: "sixty cents of it was in pennies," "pennies saved one and two at a time." The story's preoccupation with money parallels how Della herself is constantly thinking about her finances—"three times Della counted" her meager savings. Already the reader can guess that Della is having money problems. Next, think about her emotions. The first paragraph describes haggling over groceries "until one's cheeks burned"—another way to describe blushing. People tend to blush when they are embarrassed or ashamed, so readers can infer that Della is ashamed by her financial situation. This inference is also supported by the second paragraph, when she flops down and howls on her "shabby little couch." Clearly, she's in distress. Without saying, "Della has no money and is embarrassed to be poor," O. Henry is able to communicate the same impression to readers through his careful inclusion of details.

A character's *motive* is their reason for acting a certain way. Usually, characters are motivated by something that they want. In the passage above, why is Della upset about not having enough money? There's an important detail at the end of the first paragraph: "the next day would be Christmas." Why is money especially important around Christmas? Christmas is a holiday when people exchange gifts. If Della is struggling with money, she's probably also struggling to buy gifts. So a shrewd reader should be able to guess that Della's motivation is wanting to buy a gift for someone—but she's currently unable to afford it, leading to feelings of shame and frustration.

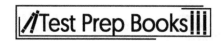

In order to understand characters in a text, readers should keep the following questions in mind:

- What words does the author use to describe the character? Are these words related to any specific emotions or personality traits (for example, characteristics like rude, friendly, unapproachable, or innocent)?

- What does the character say? Does their dialogue seem to be straightforward, or are they hiding some thoughts or emotions?

- What actions can be observed from this character? How do their actions reflect their feelings?

- What does the character want? What do they do to get it?

Interpreting Information Presented in Different Formats

Information is often presented in different formats. One of the most common ways to express data is in a table. The primary reason for plugging data into a table is to make interpretation more convenient. It's much easier to look at the table than to analyze results in a narrative paragraph. When analyzing a table, pay close attention to the title, variables, and data.

For example, the following theoretical antibiotic study can be analyzed. The study has 6 groups, named A through F, and each group receives a different dose of medicine. The results of the study are listed in the table below.

Results of Antibiotic Studies		
Group	Dosage of Antibiotics in milligrams (mg)	Efficacy (% of participants cured)
A	0 mg	20%
B	20 mg	40%
C	40 mg	75%
D	60 mg	95%
E	80 mg	100%
F	100 mg	100%

Tables generally list the title immediately above the data. The title should succinctly explain what is listed below. Here, "Results of Antibiotic Studies" informs the audience that the data pertains to the results of scientific study on antibiotics.

Identifying the variables at play is one of the most important parts of interpreting data. Remember, the independent variable is intentionally altered, and its change is independent of the other variables. Here, the dosage of antibiotics administered to the different groups is the independent variable. The study is intentionally manipulating the strength of the medicine to study the related results. Efficacy is the dependent variable since its results *depend* on a different variable, the dose of antibiotics. Generally, the independent variable will be listed before the dependent variable in tables.

Also play close attention to the variables' labels. Here, the dose is expressed in milligrams (mg) and efficacy in percentages (%). Keep an eye out for questions referencing data in a different unit measurement, or questions asking for a raw number when only the percentage is listed.

Now that the nature of the study and variables at play have been identified, the data itself needs be interpreted. Group A did not receive any of the medicine. As discussed earlier, Group A is the control, as it reflects the amount of people cured in the same timeframe without medicine. It's important to see that efficacy positively correlates with the dosage of medicine. A question using this study might ask for the lowest dose of antibiotics to achieve 100% efficacy. Although Group E and Group F both achieve 100% efficacy, it's important to note that Group E reaches 100% with a lower dose.

Interpreting Graphs

Graphs provide a visual representation of data. The variables are placed on the two axes. The bottom of the graph is referred to as the horizontal axis or X-axis. The left-hand side of the graph is known as the vertical axis or Y-axis. Typically, the independent variable is placed on the X-axis, and the dependent variable is located on the Y-axis. Sometimes, the X-axis is a timeline, and the dependent variables for different trials or groups have been measured throughout points in time; time is still an independent variable but is not always immediately thought of as the independent variable being studied.

The most common types of graphs are the bar graph and the line graph.

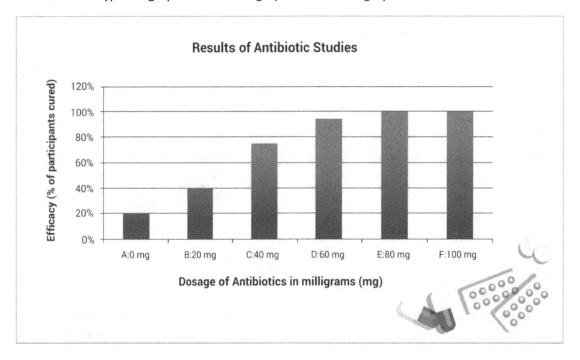

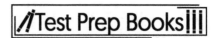

The *bar graph* above expresses the data from the table entitled "Results of Antibiotic Studies." To interpret the data for each group in the study, look at the top of their bars and read the corresponding efficacy on the Y-axis.

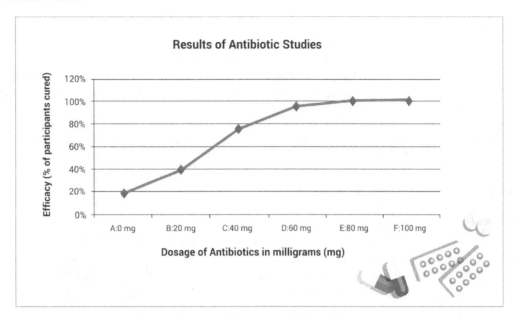

Here, the same data is expressed on a *line graph*. The points on the line correspond with each data entry. Reading the data on the line graph works like the bar graph. The data trend is measured by the slope of the line.

Interpreting Nonliteral Language

It's important to be able to recognize and interpret *figurative*, or non-literal, language. Literal statements rely directly on the denotations of words and express exactly what's happening in reality. Figurative language uses non-literal expressions to present information in a creative way. Consider the following sentences:

a. His pillow was very soft, and he fell asleep quickly.

b. His pillow was a fluffy cloud, and he floated away on it to the dream world.

Sentence *A* is literal, employing only the real meanings of each word. Sentence *B* is figurative. It employs a metaphor by stating that his pillow was a cloud. Of course, he isn't actually sleeping on a cloud, but the reader can draw on images of clouds as light, soft, fluffy, and relaxing to get a sense of how the character felt as he fell asleep. Also, in sentence *B*, the pillow becomes a vehicle that transports him to a magical dream world. The character isn't literally floating through the air—he's simply falling asleep! But by utilizing figurative language, the author creates a scene of peace, comfort, and relaxation that conveys stronger emotions and more creative imagery than the purely literal sentence. While there are countless types of figurative language, there are a few common ones that any reader should recognize.

Simile and *metaphor* are comparisons between two things, but their formats differ slightly. A simile says that two things are *similar* and makes a comparison using "like" or "as"—*A* is like *B*, or *A* is as [some characteristic] as *B*—whereas a metaphor states that two things are exactly the same—*A* is *B*. In both cases, simile and metaphor invite the reader to think more deeply about the characteristics of the two subjects and consider where they overlap. An example of metaphor can be found in the above sentence about the sleeper ("His pillow was a fluffy cloud"). For an example of simile, look at the first line of Robert Burns' famous poem:

My love is like a red, red rose

This is comparison using "like," and the two things being compared are love and a rose. Some characteristics of a rose are that it's fragrant, beautiful, blossoming, colorful, vibrant—by comparing his love to a rose, Burns asks the reader to apply these qualities to his love. In this way, he implies that his love is also fresh, blossoming, and brilliant.

Similes can also compare things that appear dissimilar. Here's a song lyric from Florence and the Machine:

Happiness hit her like a bullet in the back

"Happiness" has a very positive connotation, but getting "a bullet in the back" seems violent and aggressive, not at all related to happiness. By using an unexpected comparison, the writer forces readers to think more deeply about the comparison and ask themselves how could getting shot be similar to feeling happy. "A bullet in the back" is something that she doesn't see coming; it's sudden and forceful; and presumably, it has a strong impact on her life. So, in this way, the author seems to be saying that unexpected happiness made a sudden and powerful change in her life.

Another common form of figurative language is *personification*, when a non-human object is given human characteristics. William Blake uses personification here:

. . . the stars threw down their spears,

And watered heaven with their tears

He imagines the stars as combatants in a heavenly battle, giving them both action (throwing down their spears) and emotion (the sadness and disappointment of their tears). Personification helps to add emotion or develop relationships between characters and non-human objects. In fact, most people use personification in their everyday lives:

My alarm clock betrayed me! It didn't go off this morning!

The last piece of chocolate cake was staring at me from the refrigerator.

Next is *hyperbole*, a type of figurative language that uses extreme exaggeration. Sentences like, "I love you to the moon and back," or "I will love you for a million years," are examples of hyperbole. They aren't literally true—unfortunately, people cannot jump to outer space or live for a million years—but they're creative expressions that communicate the depth of feeling of the author.

Another way that writers add deeper meaning to their work is through *allusions*. An allusion is a reference to something from history, literature, or another cultural source. When the text is from a different culture or a time period, readers may not be familiar with every allusion. However, allusions

tend to be well-known because the author wants the reader to make a connection between what's happening in the text and what's being referenced.

> I can't believe my best friend told our professor that I was skipping class to finish my final project! What a Judas!

This sentence contains a Biblical allusion to Judas, a friend and follower of Jesus who betrayed Jesus to the Romans. In this case, the allusion to Judas is used to give a deeper impression of betrayal and disloyalty from a trusted friend. Commonly used allusions in Western texts may come from the Bible, Greek or Roman mythology, or well-known literature such as Shakespeare. By familiarizing themselves with these touchstones of history and culture, readers can be more prepared to recognize allusions.

Analysis

Topic, Main Idea, and Theme

In order to understand any text, readers first must determine the *topic*, or what the text is about. In non-fiction writing, the topic can generally be expressed in a few words. For example, a passage might be about college education, moving to a new neighborhood, or dog breeds. Slightly more specific information is found in the *main idea*, or what the writer wants readers to know about the topic. An article might be about the history of popular dog breeds; another article might tell how certain dog breeds are unfairly stereotyped. In both cases, the topic is the same—dog breeds—but the main ideas are quite different. Each writer has a distinct purpose for writing and a different set of details for what they want us to know about dog breeds. When a writer expresses their main idea in one sentence, this is known as a *thesis statement*. If a writer uses a thesis statement, it can generally be found at the beginning of the passage. Finally, the most specific information in a text is in the *supporting details*. An article about dog breed stereotyping might discuss a case study of pit bulls and provide statistics about how many dog attacks are caused by pit bulls versus other breeds.

Below is a diagram showcasing a topic with the main idea and supporting details. The topic is a single word (Cheetahs). The main idea tells us *what about* cheetahs the essay will be discussing. The supporting details offer proof that the main idea is true.

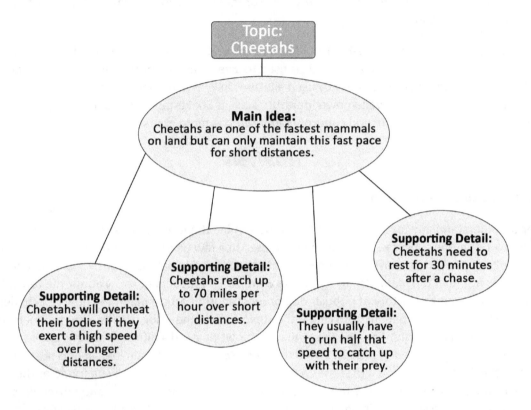

In contrast to informative writing, literary texts contain *themes*. A theme is a general way to describe the ideas and questions raised in a piece of literature. Like a topic, a theme can often be expressed in just one word or a few words rather than a full sentence. However, due to the complex nature of literature, most texts contain more than one theme. Some examples of literary themes include: isolation, sacrifice, vengeance. A text's theme might also explore the relationship between two contrasting ideas: ignorance versus knowledge, nature versus technology, science versus religion. A theme generally expresses a relatively broad and abstract idea about the text—so don't confuse a text's theme with its subject. Both theme and subject can answer the question, "What's the story about?" but the subject answers the question in a concrete way while the theme answers more abstractly. For example, the subject of *Hamlet* is Hamlet's investigation of his father's death (a concrete idea of what happens in the story). However, its themes—that is, the ideas explored through the story—include indecision and revenge, fundamental concepts that unite the events of the story. Because the theme is usually abstract, it might seem difficult to identify. Readers can ask themselves several questions to get a better idea of the theme:

- What observations does the writer make about human behavior?
- How do specific events of this story relate to society in general?
- What forces drive the characters' actions and decisions?
- How did the characters change or what did they learn during the story?

Identifying the Author's Purpose

Every story has a *narrator*, or someone who tells the story. This is sometimes also referred to as the story's *point of view*. Don't make the mistake of assuming that the narrator and the author are the same person—on the contrary, they may have completely different opinions and personalities. There are also several types of narrators commonly found in fiction, and each type serves a different purpose.

The first type of narrator is a *first-person narrator* who tells the story from an "I" perspective. A first-person narrator is almost always a character in the story. They might be a main protagonist (like Jane Eyre in the novel of the same name) or a side character (such as Nelly in *Wuthering Heights*). Also, first-person narrator has an immediate response to the events of the story. However, keep in mind that they are reporting the story through their own perspective, so this narrator may not be totally objective or reliable. Pay attention to language that shows the narrator's tone and consider how the events of the story might be interpreted differently by another character.

Another type of narrator is a *second-person narrator* who tells the story from a "you" perspective. Readers probably won't encounter this type of narrator often, although it's becoming slightly more common in modern literature. A second-person narrator forces the reader to insert themselves into the story and consider how they would respond to the action themselves.

Much more common is a *third-person narrator*. This type of narrator is further divided into two subtypes: *third-person limited* and *third-person omniscient*. In both cases, the story is told from an outside perspective ("he," "she," and "they," rather than "I," "me," or "you"). In this way, a third-person narrator is more distanced and objective than a first-person narrator. A third-person limited narrator tends to follow one main character in the story, reporting only that character's thoughts and only events that directly involve that character (think of the Harry Potter books, which are told by a third-person narrator but primarily limited to Harry's thoughts and experiences). A third-person omniscient narrator isn't limited to one character, but instead reports freely on the thoughts, actions, and events of all characters and situations in the story (omniscient means "all knowing"). Third-person omniscient is generally the most flexible and objective type of narrator. Readers can find this narrator in the novels of writers like Charles Dickens and Alexandre Dumas, whose stories are spread across dozens of different characters and locations.

By identifying the type of narrator, readers can assess a narrator's objectivity as well as how the narrator's perspective is limited to a single character's experience. When a narrator is unbiased, readers can take their account of events at face value; when a narrator is less objective, readers have to consider how the narrator's personal perspective influences the reader's understanding.

In addition to the point of view, another important factor to consider is the writer's *purpose*. This answers the question, "Why did the author write this?" Three of the most common purposes are to persuade, to inform, or to entertain. Of course, it's possible for these purposes to overlap.

Generally, literary texts are written to *entertain*. That is, the reader is simply supposed to enjoy the story! The romantic, dramatic, and adventurous elements are all designed to get readers involved with the fates of the characters. When the primary purpose of a text is to entertain, the focus is on the plot elements, the characters' thoughts and emotions, and descriptions of places, events, and characters. Examples of texts that entertain include novels, plays, poetry, and memoirs. Within the more general purpose of entertaining, literary texts might include passages to convey emotion or to describe a character's thoughts or feelings.

Another common writing purpose is to *inform*. Informative writing intends to teach the reader about a particular topic. Generally, this type of writing is objective and unbiased. Informative writing may also be descriptive. However, instead of describing thoughts and emotions, informative writing tends to describe concrete facts. After reading an informative text, the reader should be more knowledgeable about the text's subject. Examples of informative writing include textbooks, research articles, and texts about academic topics like science and history. Sometimes, informative writing is also described as writing that intends to teach or give information about a subject.

The third common writing purpose is to *persuade*. Unlike the objective writing found in informative texts, persuasive texts are more subjective. Rather than being neutral, the author expresses an opinion about the subject and tries to convince readers to agree. As with informative texts, persuasive texts often include facts and statistics—however, these are used to support the writer's perspective. Examples of persuasive writing include newspaper editorials and texts about controversial topics like politics or social issues. Sometimes, persuasive writing is also described as writing that intends to convince, argue, or express an opinion about a subject.

In order to determine the purpose of a text, readers can keep the following questions in mind:

- What type of details are included (facts, emotions, imagery, etc.)?

- Facts and statistics tend to indicate informative or persuasive writing, while emotions or imagery point to writing that entertains.

- Is the author neutral or opinionated?

- An opinionated or biased author is probably writing to persuade, while a neutral author writes to inform.

- Is this passage teaching information?

- If readers gained a lot of new knowledge about a subject after reading a text, the primary purpose was probably to inform.

Analyzing Individuals, Events, and Ideas Over the Course of a Text

Transitions are the glue that holds the writing together. They function to purposefully incorporate new topics and supporting details in a smooth and coherent way. Transitions and the corresponding structure they create can be used to determine how individuals, events, and ideas change and develop over the course of the text.

Transition words can be categorized based on the relationships they create between ideas:

- *General order*: signaling elaboration of an idea to emphasize a point—e.g., *for example, for instance, to demonstrate, including, such as, in other words, that is, in fact, also, furthermore, likewise, and, truly, so, surely, certainly, obviously, doubtless*

- *Chronological order*: referencing the time frame in which main event or idea occurs—e.g., *before, after, first, while, soon, shortly thereafter, meanwhile*

- *Numerical order/order of importance*: indicating that related ideas, supporting details, or events will be described in a sequence, possibly in order of importance—e.g., *first, second, also, finally,*

another, in addition, equally important, less importantly, most significantly, the main reason, last but not least

- *Spatial order*: referring to the space and location of something or where things are located in relation to each other—e.g., *inside, outside, above, below, within, close, under, over, far, next to, adjacent to*

- *Cause and effect order*: signaling a causal relationship between events or ideas—e.g., *thus, therefore, since, resulted in, for this reason, as a result, consequently, hence, for, so*

- *Compare and contrast order*: identifying the similarities and differences between two or more objects, ideas, or lines of thought—e.g., *like, as, similarly, equally, just as, unlike, however, but, although, conversely, on the other hand, on the contrary*

- *Summary order*: indicating that a particular idea is coming to a close—e.g., *in conclusion, to sum up, in other words, ultimately, above all*

Style, Structure, Mood, and Tone

Readers should be able to identify and analyze the components of a writer's *style*. Think about someone's fashion style—a person might dress casually or formally; they might wear trendy clothes or classic clothes; they might prefer simple looks or flashy ones. And the way a person styles their fashion often determines the impression they give to other people. Similarly, writers combine elements of structure, diction, and figurative and connotative language to create their own style.

Structure refers to how a writer organizes ideas. In literature, a text may be either *prose* or *poetry*. Poetry relies on careful word choice (especially in terms of sound and emotional meaning) and rhythm in order to communicate a special feeling or idea. Contrary to the popular assumption, poetry doesn't have to rhyme or follow a strict structure. In fact, there are two types of poetic form: *open form* and *closed form*. In closed form, the poet follows a predictable and repetitive structure, perhaps by using a fixed number of syllables in each line or repeating the same rhyme scheme. Examples of closed-form structure include sonnets and haiku, both of which require the poet to follow an established pattern of rhythm or rhyme. Open-form poetry doesn't have restrictions on length, the number of syllables or pattern of stress in each line (also known as meter), or the rhyme pattern. Open-form poetry has a structure, but it's more flexible and open to the creative whims of the poet. When a poet uses open form, changes in structure can reflect changes in emotion. For example, if a poem starts out with blunt, brief lines but then develops into long and complex lines, it might represent the speaker becoming more open and expressive of emotions that they had previously been reluctant to share.

Prose is regular written language without any meter or rhythmic form. Literary prose includes novels, short stories, and memoirs. An author may choose prose over poetry when they want to communicate in colloquial language, or when they want to convey information that is more straightforward (but of course, both poetry and prose can be emotional and creative). It's also possible to combine prose and poetry. In Shakespeare's plays, for example, some characters speak in metered lines while other characters speak in prose. This separation may indicate the topic under discussion. For example, in *Julius Caesar*, Brutus' speech is in prose, while Marc Antony's speech is written in iambic pentameter, a common poetic meter. Antony's speech begins with "Friends, Romans, countrymen, lend me your ears; / I come to bury Caesar, not to praise him." The cadence and stress of the language in Antony's speech makes for a more powerful listening device compared to Brutus' opening, "Romans, countrymen, and friends! Listen to my reasons and be silent so you can hear." In this way, employing prose or poetry can influence the impression that readers get from a text or drama. The crowd, in *Julius Caesar*, is persuaded by Marc Antony's speech in the end, for all its rhetorical glory.

There are also different story structures, or ways for the writer to present their narrative. A story can be either *linear* (told in the same order that events happened) or *non-linear* (the events are presented to the reader out of order). In a non-linear structure, the author may use flashbacks, when the timeline of the story shifts backwards to reveal earlier events. Non-linear storytelling is common in mystery or suspense writing, where the author keeps some information or events hidden from the reader until later in the story.

An author's style can also come from *diction*, or word choice. Just like a person's fashion style can be casual or formal depending on the event, an author's writing style can be anywhere from conversational to academic, elevated to colloquial, reflecting the audience and subject matter. For example, a chemistry textbook is going to contain more academic language and scientific terminology than a newspaper article, which is likely to contain common expressions and easier vocabulary. *Colloquial language* refers to the informal language of normal speech, and may include elements of non-standard pronunciation or grammar (words like "y'all," for example). Colloquialisms can often be found in "local color" pieces where the writer wants the reader to feel directly involved in the everyday lives and conversations of people or characters in the text.

Diction also contributes to the *tone* of the text. Keys to recognizing an author's tone include paying attention to any connotative or emotional language as well as what types of details and information are included (or if any important information seems to be missing). If an article about a proposal to build a new highway only includes information about how the highway will increase traffic congestion and negatively impact the environment, readers can feel the author's critical tone towards the subject. On the other hand, if the article also mentions research about how the highway could direct more customers to local businesses and boost the town's economy, the author's tone will probably seem more balanced.

A text's *mood* is the general feeling or atmosphere created by the author's descriptions and imagery (and, again, it relies on diction and selection of details). *Imagery* refers to all of the details in a text that appeal to any of the five senses; it's how the author helps draw a picture in the reader's mind. Imagine a story that starts with, "It was a dark and stormy night…" and includes descriptions of the howling wind outside, the dim flicker of candlelight, the mysterious creak of unknown footsteps coming upstairs. All of this imagery comes together to create a mood of creepiness and mystery.

Literary and Argumentative Techniques

Authors use a wide range of techniques to tell a story or communicate information. These techniques are also known as *rhetorical devices*.

In non-fiction writing, particularly persuasive writing, authors employ argumentative techniques to present their opinion in the most convincing way. Persuasive writing usually includes at least one type of appeal: an appeal to logic (logos), emotion (pathos), or authority and trustworthiness (ethos). When a writer appeals to logic, they are asking readers to agree based on research, evidence, and an established line of reasoning. An author's argument might also appeal to readers' emotions, perhaps by including personal stories and anecdotes (a short narrative of a specific event). An appeal to authority asks the

reader to agree on the basis of someone's expertise or credentials. Consider three different approaches to arguing the same opinion:

> Our school should abolish its current ban on cell phone use on campus. This rule was adopted last year as an attempt to reduce class disruptions and help students focus more on their lessons. However, since the rule was enacted, there has been no change in the number of disciplinary problems in class. Therefore, the rule is ineffective and should be done away with.

This is an appeal to logic. The author uses evidence to disprove the logic of the rule (the rule was supposed to reduce discipline problems, but the number of problems has not been reduced; therefore, the rule isn't working) and call for its repeal.

> Our school should abolish its current ban on cell phone use on campus. If they aren't able to use their phones during the school day, many students feel isolated from their loved ones. For example, last semester, one student's grandmother had a heart attack in the morning. However, because he couldn't use his cell phone, the student didn't know about his grandmother until the end of the day—when she had already passed away, and it was too late to say goodbye. By preventing students from contacting their friends and family, our school is placing undue stress and anxiety on students.

This is an appeal to emotion. By sharing the anecdote of one student and speaking about emotional topics like family relationships, the author invokes the reader's empathy in asking them to reconsider the rule.

> Our school should abolish its current ban on cell phone use on campus. According to Dr. Bartholomew Everett, a leading educational expert, "Research studies show that cell phone usage has no real impact on student attentiveness. Rather, phones provide a valuable technological resource for learning. Schools need to learn how to integrate this new technology into their curriculum." Rather than banning phones altogether, our school should follow the advice of experts and allow students to use phones as part of their learning.

This appeal to authority includes a statement from a relevant expert (in this case, a doctor in the field of education) to support the author's argument. All three examples argue the same opinion—the school's phone ban needs to change—but rely on different argumentative styles to persuade the reader.

Another argumentative technique is asking *rhetorical questions*, which don't require an answer but push the reader to consider the topic further.

> I wholly disagree with the proposal to ban restaurants from serving foods with high sugar and sodium contents. Do we really want to live in a world where the government can control what we eat? I prefer to make my own food choices.

Here, the author's rhetorical question prompts readers to put themselves in a hypothetical situation and imagine how they would feel.

Readers must also be able to tell the difference between *facts* and *opinions*. A fact doesn't simply refer to something that is true. Rather, it refers to anything that can be objectively measured and can be *proven* true or false. On the other hand, an opinion is a statement based on subjective observation. There is no measurable way to determine whether an opinion is correct—it simply depends on the perspective of the writer.

Consider the following examples:

Napoleon Bonaparte was the emperor of France from 1804 until 1814, and again in 1815.

This statement is a fact because it contains information that can be proven. By referencing historical documents and records from the nineteenth century, historians are able to objectively establish the dates of Napoleon's reign.

Napoleon was the most charismatic ruler France has ever had.

This statement is an opinion because it contains information that cannot be proven. First, there is no clear way to measure someone's charisma and compare it to that of other leaders. Even if an opinion is based on popular consensus, if it cannot be objectively proven, then it's not a fact.

Consider one more example:

a. *Beauty and the Beast* is the most important Disney movie.

b. According to a recent survey of Americans aged twenty to forty, the majority of respondents chose *Beauty and the Beast* as "the Disney movie that had the strongest impact on your life."

Sentence *A* is an opinion because it's based on a concept ("important") rather than something measurable. There is no agreed-upon definition of what "important" means, and no way of standardizing people's interpretation of the term. However, sentence *B* is a fact because it presents the statistical results of research. Anybody could look over the research results to confirm whether this statement is true, or they could even attempt to recreate the results themselves.

Remember that informative writing should contain purely factual statements without any biased opinions. Persuasive writing may combine facts and opinions, such as opinions that are supported by facts.

Three other types of statements are *observations*, *assumptions*, and *conclusions*. An observation states something the writer has noticed based on things they have seen or experienced. It's a descriptive statement that tends to be objective and unemotional. An assumption is something the author may not have observed, but is generally believed to be true. Finally, a conclusion is when the writer draws together observations and assumptions in order to make a final statement about them. For example:

a. The dining hall has started offering a new discounted lunch special for students. The dining hall has become much more crowded during lunch hours.

b. Students are usually on a budget and were probably attracted by the discounted prices.

c. The lunch special discount has been a successful way to gain diners.

Statement *A* contains observations because it describes things that can be seen—like the number of customers, or a sign advertising discounted lunch—by anyone who visits the dining hall. Statement *B* is an assumption. Without directly interviewing every student about their finances and their reasons for eating in the dining hall, it's impossible to know whether they're all looking for a low-budget meal. However, it's reasonable to believe this statement could be true based on what the author knows about student life. Finally, statement *C* is a conclusion because it combines all the information in the observations and assumptions to make a statement about the situation as a whole.

Literary texts also employ rhetorical devices like *figurative language* (including simile and metaphor, discussed earlier). In addition to rhetorical devices that play on the *meanings* of words, there are also rhetorical devices that use the *sounds* of words. These devices are most often found in poetry, but may also be found in other types of literature and in non-fiction writing like speech texts.

Alliteration and *assonance* are varieties of sound repetition. Alliteration refers to the repetition of the first sound of each word. Recall Robert Burns' opening line:

> My love is like a red, red rose

This line includes two examples of alliteration: "love" and "like" (repeated *L* sound), as well as "red" and "rose" (repeated *R* sound). Next, assonance refers to the repetition of vowel sounds, and can occur anywhere with a word (not just the opening sound). Here is the opening of a poem by John Keats:

> When I have fears that I may cease to be

> Before my pen has glean'd my teeming brain

Assonance can be found in the words "fears," "cease," "be," "glean'd," and "teeming," all of which stress the long *E* sound. Both alliteration and assonance create a harmony that unifies the writer's language.

Another sound device is *onomatopoeia*, or words whose spelling mimics the sound they describe. Words such as "crash," "bang," and "sizzle" are all examples of onomatopoeia. Use of onomatopoetic language adds auditory imagery to the text.

Readers are probably familiar with the *pun*. A pun is a play on words, taking advantage of two words that have the same or similar pronunciation. Puns can be found throughout Shakespeare's plays, for example:

> Now is the winter of our discontent

> Made glorious summer by this son of York

Here, Richard III refers to his brother, the newly crowned King Edward IV, as the "son of York," referencing their family heritage from the house of York. However, while drawing a comparison between the political climate and the weather (times of political trouble were the "winter," but now the new king brings "glorious summer"), Richard's use of the word "son" also implies another word with the same pronunciation, "sun"—so Edward IV is also like the sun, bringing light, warmth, and hope to England. Puns are a clever way for writers to suggest two meanings at once.

Synthesis and Generalization

Drawing Conclusions and Making Generalizations

As readers are presented with new information, they should organize it, make sense of it, and reflect on what they learned from the text. Readers draw conclusions at the end of a text by bringing together all of the details, descriptions, facts, and/or opinions presented by the author and asking, "What did I gain from reading this text? How have my ideas or emotions changed? What was the author's overall purpose for writing?" In this case, a *conclusion* is a unifying idea or final thought about the text that the reader can form after they are done reading. As discussed earlier, sometimes writers are very explicit in stating what conclusions should be drawn from a text and what readers are meant to have learned. However, more often than not, writers simply present descriptions or information and then leave it up to readers to draw their own conclusions. As with making inferences, though, readers always need to base their conclusions on textual evidence rather than simply guessing or making random statements.

> When the school district's uniform policy was first introduced fifteen years ago, parents and students alike were incredibly enthusiastic about it. Some of the most appealing arguments in favor of enforcing school uniforms was to create an equal learning environment for all students, to eliminate the focus on fashion and appearance, and to simplify students' morning routine by removing the need to pick a different outfit every day. However, despite this promising beginning, the uniform policy has steadily lost favor over the years. First of all, schools did not notice a significant drop in examples of bullying at school, and students continue to report that they feel judged on their appearance based on things like weight and hairstyle. This seems to indicate that uniforms have not been particularly effective at removing the social pressure that teens feel to appear a certain way in front of their peers. Also, many parents have complained that the school's required uniform pieces like jackets, sweaters, and neckties can only be purchased from one specific clothing shop. Because this retailer has cornered the market on school uniforms, they are operating under a total monopoly, and disgruntled parents feel that they are being grossly overcharged for school clothing for their children. The uniform policy is set to be debated at the upcoming school board meeting, and many expect it to be overturned.

After reading this article, a reader might conclude any of the following: that ideas that start with popular support might become unpopular over time; or that there are several compelling counterarguments to the benefits of school uniforms; or that this school district is open to new ideas but also open to criticism. While each conclusion is slightly different, they are all based on information and evidence from the article, and therefore all are plausible. Each conclusion sums up what the reader learned from the passage and what overall idea the writer seems to be communicating.

Another way for readers to make sense of information in a text is to make *generalizations*. This is somewhat related to the concept of inductive reasoning, by which readers move from specific evidence to a more general idea. When readers generalize, they take the specific content of a text and apply it to a larger context or to a different situation. Let's make a generalization from the topic, the bystander effect:

> A bystander is simply a person who watches something happen. Paradoxically, the more people who witness an accident happen, the less likely each individual is to actually intervene and offer assistance. This is known as the bystander effect. Psychologists attribute the bystander effect to something called "diffusion of responsibility." If one individual witnesses an accident, that single person feels the whole burden of responsibility to respond to the accident. However, if there are many witnesses, each person feels that responsibility has been divided amongst many people, so their individual sense of responsibility is much lower and they are less likely to offer help.

This article describes one very specific psychological phenomenon known as the "bystander effect." However, based on this specific information, a reader could form a more general psychological statement such as, "Humans sometimes behave differently when they are alone and when they are in a group."

It's also possible to make generalizations from literary texts. This is a particularly useful reading skill when evaluating the collective works of a particular writer or when forming a general characterization of texts from a particular genre or time period. For example, after reading a handful of novels by Jane Austen, all of which feature clever female protagonists and contain several examples of cynical or unflattering depictions of marriage, a reader can form a general impression of Jane Austen's thoughts on women's social roles. An overarching generalization from these novels might be that, "Education is just as important for women as marriage," or "Marriage isn't a guarantee of happiness or satisfaction." Being able to form generalizations is an important step in drawing connections and establishing relationships between texts.

Making Predictions

When readers make *predictions*, they try to anticipate what will happen next in the text. Think about how weather forecasters make predictions for future weather conditions. It's not purely guesswork. Rather, they gather a wide variety of relevant data, analyze the information they've collected, and also compare it with previous weather patterns. Finally, they're able to make a well-researched prediction with a high rate of probability. Readers must do the same when making predictions in a text—rather than simply guessing, they draw information from earlier in the text and use preexisting knowledge to form a prediction that's reasonably likely to be true.

In literary texts, authors tend to give clues that guide readers through the narrative. One method is *foreshadowing*, where the author hints at what will happen next in the story. Consider this line from *Romeo and Juliet* in which Juliet desires to learn more about Romeo.

She says to her nurse:

> Go ask his name.—If he be married.

> My grave is like to be my wedding bed.

At this point in the play, Juliet's lines simply mean that she would be sorely disappointed to learn that Romeo was already attached to another woman. However, for readers who already know how the play ends, this sentence carries another level of meaning—Juliet's marital choices will go hand in hand with her death. In this way, Shakespeare foreshadows the tragic end of Juliet's love story at the very moment it begins.

Of course, authors sometimes give false hints that lead readers to dead ends. This type of misdirection is especially commonplace in genres like mystery and suspense, where the author wants to keep the reader guessing until the very end. A distracting hint that turns out to be false is known as a *red herring*. However, if readers are aware that certain genres are likely to contain red herrings, then readers can be more cautious in evaluating hints. If a clue seems too obvious, it might be a red herring! In a roundabout way, then, red herrings can actually *help* readers make predictions by forcing them to look beyond the most obvious details.

In addition to clues sprinkled throughout the text, readers can also make predictions by considering the tone and mood of a text. For example, if a story has an overall gloomy mood or its diction creates a tone that is melancholy and foreboding, readers will expect that dark or depressing events will follow. On the other hand, if the tone is playful and lighthearted, readers are less likely to expect tragedy and might instead predict a comic or happy outcome. If the outcome of the story is vastly different from what either readers or the story's characters themselves are expecting, the author is probably using irony.

It's also possible to make predictions in non-literary texts. Consider a persuasive article that opens with the following thesis statement:

> There are countless reasons why closing down the city's only public dog park is a bad idea for local citizens.

Readers can expect that the rest of the essay will contain evidence that supports the author's opinion. The same is also true of informative texts. Imagine a scientific article that contains this sentence:

> Surprising new evidence challenges long-held beliefs about the cognitive capabilities of non-human animals.

In the paragraphs that follow, readers might expect to find any of the following: background on previously accepted theories of animal cognition, description of new scientific research or experimentation, and interpretation and discussion of the experiment's results. By making these predictions whenever they encounter a new text, readers will be more prepared to understand new information and opinions. Also, making predictions about information is especially useful when readers have a limited amount of time to read a text for relevant details. By reading either the thesis statement of the article or the topic sentences of each paragraph, readers can then make logical predictions about what information might be discussed later without having to read the entire text first.

Compare and Contrast

In order to understand the relationship between ideas, readers should be able to *compare* and *contrast*. Comparing two things means identifying their similarities, while contrasting two things means finding their differences. Recall the excerpt from "The Lamb" by William Blake:

> Little lamb, who made thee?
> Dost thou know who made thee,
> Gave thee life, and bid thee feed
> By the stream and o'er the mead;
> Gave thee clothing of delight,
> Softest clothing, woolly, bright;
> Gave thee such a tender voice,
> Making all the vales rejoice?
> Little lamb, who made thee?
> Dost thou know who made thee?

Consider that poem alongside an excerpt from another work by Blake called "The Tyger."

> Tyger! Tyger! burning bright
> In the forests of the night,
> What immortal hand or eye
> Could frame thy fearful symmetry?
> [...]
> What the hammer? what the chain?
> In what furnace was thy brain?
> What the anvil? what dread grasp
> Dare its deadly terrors clasp?
> When the stars threw down their spears,

And watered heaven with their tears,
Did he smile his work to see?
Did he who made the Lamb make thee?

These poems have quite a few things in common. Each poem's subject is an animal—a lamb and a tiger, respectively—and each poem addresses the same question to the animal: "Who created you?" In fact, both poems are formed primarily of questions.

However, the poems also exhibit many differences. For example, it's easy to contrast the tone and word choice in each poem. Whereas "The Lamb" uses words with positive and gentle connotations to create a tone of innocence and serenity, "The Tyger" gives a completely different impression. Some strongly connotative words that stand out include "night," "fearful," and "deadly terrors," all of which contribute to a tone that's tense and full of danger.

When taken together, then, the two poems address the same question—who created the world and all of its creatures?—from two different perspectives. "The Lamb" considers all of the sweet and delightful things that exist, leaving "The Tyger" to ponder the problem of why evil exists. In fact, Blake relies on the contrast between the two poems to fully communicate his dilemma over the paradox of creation—"Did he who make the Lamb make thee?" Although the poems present a strong contrast to one another, it's also possible to find similarities in their subject matter.

Authors often intentionally use contrast in order to ask readers to delve deeper into the qualities of the two things being compared. When an author deliberately places two things (characters, settings, etc.) side-by-side for readers to compare, it's known as *juxtaposition*. An example of juxtaposition can be found in Emily Bronte's *Wuthering Heights*, a novel in which the protagonist Cathy is caught in a love triangle between two romantic interests, Heathcliff and Edgar Linton, who are complete opposites. Cathy compares her feelings for each man in her memorable speech:

> My love for Linton is like the foliage in the woods: time will change it, I'm well aware, as winter changes the trees. My love for Heathcliff resembles the eternal rocks beneath: a source of little visible delight, but necessary. Nelly, I *am* Heathcliff! He's always, always in my mind: not as a pleasure, any more than I am always a pleasure to myself, but as my own being.

When these two characters are placed next to each other, it's easier for readers to grasp their notable characteristics. Edgar is gentle and sophisticated in comparison to Heathcliff, who is rough and wild. Here, Cathy also juxtaposes her feelings about each character. Her love for Edgar is fresh and harmless, like the new spring leaves on trees; but come winter, it will fade away. Her love for Heathcliff might be less conventionally appealing, like the rocks that form the earth; but, just like those rocks, that love forms the foundation of Cathy's being and is essential to her life. By juxtaposing these two men, Cathy is better able to express her thoughts about them.

Analyzing Information From Multiple Sources

When professors ask students to write a thesis, they expect students to base their essay on more than one source of information. When scholars delve into a research project, they too consult more than one source. In any academic endeavor, it's essential to look to *many* sources of information to get a comprehensive and well-rounded view of the subject matter. Getting information from multiple texts, though, requires readers to synthesize their content—that is, to combine ideas from various sources and express it in an organized way.

In order to synthetize information, readers first need to understand the relationship between the different sources. One way to do so is comparing and contrasting, as described above. Comparison and contrast is also useful in evaluating non-literary sources. For example, if readers want to learn more about a controversial issue, they might decide to read articles from both sides of the argument, compare differences in the arguments on each side, and identify any areas of overlap or agreement. This will allow readers to arrive at a more balanced conclusion.

In addition to synthesizing information from persuasive sources with different opinions, readers can also combine information from different types of texts—for example, from entertaining and informative texts. Readers who are interested in medieval religious life in Europe, for example, might read a text on medieval history by modern academics, a sociological research article about the role of religion in society, and a piece of literature from the Middle Ages such as Chaucer's *The Canterbury Tales*. By reading fiction from that time period, readers can look at one writer's perspective on religious activities in their world; and by reading non-fiction texts by modern researchers, readers can further enhance their background knowledge of the subject.

Practice Questions

The poem below, "The Human Seasons," was written by John Keats. Read it and answer questions 1 – 7.

> Four Seasons fill the measure of the year;
> There are four seasons in the mind of man:
> He has his lusty Spring, when fancy clear
> Takes in all beauty with an easy span:
> 5   He has his Summer, when luxuriously
> Spring's honied cud of youthful thought he loves
> To ruminate, and by such dreaming high
> Is nearest unto heaven: quiet coves
> His soul has in its Autumn, when his wings
> 10   He furleth close; contented so to look
> On mists in idleness—to let fair things
> Pass by unheeded as a threshold brook.
> He has his Winter too of pale misfeature,
> Or else he would forego his mortal nature.

1. What literary device does Keats primarily use in this poem?
   a. Simile
   b. Soliloquy
   c. Hyperbole
   d. Extended metaphor

2. The meaning of the word "ruminate" in line 7 is closest to:
   a. Ponder
   b. Unwind
   c. Respond
   d. Incorporate

3. According to the poem, how does a man change between Spring and Autumn?
   a. He starts preparing for his future.
   b. He feels more deeply connected to nature.
   c. He spends less time thinking about beautiful things.
   d. He becomes more sensible about how he spends his time.

4. Why does Keats end the poem with Winter?
   a. Winter represents the end of man's life.
   b. The narrator's least favorite season is winter.
   c. Winter is the final season of the calendar year.
   d. The poem is organized from the hottest season to the coldest.

5. Which statement would the narrator probably agree with?
   a. People are most content when they are young.
   b. People should appreciate the beauty of everyday life more.
   c. People change as they move through different stages of life.
   d. People spend too much time on daydreaming instead of being active.

6. What does "he would forego his mortal nature" mean in the final line?
  a. He would take a break.
  b. He would postpone or avoid death.
  c. He would give up nature for technology.
  d. He would move away from the countryside.

7. Which of the following is an example of alliteration in this poem?
  a. "in the mind of man"
  b. "On mists of idleness"
  c. "his wings / He furleth closed"
  d. "unheeded as a threshold brook"

In this excerpt from a novel set in nineteenth-century France, two friends, Albert de Morcerf and the Count of Monte Cristo, discuss Parisian social life. Read it and answer questions 8 – 14.

"Mademoiselle Eugénie is pretty—I think I remember that to be her name."

"Very pretty, or rather, very beautiful," replied Albert, "but of that style of beauty which I don't appreciate; I am an ungrateful fellow."

"Really," said Monte Cristo, lowering his voice, "you don't appear to me to be very enthusiastic on the subject of this marriage."

"Mademoiselle Danglars is too rich for me," replied Morcerf, "and that frightens me."

"Bah," exclaimed Monte Cristo, "that's a fine reason to give. Are you not rich yourself?"

"My father's income is about 50,000 francs per annum; and he will give me, perhaps, ten or twelve thousand when I marry."

"That, perhaps, might not be considered a large sum, in Paris especially," said the count; "but everything doesn't depend on wealth, and it's a fine thing to have a good name, and to occupy a high station in society. Your name is celebrated, your position magnificent; and then the Comte de Morcerf is a soldier, and it's pleasing to see the integrity of a Bayard united to the poverty of a Duguesclin; disinterestedness is the brightest ray in which a noble sword can shine. As for me, I consider the union with Mademoiselle Danglars a most suitable one; she will enrich you, and you will ennoble her."

Albert shook his head, and looked thoughtful. "There is still something else," said he.

"I confess," observed Monte Cristo, "that I have some difficulty in comprehending your objection to a young lady who is both rich and beautiful."

"Oh," said Morcerf, "this repugnance, if repugnance it may be called, isn't all on my side."

"Whence can it arise, then? for you told me your father desired the marriage."

"It's my mother who dissents; she has a clear and penetrating judgment, and doesn't smile on the proposed union. I cannot account for it, but she seems to entertain some prejudice against the Danglars."

"Ah," said the count, in a somewhat forced tone, "that may be easily explained; the Comtesse de Morcerf, who is aristocracy and refinement itself, doesn't relish the idea of being allied by your marriage with one of ignoble birth; that is natural enough."

8. The meaning of the word "repugnance" is closest to:
    a. Strong resemblance
    b. Strong dislike
    c. Extreme shyness
    d. Extreme dissimilarity

9. What can be inferred about Albert's family?
    a. Their finances are uncertain.
    b. Albert is the only son in his family.
    c. Their name is more respected than the Danglars'.
    d. Albert's mother and father both agree on their decisions.

10. What is Albert's attitude towards his impending marriage?
    a. Pragmatic
    b. Romantic
    c. Indifferent
    d. Apprehensive

11. What is the best description of the Count's relationship with Albert?
    a. He's like a strict parent, criticizing Albert's choices.
    b. He's like a wise uncle, giving practical advice to Albert.
    c. He's like a close friend, supporting all of Albert's opinions.
    d. He's like a suspicious investigator, asking many probing questions.

12. Which sentence is true of Albert's mother?
    a. She belongs to a noble family.
    b. She often makes poor choices.
    c. She is primarily occupied with money.
    d. She is unconcerned about her son's future.

13. Based on this passage, what is probably NOT true about French society in the 1800s?
    a. Children often received money from their parents.
    b. Marriages were sometimes arranged between families.
    c. The richest people in society were also the most respected.
    d. People were often expected to marry within their same social class.

14. Why is the Count puzzled by Albert's attitude toward his marriage?
    a. He seems reluctant to marry Eugénie, despite her wealth and beauty.
    b. He is marrying against his father's wishes, despite usually following his advice.
    c. He appears excited to marry someone he doesn't love, despite being a hopeless romantic.
    d. He expresses reverence towards Eugénie, despite being from a higher social class than her.

Read this article about NASA technology and answer questions 15 – 20.

When researchers and engineers undertake a large-scale scientific project, they may end up making discoveries and developing technologies that have far wider uses than originally intended. This is especially true in NASA, one of the most influential and innovative scientific organizations in America. NASA *spinoff technology* refers to innovations originally developed for NASA space projects that are now used in a wide range of different commercial fields. Many consumers are unaware that products they are buying are based on NASA research! Spinoff technology proves that it's worthwhile to invest in science research because it could enrich people's lives in unexpected ways.

The first spinoff technology worth mentioning is baby food. In space, where astronauts have limited access to fresh food and fewer options about their daily meals, malnutrition is a serious concern. Consequently, NASA researchers were looking for ways to enhance the nutritional value of astronauts' food. Scientists found that a certain type of algae could be added to food, improving the food's neurological benefits. When experts in the commercial food industry learned of this algae's potential to boost brain health, they were quick to begin their own research. The nutritional substance from algae then developed into a product called life's DHA, which can be found in over 90 percent of infant food sold in America.

Another intriguing example of a spinoff technology can be found in fashion. People who are always dropping their sunglasses may have invested in a pair of sunglasses with scratch resistant lenses—that is, it's impossible to scratch the glass, even if the glasses are dropped on an abrasive surface. This innovation is incredibly advantageous for people who are clumsy, but most shoppers don't know that this technology was originally developed by NASA. Scientists first created scratch resistant glass to help protect costly and crucial equipment from getting scratched in space, especially the helmet visors in space suits. However, sunglasses companies later realized that this technology could be profitable for their products, and they licensed the technology from NASA.

15. What is the main purpose of this article?
    a. To advise consumers to do more research before making a purchase
    b. To persuade readers to support NASA research
    c. To tell a narrative about the history of space technology
    d. To define and describe examples of spinoff technology

16. What is the organizational structure of this article?
    a. A general definition followed by more specific examples
    b. A general opinion followed by supporting arguments
    c. An important moment in history followed by chronological details
    d. A popular misconception followed by counterevidence

17. Why did NASA scientists research algae?
    a. They already knew algae was healthy for babies.
    b. They were interested in how to grow food in space.
    c. They were looking for ways to add health benefits to food.
    d. They hoped to use it to protect expensive research equipment.

18. What does the word "neurological" mean in the second paragraph?
    a. Related to the body
    b. Related to the brain
    c. Related to vitamins
    d. Related to technology

19. Why does the author mention space suit helmets?
    a. To give an example of astronaut fashion
    b. To explain where sunglasses got their shape
    c. To explain how astronauts protect their eyes
    d. To give an example of valuable space equipment

20. Which statement would the author probably NOT agree with?
    a. Consumers don't always know the history of the products they are buying.
    b. Sometimes new innovations have unexpected applications.
    c. It's difficult to make money from scientific research.
    d. Space equipment is often very expensive.

Read the following poem called "The Lady's Yes" by Elizabeth Barrett Browning and answer questions 21 – 26.

"Yes!" I answered you last night;
"No!" this morning, Sir, I say!
Colours, seen by candle-light,
Will not look the same by day.
5   When the tabors* played their best,
Lamps above, and laughs below
Love me sounded like a jest,
Fit for Yes or fit for No!
Call me false, or call me free
10   Vow, whatever light may shine,
No man on your face shall see
Any grief for change on mine.
Yet the sin is on us both
Time to dance isn't to woo
15   Wooer light makes fickle troth*
Scorn of me recoils on you!
Learn to win a lady's faith
Nobly, as the thing is high;
Bravely, as for life and death
20   With a loyal gravity.
Lead her from the festive boards,
Point her to the starry skies,
Guard her, by your truthful words,
Pure from courtship's flatteries.
25   By your truth she shall be true
Ever true, as wives of yore
And her Yes, once said to you,
SHALL be Yes for evermore.

*Tabors: small drums
*Troth: promise or loyalty

21. In this poem, the speaker mainly contrasts:
    a. Life and death
    b. Honesty and deceit
    c. Husbands and wives
    d. Shallow flirtation and true love

22. What is the speaker's tone in this poem?
    a. Reproachful
    b. Intrigued
    c. Hopeless
    d. Lovesick

23. Which of the following is the best paraphrase for the third stanza (lines 9 – 12)?
    a. Promise that you will never change your feelings for me, even if I change mine for you.
    b. Whether or not you criticize me for changing my mind, don't appear too upset about it.
    c. It's your fault that our relationship ended, so don't feel sorry for yourself.
    d. Don't talk to other people about our relationship.

24. What's the speaker's purpose in the final three stanzas of the poem (lines 17 – 28)?
    a. To describe relationships that don't traditionally exist in modern romance
    b. To give advice about how to win a woman's true love
    c. To contrast the roles of men and women in society
    d. To lament the hardships of finding a husband

25. How would the speaker probably feel about modern nightclubs?
    a. They are too noisy to be enjoyable.
    b. They should have more drum music.
    c. They aren't suitable places to find true love.
    d. They allow people to speak honestly about their feelings.

26. Which word is closest in meaning to "jest" in line 7?
    a. Joke
    b. Melody
    c. Proposal
    d. Confession

This article discusses the famous poet and playwright William Shakespeare. Read it and answer questions 27 – 34.

People who argue that William Shakespeare isn't responsible for the plays attributed to his name are known as anti-Stratfordians (from the name of Shakespeare's birthplace, Stratford-upon-Avon). The most common anti-Stratfordian claim is that William Shakespeare simply was not educated enough or from a high enough social class to have written plays overflowing with references to such a wide range of subjects like history, the classics, religion, and international culture. William Shakespeare was the son of a glove-maker, he only had a basic grade school education, and he never set foot outside of England—so how could he have produced plays of such sophistication and imagination? How could he have written in such detail about historical figures and events, or about different cultures and locations around Europe? According to anti-Stratfordians, the depth of knowledge contained in Shakespeare's plays suggests a well-traveled writer from a wealthy background with a university education, not a countryside writer like Shakespeare. But in fact, there isn't much substance to such speculation, and most anti-Stratfordian arguments can be refuted with a little background about Shakespeare's time and upbringing.

First of all, those who doubt Shakespeare's authorship often point to his common birth and brief education as stumbling blocks to his writerly genius. Although it's true that Shakespeare did not come from a noble class, his father was a very *successful* glove-maker and his mother was from a very wealthy land-owning family—so while Shakespeare may have had a country upbringing, he was certainly from a well-off family and would have been educated accordingly. Also, even though he did not attend university, grade school education in Shakespeare's time was actually quite rigorous and exposed students to classic drama through writers like Seneca and Ovid. It's

not unreasonable to believe that Shakespeare received a very solid foundation in poetry and literature from his early schooling.

Next, anti-Stratfordians tend to question how Shakespeare could write so extensively about countries and cultures he had never visited before (for example, several of his most famous works like *Romeo and Juliet* and *The Merchant of Venice* were set in Italy, on the opposite side of Europe). But again, this criticism doesn't hold up under scrutiny. For one thing, Shakespeare was living in London, a bustling metropolis of international trade, the most populous city in England, and a political and cultural hub of Europe. In the daily crowds of people, Shakespeare would certainly have been able to meet travelers from other countries and hear firsthand accounts of life in their home country. And, in addition to the influx of information from world travelers, this was also the age of the printing press, a jump in technology that made it possible to print and circulate books much more easily than in the past. This also allowed for a freer flow of information across different countries, allowing people to read about life and ideas throughout Europe. One needn't travel the continent in order to learn and write about its culture.

27. The main purpose of this article is to:
    a. Explain two sides of an argument and allow readers to choose which side they agree with
    b. Encourage readers to be skeptical about the authorship of famous poems and plays
    c. Give historical background about an important literary figure
    d. Criticize a theory by presenting counterevidence

28. Which sentence contains the author's thesis?
    a. "People who argue that William Shakespeare isn't responsible for the plays attributed to his name are known as anti-Stratfordians."
    b. "But in fact, there isn't much substance to such speculation, and most anti-Stratfordian arguments can be refuted with a little background about Shakespeare's time and upbringing."
    c. "It's not unreasonable to believe that Shakespeare received a very solid foundation in poetry and literature from his early schooling."
    d. "Next, anti-Stratfordians tend to question how Shakespeare could write so extensively about countries and cultures he had never visited before."

29. In the first paragraph, "How could he have written in such detail about historical figures and events, or about different cultures and locations around Europe?" is an example of which of the following?
    a. Hyperbole
    b. Onomatopoeia
    c. Rhetorical question
    d. Appeal to authority

30. How does the author respond to the claim that Shakespeare was not well-educated because he didn't attend university?
    a. By insisting upon Shakespeare's natural genius
    b. By explaining grade school curriculum in Shakespeare's time
    c. By comparing Shakespeare with other uneducated writers of his time
    d. By pointing out that Shakespeare's wealthy parents probably paid for private tutors

31. The word "bustling" in the third paragraph most nearly means:
    a. Busy
    b. Foreign
    c. Expensive
    d. Undeveloped

32. What can be inferred from the article?
    a. Shakespeare's peers were jealous of his success and wanted to attack his reputation.
    b. Until recently, classical drama was only taught in universities.
    c. International travel was extremely rare in Shakespeare's time.
    d. In Shakespeare's time, glove-makers weren't part of the upper class.

33. Why does the author mention *Romeo and Juliet*?
    a. It's Shakespeare's most famous play.
    b. It was inspired by Shakespeare's trip to Italy.
    c. It's an example of a play set outside of England.
    d. It was unpopular when Shakespeare first wrote it.

34. Which statement would the author probably agree with?
    a. It's possible to learn things from reading rather than firsthand experience.
    b. If you want to be truly cultured, you need to travel the world.
    c. People never become successful without a university education.
    d. All of the world's great art comes from Italy.

A traveler prepares for a journey in this excerpt from a novel. Read it and answer questions 35 – 40.

When I got on the coach the driver had not taken his seat, and I saw him talking with the landlady. They were evidently talking of me, for every now and then they looked at me, and some of the people who were sitting on the bench outside the door came and listened, and then looked at me, most of them pityingly. I could hear a lot of words often repeated, queer words, for there were many nationalities in the crowd; so I quietly got my polyglot dictionary from my bag and looked them out. I must say they weren't cheering to me, for amongst them were "Ordog"—Satan, "pokol"—hell, "stregoica"—witch, "vrolok" and "vlkoslak"—both of which mean the same thing, one being Slovak and the other Servian for something that is either were-wolf or vampire.

When we started, the crowd round the inn door, which had by this time swelled to a considerable size, all made the sign of the cross and pointed two fingers towards me. With some difficulty I got a fellow-passenger to tell me what they meant; he wouldn't answer at first, but on learning that I was English, he explained that it was a charm or guard against the evil eye. This was not very pleasant for me, just starting for an unknown place to meet an unknown man; but everyone seemed so kind-hearted, and so sorrowful, and so sympathetic that I couldn't but be touched. I shall never forget the last glimpse which I had of the inn-yard and its crowd of picturesque figures, all crossing themselves, as they stood round the wide archway, with its background of rich foliage of oleander and orange trees in green tubs clustered in the centre of the yard. Then our driver cracked his big whip over his four small horses, which ran abreast, and we set off on our journey.

I soon lost sight and recollection of ghostly fears in the beauty of the scene as we drove along, although had I known the language, or rather languages, which my fellow-passengers were speaking, I might not have been able to throw them off so easily. Before us lay a green sloping land full of forests and woods, with here and there steep hills, crowned with clumps of trees or with farmhouses, the blank gable end to the road. There was everywhere a bewildering mass of fruit blossom—apple, plum, pear, cherry; and as we drove by I could see the green grass under the trees spangled with the fallen petals. In and out amongst these green hills of what they call here the "Mittel Land" ran the road, losing itself as it swept round the grassy curve, or was shut out by the straggling ends of pine woods, which here and there ran down the hillsides like tongues of flame. The road was rugged, but still we seemed to fly over it with a feverish haste. I couldn't understand then what the haste meant, but the driver was evidently bent on losing no time in reaching Borgo Prund.

35. What type of narrator is found in this passage?
    a. First person
    b. Second person
    c. Third-person limited
    d. Third-person omniscient

36. Which of the following is true of the traveler?
    a. He wishes the driver would go faster.
    b. He's returning to the country of his birth.
    c. He has some familiarity with the local customs.
    d. He doesn't understand all of the languages being used.

37. How does the traveler's mood change between the second and third paragraphs?
    a. From relaxed to rushed
    b. From fearful to charmed
    c. From confused to enlightened
    d. From comfortable to exhausted

38. Who is the traveler going to meet?
    a. A kind landlady
    b. A distant relative
    c. A friendly villager
    d. A complete stranger

39. Based on the details in this passage, what can readers probably expect to happen in the story?
    a. The traveler will become a farmer.
    b. The traveler will arrive late at his destination.
    c. The traveler will soon encounter danger or evil.
    d. The traveler will have a pleasant journey and make many new friends.

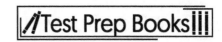

40. Which sentence from the passage provides a clue for question 39?
    a. "I must say they weren't cheering to me, for amongst them were "Ordog"—Satan, "pokol"—hell, "stregoica"—witch, "vrolok" and "vlkoslak"—both of which mean the same thing, one being Slovak and the other Servian for something that is either were-wolf or vampire."
    b. "When I got on the coach the driver had not taken his seat, and I saw him talking with the landlady."
    c. "Then our driver cracked his big whip over his four small horses, which ran abreast, and we set off on our journey."
    d. "There was everywhere a bewildering mass of fruit blossom—apple, plum, pear, cherry; and as we drove by I could see the green grass under the trees spangled with the fallen petals."

41. Which phrase below best defines *inference*?
    a. Reading between the lines
    b. Skimming a text for context clues
    c. Writing notes or questions that need answers during the reading experience
    d. Summarizing the text

42. Which phrase best defines *connotation*?
    a. An author's use of footnotes in his or her informational text
    b. Words or phrases that mean exactly what they say
    c. The author's use of allusion
    d. When an author chooses words or phrases that invoke feelings rather than a literal meaning

Questions 43-50 are based on the following passages:

Passage I

Lethal force, or deadly force, is defined as the physical means to cause death or serious harm to another individual. The law holds that lethal force is only accepted when you or another person are in immediate and unavoidable danger of death or severe bodily harm. For example, a person could be beating a weaker person in such a way that they are suffering severe enough trauma that could result in death or serious harm. This would be an instance where lethal force would be acceptable and possibly the only way to save that person from irrevocable damage.

Another example of when to use lethal force would be when someone enters your home with a deadly weapon. The intruder's presence and possession of the weapon indicate mal-intent and the ability to inflict death or severe injury to you and your loved ones. Again, lethal force can be used in this situation. Lethal force can also be applied to prevent the harm of another individual. If a woman is being brutally assaulted and is unable to fend off an attacker, lethal force can be used to defend her as a last-ditch effort. If she is in immediate jeopardy of rape, harm, and/or death, lethal force could be the only response that could effectively deter the assailant.

The key to understanding the concept of lethal force is the term *last resort*. Deadly force cannot be taken back; it should be used only to prevent severe harm or death. The law does distinguish whether the means of one's self-defense is fully warranted, or if the individual goes out of control in the process. If you continually attack the assailant after they are rendered incapacitated, this would be causing unnecessary harm, and the law can bring charges against you. Likewise, if you kill an attacker unnecessarily after

defending yourself, you can be charged with murder. This would move lethal force beyond necessary defense, making it no longer a last resort but rather a use of excessive force.

Passage II

Assault is the unlawful attempt of one person to apply apprehension on another individual by an imminent threat or by initiating offensive contact. Assaults can vary, encompassing physical strikes, threatening body language, and even provocative language. In the case of the latter, even if a hand has not been laid, it is still considered an assault because of its threatening nature.

Let's look at an example: A homeowner is angered because his neighbor blows fallen leaves into his freshly mowed lawn. Irate, the homeowner gestures a fist to his fellow neighbor and threatens to bash his head in for littering on his lawn. The homeowner's physical motions and verbal threat heralds a physical threat against the other neighbor. These factors classify the homeowner's reaction as an assault. If the angry neighbor hits the threatening homeowner in retaliation, that would constitute an assault as well because he physically hit the homeowner.

Assault also centers on the involvement of weapons in a conflict. If someone fires a gun at another person, this could be interpreted as an assault unless the shooter acted in self-defense. If an individual drew a gun or a knife on someone with the intent to harm them, that would be considered assault. However, it's also considered an assault if someone simply aimed a weapon, loaded or not, at another person in a threatening manner.

43. What is the purpose of the second passage?
    a. To inform the reader about what assault is and how it is committed
    b. To inform the reader about how assault is a minor example of lethal force
    c. To disprove the previous passage concerning lethal force
    d. The author is recounting an incident in which they were assaulted

44. Which of the following situations, according to the passages, would not constitute an illegal use of lethal force?
    a. A disgruntled cashier yells obscenities at a customer.
    b. A thief is seen running away with stolen cash.
    c. A man is attacked in an alley by another man with a knife.
    d. A woman punches another woman in a bar.

45. Given the information in the passages, which of the following must be true about assault?
    a. Assault charges are more severe than unnecessary use of force charges.
    b. There are various forms of assault.
    c. Smaller, weaker people cannot commit assaults.
    d. Assault is justified only as a last resort.

46. Which of the following, if true, would most seriously undermine the explanation proposed by the author of Passage I in the third paragraph?

a. An instance of lethal force in self-defense is not absolutely absolved from blame. The law considers the necessary use of force at the time it is committed.

b. An individual who uses lethal force under necessary defense is in direct compliance of the law under most circumstances.

c. Lethal force in self-defense should be forgiven in all cases for the peace of mind of the primary victim.

d. The use of lethal force is not evaluated on the intent of the user but rather the severity of the primary attack that warranted self-defense.

47. Based on the passages, what can be inferred about the relationship between assault and lethal force?

a. An act of lethal force always leads to a type of assault.

b. An assault will result in someone using lethal force.

c. An assault with deadly intent can lead to an individual using lethal force to preserve their well-being.

d. If someone uses self-defense in a conflict, it is called deadly force; if actions or threats are intended, it is called assault.

48. Which of the following best describes the way the passages are structured?

a. Both passages open by defining a legal concept and then continue to describe situations that further explain the concept.

b. Both passages begin with situations, introduce accepted definitions, and then cite legal ramifications.

c. Passage I presents a long definition while the Passage II begins by showing an example of assault.

d. Both cite specific legal doctrines, then proceed to explain the rulings.

49. What can be inferred about the role of intent in lethal force and assault?

a. Intent is irrelevant. The law does not take intent into account.

b. Intent is vital for determining the lawfulness of using lethal force.

c. Intent is very important for determining both lethal force and assault; intent is examined in both parties and helps determine the severity of the issue.

d. The intent of the assailant is the main focus for determining legal ramifications; it is used to determine if the defender was justified in using force to respond.

50. The author uses the example in the second paragraph of Passage II in order to do what?

a. To demonstrate two different types of assault by showing how each specifically relates to the other

b. To demonstrate a single example of two different types of assault, then adding in the third type of assault in the example's conclusion

c. To prove that the definition of lethal force is altered when the victim in question is a homeowner and his property is threatened

d. To suggest that verbal assault can be an exaggerated crime by the law and does not necessarily lead to physical violence

Answer Explanations

1. D: Extended metaphor. Metaphor is a direct comparison between two things, and extended metaphor is a lengthy, well-developed metaphor that usually extends over the length of the poem. In this poem, Keats forms an extended metaphor by drawing a comparison between the four seasons of nature and the "seasons" that humans experience from youth to old age.

2. A: Ponder. This question can be answered using context clues from the sentence: "Spring's honied cud of youthful thought he loves / To ruminate, and by such dreaming high / Is nearest unto heaven." Following the word "ruminate," it's restated as "such dreaming"; also, immediately before is the expression "youthful thought." Together, this sentence describes a young man pleasantly daydreaming. The only word related to thinking and daydreaming is "ponder," Choice *A*.

3. C: He spends less time thinking about beautiful things. This is a general comprehension question. The narrator describes a man in Autumn "contented so . . . to let fair things / Pass by unheeded." In this case, "fair" is another word for "beautiful," and letting things "pass by unheeded" means "he doesn't pay attention to them." In contrast, a man in the Spring and Summer of life spends time appreciating and daydreaming about beautiful things.

4. A: Winter represents the end of man's life. This is a purpose question, but it also requires readers to understand that this poem is an extended metaphor. Since the narrator is developing an extended comparison between seasons and life, it's natural that winter should come last because it's the season of death, dormancy, and "pale" nature (unlike, say, Spring, which is a season of life and rebirth in nature).

5. C: People change as they move through different stages of life. This is an inference question asking readers to understand the narrator's perspective. Choices *B* and *D* both include an opinion or advice to the reader, while the tone of the poem is more neutral or purely descriptive (the narrator is simply describing the stages of life, rather than advising readers on how to behave). Choice *C* more closely agrees with the comparison that the narrator sets up in the poem; just as seasons change in nature, people also change throughout their lives.

6. B: He would postpone or avoid death. This is both a vocabulary and a comprehension question. Based on the poem's extended metaphor, readers can gather that Winter is a metaphor for the end of life; all people must pass through Winter or else they would never die. Looking at the poem's vocabulary, "mortal" refers to human's limited life span (the opposite of "immortal"), and "forego" means to turn something down.

7. A: "in the mind of man" (2). This is a fairly straightforward question about literary devices. Alliteration refers to repetition of a word's beginning sound, and Choice *A* is the only example of that ("mind" and "man" both start with the letter M).

8. B: Strong dislike. This vocabulary question can be answered using context clues. Based on the rest of the conversation, the reader can gather that Albert isn't looking forward to his marriage. As the Count notes that "you don't appear to me to be very enthusiastic on the subject of this marriage," and also remarks on Albert's "objection to a young lady who is both rich and beautiful," readers can guess Albert's feelings. The answer choice that most closely matches "objection" and "not . . . very enthusiastic" is *B*, "strong dislike."

9. C: Their name is more respected than the Danglars'. This inference question can be answered by eliminating incorrect answers. Choice *A* is tempting, considering that Albert mentions money as a concern in his marriage. However, although he may not be as rich as his fiancée, his father still has a stable income of 50,000 francs a year. Choice *B* isn't mentioned at all in the passage, so it's impossible to make an inference. Finally, Choice *D* is false because Albert's father arranged his marriage, but his mother doesn't approve of it. Evidence for Choice *C* can be found in the Count's comparison of Albert and Eugénie: "she will enrich you, and you will ennoble her." In other words, the Danglars are wealthier but the Morcerf family has a more noble background.

10. D: Apprehensive. There are many clues in the passage that indicate Albert's attitude towards his marriage—far from enthusiastic, he has many reservations. This question requires test takers to understand the vocabulary in the answer choices. "Pragmatic" is closest in meaning to "realistic," and "indifferent" means "uninterested." The only word related to feeling worried, uncertain, or unfavorable about the future is "apprehensive."

11. B: He is like a wise uncle, giving practical advice to Albert. Choice *A* is incorrect because the Count's tone is friendly and conversational. Choice *C* is also incorrect because the Count questions why Albert doesn't want to marry a young, beautiful, and rich girl. While the Count asks many questions, he isn't particularly "probing" or "suspicious"—instead, he's asking to find out more about Albert's situation and then give him advice about marriage.

12. A: She belongs to a noble family. Though Albert's mother doesn't appear in the scene, there's more than enough information to answer this question. More than once is his family's noble background mentioned (not to mention that Albert's mother is the Comtesse de Morcerf, a noble title). The other answer choices can be eliminated—she is deeply concerned about her son's future; money isn't her highest priority because otherwise she would favor a marriage with the wealthy Danglars; and Albert describes her "clear and penetrating judgment," meaning she makes good decisions.

13. C: The richest people in society were also the most respected. The Danglars family is wealthier but the Morcerf family has a more aristocratic name, which gives them a higher social standing. Evidence for the other answer choices can be found throughout the passage: Albert mentioned receiving money from his father's fortune after his marriage; Albert's father has arranged this marriage for him; and the Count speculates that Albert's mother disapproves of this marriage because Eugénie isn't from a noble background like the Morcerf family, implying that she would prefer a match with a girl from aristocratic society.

14. A: He seems reluctant to marry Eugénie, despite her wealth and beauty. This is a reading comprehension question, and the answer can be found in the following lines: "'I confess,' observed Monte Cristo, 'that I have some difficulty in comprehending your objection to a young lady who is both rich and beautiful.'" Choice *B* is the opposite (Albert's father is the one who insists on the marriage), Choice *C* incorrectly represents Albert's eagerness to marry, and Choice *D* describes a more positive attitude than Albert actually feels ("repugnance").

15. D: To define and describe examples of spinoff technology. This is a purpose question—*why* did the author write this? The article contains facts, definitions, and other objective information without telling a story or arguing an opinion. In this case, the purpose of the article is to inform the reader. The only answer choice related to giving information is Choice *D*: to define and describe.

16. A: A general definition followed by more specific examples. This organization question asks readers to analyze the structure of the essay. The topic of the essay is spinoff technology; the first paragraph gives a general definition of the concept, while the following two paragraphs offer more detailed examples to help illustrate this idea.

17. C: They were looking for ways to add health benefits to food. This reading comprehension question can be answered based on the second paragraph—scientists were concerned about astronauts' nutrition and began researching nutritional supplements. Choice *A* isn't true because it reverses the order of discovery (first NASA identified algae for astronaut use, and then it was further developed for use in baby food).

18. B: Related to the brain. This vocabulary question could be answered based on the reader's prior knowledge, but the passage provides context clues for readers who've never encountered the word "neurological." The next sentence talks about "this algae's potential to boost brain health," which is a paraphrase of "neurological benefits." From this context, readers should be able to infer that "neurological" relates to the brain.

19. D: To give an example of valuable space equipment. This purpose question requires readers to understand the relevance of the given detail. In this case, the author mentions "costly and crucial equipment" before space suit visors, which are given as an example of something valuable. Choice *A* isn't correct because fashion is only related to sunglasses, not to NASA equipment. Choice *B* can be eliminated because it's simply not mentioned. While Choice *C* seems like it could be true, it's not relevant.

20. C: It's difficult to make money from scientific research. The article gives several examples of how businesses have capitalized on NASA research, so it's unlikely that the author would agree with this statement. Evidence for the other answer choices can be found in the article: In Choice *A*, the author mentions that "many consumers are unaware that products they are buying are based on NASA research"; Choice *B* is a general definition of spinoff technology; and Choice *D* is mentioned in the final paragraph.

21. D: Shallow flirtation and true love. This is a general comprehension question. The speaker is a woman who exchanged romantic words with another at an exciting party one evening, but changed her mind after deeper consideration the next morning. Their flirtation was influenced by the ambience of music, laughter, and candlelight, but she spends the latter part of the poem describing the course of true love. The poem, then, focuses on contrasting short-term flirtation with lasting love. Readers might be tempted by Choice *B* because the final few stanzas of the poem focus on "truth" and "true," but the speaker is using this to specifically describe love, not to talk about truth and lies in general.

22. A: Reproachful. In the poem, the speaker chastises herself and her partner for their "sin" and tries to direct them toward "truth"—she is criticizing their past actions and outlining a new way forward, so her tone is "reproachful," or critical and judgmental. For test takers who don't know the word "reproachful," this question can also be answered through process of elimination. The speaker is clearly not "lovesick" because she's rejecting someone's love. Nor is she "hopeless," because at the end of the poem she explains how true love is possible. Finally, she feels no curiosity or temptation about continuing this flirtation, so her tone isn't "intrigued."

23. B: Whether or not you criticize me for changing my mind, don't appear too upset about it. A paraphrase is a restatement of information without losing meaning. Choice *B* is the only sentence that retains the speaker's message ("call me false, or call me free" can be paraphrased as "what you think of me is up to you"). The other answer choices don't keep the original meaning intact.

24. B: To give advice about how to win a woman's true love. In the first half of the poem, the speaker criticizes shallow flirtation at a party; in the second half of the poem, she gives advice about how to develop a deeper kind of love. In these stanzas, her sentences are also written as commands ("Learn," "Lead," "Point," and "Guard"), directly instructing the reader how to behave. Choice *A* isn't a good answer choice because she tells the reader to use these romantic tips—clearly she thinks they are still effective. Choices *C* and *D* are also incorrect because they aren't directly related to the poem's content.

25. C: They aren't suitable places to find true love. This is an inference question, and hints can be found throughout the poem. For example, in the second stanza, the speaker describes the noisy, musical atmosphere and says that "Love me sounded like a jest, / Fit for yes or fir for No!"—in other words, it wasn't a serious or sincere atmosphere. Also, in the fourth stanza, she mentions "Time to dance isn't to woo," indicating her belief that dancing and finding true love don't really overlap.

26. A: Joke. Readers may already have enough vocabulary to answer this question. For readers who need context clues, the line that mentions "laughs" hints at the playful situation at hand.

27. D: Criticize a theory by presenting counterevidence. The author mentions anti-Stratfordian arguments in the first paragraph, but then goes on to debunk these theories with facts about Shakespeare's life in the second and third paragraphs. Choice *A* is incorrect because the author is far from unbiased; in fact, the author clearly disagrees with anti-Stratfordians. Choice *B* is also incorrect because it's more closely aligned with the beliefs of anti-Stratfordians. Choice *C* can be eliminated because, while it's true that the author gives historical background, the purpose is using that information to disprove a theory.

28. B: "But in fact, there isn't much substance to such speculation, and most anti-Stratfordian arguments can be refuted with a little background about Shakespeare's time and upbringing." The thesis is a statement that contains the author's topic and main idea. As seen in question 27, the purpose of this article is to use historical evidence to provide counterarguments to anti-Stratfordians. Choice *A* is simply a definition; Choice *C* is a supporting detail, not a main idea; and Choice *D* represents an idea of anti-Stratfordians, not the author's opinion.

29. C: Rhetorical question. A rhetorical question is asked not to obtain an answer but to encourage readers to more deeply consider an issue.

30. B: By explaining grade school curriculum in Shakespeare's time. This question asks readers to refer to the organizational structure of the article and demonstrate understanding of how the author provides details to support the argument. This particular detail can be found in the second paragraph: "even though he did not attend university, grade school education in Shakespeare's time was actually quite rigorous."

31. A: Busy. This is a vocabulary question that can be answered using context clues. Other sentences in the paragraph describe London as "the most populous city in England" filled with "crowds of people." Choice *B* is incorrect because London was in Shakespeare's home country, not a foreign one. Choice *C* isn't mentioned in the passage. Choice *D* isn't a good answer because the passage describes London as an important city, not underdeveloped.

32. D: In Shakespeare's time, glove-makers weren't part of the upper class. Anti-Stratfordians doubt Shakespeare's ability because he wasn't from the upper class; his father was a glove-maker; therefore, in at least this example, glove-makers weren't included in the upper class. This is an example of inductive reasoning, using two specific pieces of information to draw a more general conclusion.

33. C: It's an example of a play set outside of England. This detail comes from the third paragraph, where the author responds to skeptics who claim that Shakespeare wrote too much about places he never visited, so *Romeo and Juliet* is mentioned as a famous example of a play with a foreign setting. In order to answer this question, readers need to understand the author's purpose in the third paragraph and how the author uses details to support this purpose. Choices *A* and *D* aren't mentioned, and Choice *B* is clearly false because the passage mentions more than once that Shakespeare never left England.

34. A: It's possible to learn things from reading rather than firsthand experience. This inference can be made from the final paragraph, where the author refutes anti-Stratfordian skepticism by noting that books about life in Europe could circulate throughout London. From this statement, readers can conclude the author believes it's possible that Shakespeare learned about European culture from books. Choice *B* isn't true because the author believes that Shakespeare contributed to English literature without traveling extensively. Similarly, Choice *C* isn't a good answer because the author explains how Shakespeare got his education without attending a university. Choice *D* can also be eliminated because the author describes Shakespeare's genius, and Shakespeare clearly isn't from Italy.

35. A: First person. This is a straightforward question that requires readers to know that a first-person narrator speaks from an "I" point of view.

36. D: He doesn't understand all of the languages being used. This can be inferred from the fact that the traveler must refer to his dictionary to understand those around him. Choice *A* isn't a good choice because the traveler seems to wonder why the driver needs to drive so fast. Choice B isn't mentioned in the passage and doesn't seem like a good answer choice because he seems wholly unfamiliar with his surroundings. This is why Choice C can also be eliminated.

37. B: From fearful to charmed. This can be found in the first sentence of the third paragraph, which states, "I soon lost sight and recollection of ghostly fears in the beauty of the scene as we drove along." Also, readers should get a sense of foreboding from the first two paragraphs, where superstitious villagers seem frightened on the traveler's behalf. However, the final paragraph changes to delighted descriptions of the landscape's natural beauty. Choices *A* and *D* can be eliminated because the traveler is anxious, not relaxed or comfortable at the beginning of the passage. Choice *C* can also be eliminated because the traveler doesn't gain any particular insights in the last paragraph, and in fact continues to lament that he cannot understand the speech of those around him.

38. D: A complete stranger. The answer to this reading comprehension question can be found in the second paragraph, when the traveler is "just starting for an unknown place to meet an unknown man"—in other words, a complete stranger.

39. C: The traveler will soon encounter danger or evil. Answering this prediction question requires readers to understand foreshadowing, or hints that the author gives about what will happen next. There are numerous hints scattered throughout this passage: the villager's sorrow and sympathy for the traveler and their superstitious actions; the spooky words that the traveler overhears; the driver's unexplained haste. All of these point to a danger that awaits the protagonist.

40. A: "I must say they weren't cheering to me, for amongst them were "Ordog"—Satan, "pokol"—hell, "stregoica"—witch, "vrolok" and "vlkoslak"—both of which mean the same thing, one being Slovak and the other Servian for something that is either were-wolf or vampire." As mentioned in question 39, this sentence is an example of how the author hints at evil to come for the traveler. The other answer choices aren't related to the passage's grim foreshadowing.

41. A: Inferring is reading between the lines. Choice *B* describes the skimming technique. Choice *C* describes a questioning technique readers should employ, and Choice *D* is a simple statement regarding summary. It's an incomplete answer and not applicable to inference.

42. D: The correct answer is when an author chooses words or phrases that invoke feelings other than their literal meaning. Choice *A* refers to footnoting, which isn't applicable, and Choice *C* refers to a literary device. Choice *B* defines denotation, which is conceptually the opposite of connotation.

43. A: The purpose is to inform the reader about what assault is and how it is committed. Choice *B* is incorrect because the passage does not state that assault is a lesser form of lethal force, only that an assault can use lethal force, or alternatively, lethal force can be utilized to counter a dangerous assault. Choice *C* is incorrect because the passage is informative and does not have a set agenda. Finally, Choice *D* is incorrect because although the author uses an example in order to explain assault, it is not indicated that this is the author's personal account.

44. C: If the man being attacked in an alley by another man with a knife used self-defense by lethal force, it would not be considered illegal. The presence of a deadly weapon indicates mal-intent and because the individual is isolated in an alley, lethal force in self-defense may be the only way to preserve his life. Choices *A* and *B* can be ruled out because in these situations, no one is in danger of immediate death or bodily harm by someone else. Choice *D* is an assault and does exhibit intent to harm, but this situation isn't severe enough to merit lethal force; there is no intent to kill.

45. B: As discussed in the second passage, there are several forms of assault, like assault with a deadly weapon, verbal assault, or threatening posture or language. Choice *A* is incorrect because the author does mention what the charges are on assaults; therefore, we cannot assume that they are more or less than unnecessary use of force charges. Choice *C* is incorrect because anyone is capable of assault; the author does not state that one group of people cannot commit assault. Choice *D* is incorrect because assault is never justified. Self-defense resulting in lethal force can be justified.

46. D: The use of lethal force is not evaluated on the intent of the user, but rather on the severity of the primary attack that warranted self-defense. This statement most undermines the last part of the passage because it directly contradicts how the law evaluates the use of lethal force. Choices *A* and *B* are stated in the paragraph, so they do not undermine the explanation from the author. Choice *C* does not necessarily undermine the passage, but it does not support the passage either. It is more of an opinion that does not offer strength or weakness to the explanation.

47. C: An assault with deadly intent can lead to an individual using lethal force to preserve their well-being. Choice *C* is correct because it clearly establishes what both assault and lethal force are and gives the specific way in which the two concepts meet. Choice *A* is incorrect because lethal force doesn't necessarily result in assault. This is also why Choice *B* is incorrect. Not all assaults would necessarily be life-threatening to the point where lethal force is needed for self-defense. Choice *D* is compelling but ultimately too vague; the statement touches on aspects of the two ideas but fails to present the concrete way in which the two are connected to each other.

48. A: Both passages open by defining a legal concept and then continue to describe situations in order to further explain the concept. Choice *D* is incorrect because while the passages utilize examples to help explain the concepts discussed, the author doesn't indicate that they are specific court cases. It's also clear that the passages don't open with examples, but instead, they begin by defining the terms addressed in each passage. This eliminates Choice *B*, and ultimately reveals Choice *A* to be the correct answer. Choice *A* accurately outlines the way both passages are structured. Because the passages follow a nearly identical structure, the Choice *C* can easily be ruled out.

49. C: Intent is very important for determining both lethal force and assault; intent is examined in both parties and helps determine the severity of the issue. Choices *A* and *B* are incorrect because it is clear in both passages that intent is a prevailing theme in both lethal force and assault. Choice *D* is compelling, but if a person uses lethal force to defend himself or herself, the intent of the defender is also examined in order to help determine if there was excessive force used. Choice *C* is correct because it states that intent is important for determining both lethal force and assault, and that intent is used to gauge the severity of the issues. Remember, just as lethal force can escalate to excessive use of force, there are different kinds of assault. Intent dictates several different forms of assault.

50. B: The example is used to demonstrate a single example of two different types of assault, then adding in a third type of assault to the example's conclusion. The example mainly provides an instance of "threatening body language" and "provocative language" with the homeowner gesturing threats to his neighbor. It ends the example by adding a third type of assault: physical strikes. This example is used to show the variant nature of assaults. Choice *A* is incorrect because it doesn't mention the "physical strike" assault at the end and is not specific enough. Choice *C* is incorrect because the example does not say anything about the definition of lethal force or how it might be altered. Choice *D* is incorrect, as the example mentions nothing about cause and effect.

Language Arts: Writing

Organization of Ideas

Good writing is not merely a random collection of sentences. No matter how well written, sentences must relate and coordinate appropriately with one another. If not, the writing seems random, haphazard, and disorganized. Therefore, good writing must be organized, where each sentence fits a larger context and relates to the sentences around it.

Transition Words

The writer should act as a guide, showing the reader how all the sentences fit together. Consider this example concerning seat belts:

> Seat belts save more lives than any other automobile safety feature. Many studies show that airbags save lives as well. Not all cars have airbags. Many older cars don't. Air bags aren't entirely reliable. Studies show that in 15 percent of accidents, airbags don't deploy as designed. Seat belt malfunctions are extremely rare.

There's nothing wrong with any of these sentences individually, but together they're disjointed and difficult to follow. The best way for the writer to communicate information is through the use of transition words. Here are examples of transition words and phrases that tie sentences together, enabling a more natural flow:

- To show causality: as a result, therefore, and consequently
- To compare and contrast: *however, but*, and *on the other hand*
- To introduce examples: *for example, namely*, and *including*
- To show order of importance: *foremost, primarily, secondly*, and *lastly*

Note that this is not a complete list of transitions. There are many more that can be used; however, most fit into these or similar categories. The important point is that the words should clearly show the relationship between sentences, supporting information, and the main idea.

Here is an update to the previous example using transition words. These changes make it easier to read and bring clarity to the writer's points:

> Seat belts save more lives than any other automobile safety feature. Many studies show that airbags save lives as well; however, not all cars have airbags. For example, some older cars don't. Furthermore, air bags aren't entirely reliable. For example, studies show that in 15 percent of accidents, airbags don't deploy as designed, but, on the other hand, seat belt malfunctions are extremely rare.

Also, be prepared to analyze whether the writer is using the best transition word or phrase for the situation. Take this sentence for example: "As a result, seat belt malfunctions are extremely rare." This sentence doesn't make sense in the context above because the writer is trying to show the contrast between seat belts and airbags, not the causality.

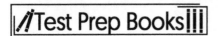

Relevance of Content

A reader must be able to evaluate the argument or point the author is trying to make and determine if it is adequately supported. The first step is to determine the main idea. The main idea is what the author wants to say about a specific topic. The next step is to locate the supporting details. An author uses supporting details to illustrate the main idea. These are the details that provide evidence or examples to help make a point. Supporting details often appear in the form of quotations, paraphrasing, or analysis. Test takers should then examine the text to make sure the author connects details and analysis to the main point. These steps are crucial to understanding the text and evaluating how well the author presents his or her argument and evidence. The following graphic demonstrates the connection between the main idea and the supporting details.

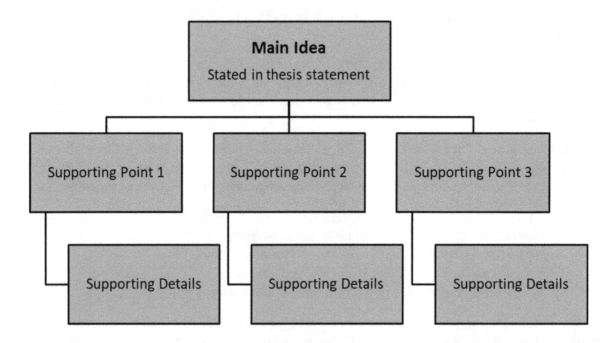

It is important to evaluate the author's supporting details to be sure that they are credible, provide evidence of the author's point, and directly support the main idea. Critical readers examine the facts used to support an author's argument and check those facts against other sources to be sure they are correct. They also check the validity of the sources used to be sure those sources are credible, academic, and/or peer-reviewed. A strong argument uses valid, measurable facts to support ideas.

Analyzing Organizational Structure

Depending on what the author is attempting to accomplish, certain formats or text structures work better than others. For example, a sequence structure might work for narration but not when identifying similarities and differences between dissimilar concepts. Similarly, a comparison-contrast structure is not useful for narration. It's the author's job to put the right information in the correct format.

Readers should be familiar with the five main literary structures:

1. *Sequence* structure (sometimes referred to as the order structure) is when the order of events proceed in a predictable order. In many cases, this means the text goes through the plot elements: exposition, rising action, climax, falling action, and resolution. Readers are introduced to characters, setting, and conflict in the exposition. In the rising action, there's an increase in tension and suspense. The climax is the height of tension and the point of no return. Tension decreases during the falling action. In the resolution, any conflicts presented in the exposition are solved, and the story concludes. An informative text that is structured sequentially will often go in order from one step to the next.

2. In the *problem-solution* structure, authors identify a potential problem and suggest a solution. This form of writing is usually divided into two paragraphs and can be found in informational texts. For example, cell phone, cable, and satellite providers use this structure in manuals to help customers troubleshoot or identify problems with services or products.

3. When authors want to discuss similarities and differences between separate concepts, they arrange thoughts in a *comparison-contrast* paragraph structure. Venn diagrams are an effective graphic organizer for comparison-contrast structures because they feature two overlapping circles that can be used to organize similarities and differences. A comparison-contrast essay organizes one paragraph based on similarities and another based on differences. A comparison-contrast essay can also be arranged with the similarities and differences of individual traits addressed within individual paragraphs. Words such as *however*, *but*, and *nevertheless* help signal a contrast in ideas.

4. *Descriptive* writing structure is designed to appeal to your senses. Much like an artist who constructs a painting, good descriptive writing builds an image in the reader's mind by appealing to the five senses: sight, hearing, taste, touch, and smell. However, overly descriptive writing can become tedious; whereas sparse descriptions can make settings and characters seem flat. Good authors strike a balance by applying descriptions only to passages, characters, and settings that are integral to the plot.

5. Passages that use the *cause and effect* structure are simply asking *why* by demonstrating some type of connection between ideas. Words such as *if*, *since*, *because*, *then*, or *consequently* indicate relationship. By switching the order of a complex sentence, the writer can rearrange the emphasis on different clauses. Saying *If Sheryl is late, we'll miss the dance* is different from saying *We'll miss the dance if Sheryl is late*. One emphasizes Sheryl's tardiness while the other emphasizes missing the dance. Paragraphs can also be arranged in a cause and effect format. Since the format—before and after—is sequential, it is useful when authors wish to discuss the impact of choices. Researchers often apply this paragraph structure to the scientific method.

Forming Paragraphs

A good *paragraph* should have the following characteristics:

- Be logical with organized sentences
- Have a *unified* purpose within itself
- Use sentences as *building blocks*
- Be a *distinct section* of a piece of writing
- Present a *single theme* introduced by a *topic sentence*
- Maintain a *consistent flow* through subsequent, relevant, well-placed sentences
- *Tell a story* of its own or have its own purpose, yet connect with what is written before and after
- Enlighten, entertain, and/or inform

Though certainly not set in stone, the length should be a consideration for the reader's sake, not merely for the sake of the topic. When paragraphs are especially short, the reader might experience an irregular, uneven effect; when they're much longer than 250 words, the reader's attention span, and probably their retention, is challenged. While a paragraph can technically be a sentence long, a good rule of thumb is for paragraphs to be at least three sentences long and no more than ten sentence long. An optimal word length is 100 to 250 words.

Recognizing Logical Transitions

Even if the writer includes plenty of information to support their point, the writing is only coherent when the information is in a logical order. First, the writer should introduce the main idea, whether for a paragraph, a section, or the entire piece. Second, they should present evidence to support the main idea by using transitional language. This shows the reader how the information relates to the main idea and to the sentences around it. The writer should then take time to interpret the information, making sure necessary connections are obvious to the reader. Finally, the writer can summarize the information in a closing section.

The logical order of a piece is supported by certain transitions that signal information is related or linked in a logical way. These transitions can include words like however, consequently, and likewise.

Though most writing follows this pattern, it isn't a set rule. Sometimes writers change the order for effect. For example, the writer can begin with a surprising piece of supporting information to grab the reader's attention, and then transition to the main idea. Thus, if a passage doesn't follow the logical order, don't immediately assume it's wrong. However, most writing usually settles into a logical sequence after a nontraditional beginning.

Language Facility

Sentence Fluency

Learning and utilizing the mechanics of structure will encourage effective, professional results, and adding some creativity will elevate one's writing to a higher level.

First, let's review the basic elements of sentences.

A *sentence* is a set of words that make up a grammatical unit. The words must have certain elements and be spoken or written in a specific order to constitute a complete sentence that makes sense.

> 1. A sentence must have a *subject* (a noun or noun phrase). The subject tells whom or what the sentence is addressing (i.e. what it is about).

> 2. A sentence must have an *action* or *state of being* (*a verb*). To reiterate: A verb forms the main part of the predicate of a sentence. This means that it explains what the noun is doing.

> 3. A sentence must convey a complete thought.

Sometimes a sentence has two ideas that work together. For example, say the writer wants to make the following points:

> Seat belt laws have saved an estimated 50,000 lives.

More lives are saved by seat belts every year.

These two ideas are directly related and appear to be of equal importance. Therefore they can be joined with a simple "and" as follows:

Seat belt laws have saved an estimated 50,000 lives, and more lives are saved by seat belts every year.

The word *and* in the sentence helps the two ideas work together or, in other words, it "coordinates" them. It also serves as a junction where the two ideas come together, better known as a *conjunction*. Therefore, the word *and* is known as a *coordinating conjunction* (a word that helps bring two equal ideas together). Now that the ideas are joined together by a conjunction, they are known as *clauses*. Other coordinating conjunctions include *or*, *but*, and *so*.

Sometimes, however, two ideas in a sentence are *not* of equal importance:

Seat belt laws have saved an estimated 50,000 lives.

Many more lives could be saved with stronger federal seat belt laws.

In this case, combining the two with a coordinating conjunction (*and*) creates an awkward sentence:

Seat belt laws have saved an estimated 50,000 lives, and many more lives could be saved with stronger federal seat belt laws.

Now the writer uses a word to show the reader which clause is the most important (or the "boss") of the sentence:

Although seat belt laws have saved an estimated 50,000 lives, many more lives could be saved with stronger federal seat belt laws.

In this example, the second clause is the key point that the writer wants to make, and the first clause works to set up that point. Since the first clause "works for" the second, it's called the *subordinate clause*. The word *although* tells the reader that this idea isn't as important as the clause that follows. This word is called the *subordinating conjunction*. Other subordinating conjunctions include *after*, *because*, *if*, *since*, *unless*, and many more. As mentioned before, it's easy to spot subordinate clauses because they don't stand on their own (as shown in this previous example):

Although seat belt laws have saved an estimated 50,000 lives.

This is not a complete thought. It needs the other clause (called the *independent clause*) to make sense. On the test, when asked to choose the best subordinating conjunction for a sentence, look at the surrounding text. Choose the word that best allows the sentence to support the writer's argument.

Conjunctions are vital words that connect words, phrases, thoughts, and ideas. Conjunctions show relationships between components. There are two types: coordinating and subordinating.

Coordinating conjunctions are the primary class of conjunctions placed between words, phrases, clauses, and sentences that are of equal grammatical rank; the coordinating conjunctions are *for*, *and*, *nor*, *but*, *or*, *yet*, and *so*. A useful memorization trick is to remember that all the first letters of these conjunctions collectively spell the word fanboys.

I need to go shopping, *but* I must be careful to leave enough money in the bank.
She wore a black, red, *and* white shirt.

Subordinating conjunctions are the secondary class of conjunctions. They connect two unequal parts, one main (or independent) and the other subordinate (or dependent). I must go to the store *even though* I do not have enough money in the bank.

Because I read the review, I do not want to go to the movie.

Notice that the presence of subordinating conjunctions makes clauses dependent. *I read the review* is an independent clause, but *because* makes the clause dependent. Thus, it needs an independent clause to complete the sentence.

Parallel Structure

Parallel structure usually has to do with lists. Look at the following sentence and spot the mistake:

Increased seat belt legislation has been supported by the automotive industry, the insurance industry, and doctors.

Many people don't see anything wrong, but the word *doctors* breaks the sentence's parallel structure. The previous items in the list refer to an industry as a singular noun, so every item in the list must follow that same format:

Increased seat belt legislation has been supported by the automotive industry, the insurance industry, and the healthcare industry.

Another common mistake in parallel structure might look like this:

Before the accident, Maria enjoyed swimming, running, and played soccer.

Here, the words "played soccer" break the parallel structure. To correct it, the writer must change the final item in the list to match the format of the previous two:

Before the accident, Maria enjoyed swimming, running, and playing soccer.

Types of Sentences

All sentences contain the same basic elements: a subject and a verb. The *subject* is who or what the sentence is about; the *verb* describes the subject's action or condition. However, these elements, subjects and verbs, can be combined in different ways. The following graphic describes the different types of sentence structures.

Sentence Structure	Independent Clauses	Dependent Clauses
Simple	1	0
Compound	2 or more	0
Complex	1	1 or more
Compound-Complex	2 or more	1 or more

A *simple sentence* expresses a complete thought and consists of one subject and verb combination:

The children ate pizza.

The subject is *children*. The verb is *ate*.

Either the subject or the verb may be *compound*—that is, it could have more than one element:

> *The children and their parents* ate pizza.

> The children *ate pizza and watched a movie.*

All of these are still simple sentences. Despite having either compound subjects or compound verbs, each sentence still has only one subject and verb combination.

Compound sentences combine two or more simple sentences to form one sentence that has multiple subject-verb combinations:

> *The children ate pizza,* and *their parents watched a movie.*

This structure is comprised of two independent clauses: (1) *the children ate pizza* and (2) *their parents watched a movie.* Compound sentences join different subject-verb combinations using a comma and a coordinating conjunction.

> I called my mom, *but* she didn't answer the phone.

> The weather was stormy, *so* we canceled our trip to the beach.

A *complex sentence* consists of an independent clause and one or more dependent clauses. Dependent clauses join a sentence using *subordinating conjunctions*. Some examples of subordinating conjunctions are *although*, *unless*, *as soon as*, *since*, *while*, *when*, *because*, *if*, and *before*.

> I missed class yesterday *because* my mother was ill.

> *Before* traveling to a new country, you need to exchange your money to the local currency.

The order of clauses determines their punctuation. If the dependent clause comes first, it should be separated from the independent clause with a comma. However, if the complex sentence consists of an independent clause followed by a dependent clause, then a comma is not always necessary.

A *compound-complex sentence* can be created by joining two or more independent clauses with at least one dependent clause:

> After the earthquake struck, thousands of homes were destroyed, and many families were left without a place to live.

The first independent clause in the compound structure includes a dependent clause—*after the earthquake struck*. Thus, the structure is both complex and compound.

Other possible components of a sentence include descriptive words (adjectives or adverbs) that provide additional information called phrases or dependent clauses to the sentence but are not themselves complete sentences; independent clauses add more information to the sentence but could stand alone as their own sentence.

There isn't an overabundance of absolutes in grammar, but here is one: every sentence in the English language falls into one of four categories.

- Declarative: a simple statement that ends with a period

  The price of milk per gallon is the same as the price of gasoline.

- Imperative: a command, instruction, or request that ends with a period

  Buy milk when you stop to fill up your car with gas.

- Interrogative: a question that ends with a question mark

  Will you buy the milk?

- Exclamatory: a statement or command that expresses emotions like anger, urgency, or surprise and ends with an exclamation mark

  Buy the milk now!

Idiomatic Usage

A figure of speech (sometimes called an idiom) is a rhetorical device. It's a phrase that is not intended to be taken literally.

When the writer uses a figure of speech, their intention must be clear if it's to be used effectively. Some phrases can be interpreted in a number of ways, causing confusion for the reader. Look for clues to the writer's true intention to determine the best replacement. Likewise, some figures of speech may seem out of place in a more formal piece of writing. To show this, here is another example involving seat belts:

> Seat belts save more lives than any other automobile safety feature. Many studies show that airbags save lives as well, however not all cars have airbags. For example, some older cars don't. In addition, air bags aren't entirely reliable. For example, studies show that in 15 percent of accidents, airbags don't deploy as designed, but, on the other hand, seat belt malfunctions happen once in a blue moon.

Most people know that "once in a blue moon" refers to something that rarely happens. However, because the rest of the paragraph is straightforward and direct, using this figurative phrase distracts the reader. In this example, the earlier version is much more effective.

Now it's important to take a moment and review the meaning of the word *literally*. This is because it's one of the most misunderstood and misused words in the English language. *Literally* means that something is exactly what it says it is, and there can be no interpretation or exaggeration. Unfortunately, *literally* is often used for emphasis as in the following example:

> This morning, I literally couldn't get out of bed.

This sentence meant to say that the person was extremely tired and wasn't able to get up. However, the sentence can't *literally* be true unless that person was tied down to the bed, paralyzed, or affected by a strange situation that the writer (most likely) didn't intend. Here's another example:

> I literally died laughing.

The writer tried to say that something was very funny. However, unless they're writing this from beyond the grave, it can't *literally* be true.

Note that this doesn't mean that writers can't use figures of speech. The colorful use of language and idioms make writing more interesting and draw in the reader. However, for these kinds of expressions to be used correctly, they cannot include the word *literally*.

Style, Tone, and Mood

Style, tone, and mood are often thought to be the same thing. Though they're closely related, there are important differences to keep in mind. The easiest way to do this is to remember that style "creates and affects" tone and mood. More specifically, style is how the writer uses words to create the desired tone and mood for their writing.

Style

Style can include any number of technical writing choices. A few examples of style choices include:

- Sentence Construction: When presenting facts, does the writer use shorter sentences to create a quicker sense of the supporting evidence, or do they use longer sentences to elaborate and explain the information?

- Technical Language: Does the writer use jargon to demonstrate their expertise in the subject, or do they use ordinary language to help the reader understand things in simple terms?

- Formal Language: Does the writer refrain from using contractions such as *won't* or *can't* to create a more formal tone, or do they use a colloquial, conversational style to connect to the reader?

- Formatting: Does the writer use a series of shorter paragraphs to help the reader follow a line of argument, or do they use longer paragraphs to examine an issue in great detail and demonstrate their knowledge of the topic?

On the test, examine the writer's style and how their writing choices affect the way the text comes across.

Tone

Tone refers to the writer's attitude toward the subject matter. Tone is usually explained in terms of a work of fiction. For example, the tone conveys how the writer feels about their characters and the situations in which they're involved. Nonfiction writing is sometimes thought to have no tone at all; however, this is incorrect.

A lot of nonfiction writing has a neutral tone, which is an important tone for the writer to take. A neutral tone demonstrates that the writer is presenting a topic impartially and letting the information speak for itself. On the other hand, nonfiction writing can be just as effective and appropriate if the tone isn't neutral. For example, let's look at the seat belt example again:

> Seat belts save more lives than any other automobile safety feature. Many studies show that airbags save lives as well; however, not all cars have airbags. For example, some older cars don't. Furthermore, air bags aren't entirely reliable. For example, studies show that in 15 percent of accidents airbags don't deploy as designed, but, on the other hand, seat belt

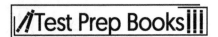

malfunctions are extremely rare. The number of highway fatalities has plummeted since laws requiring seat belt usage were enacted.

In this passage, the writer mostly chooses to retain a neutral tone when presenting information. If the writer would instead include their own personal experience of losing a friend or family member in a car accident, the tone would change dramatically. The tone would no longer be neutral and would show that the writer has a personal stake in the content, allowing them to interpret the information in a different way. When analyzing tone, consider what the writer is trying to achieve in the text and how they *create* the tone using style.

Mood

Mood refers to the feelings and atmosphere that the writer's words create for the reader. Like tone, many nonfiction texts can have a neutral mood. To return to the previous example, if the writer would choose to include information about a person they know being killed in a car accident, the text would suddenly carry an emotional component that is absent in the previous example. Depending on how they present the information, the writer can create a sad, angry, or even hopeful mood. When analyzing the mood, consider what the writer wants to accomplish and whether the best choice was made to achieve that end.

Analyzing Nuances in Words

Language is not as simple as one word directly correlated to one meaning. Rather, one word can express a vast array of diverse meanings, and similar meanings can be expressed through different words. However, there are very few words that express exactly the same meaning. For this reason, it is important to be able to pick up on the nuances of word meaning.

Many words contain two levels of meaning: connotation and denotation as discussed previously in the informational texts and rhetoric section. A word's *denotation* is its most literal meaning—the definition that can readily be found in the dictionary. A word's *connotation* includes all of its emotional and cultural associations.

In literary writing, authors rely heavily on connotative meaning to create mood and characterization. The following are two descriptions of a rainstorm:

A. The rain slammed against the windowpane, and the wind howled through the fireplace. A pair of hulking oaks next to the house cast eerie shadows as their branches trembled in the wind.

B. The rain pattered against the windowpane, and the wind whistled through the fireplace. A pair of stately oaks next to the house cast curious shadows as their branches swayed in the wind.

Description A paints a creepy picture for readers with strongly emotional words like *slammed*, connoting force and violence. *Howled* connotes pain or wildness, and *eerie* and *trembled* connote fear. Overall, the connotative language in this description serves to inspire fear and anxiety.

However, as can be seen in description B, swapping out a few key words for those with different connotations completely changes the feeling of the passage. *Slammed* is replaced with the more cheerful *pattered*, and *hulking* has been swapped out for *stately*. Both words imply something large, but *hulking* is more intimidating whereas *stately* is more respectable. *Curious* and *swayed* seem more playful than the language used in the earlier description. Although both descriptions represent roughly the

same situation, the nuances of the emotional language used throughout the passages create a very different sense for readers.

Selective choice of connotative language can also be extremely impactful in other forms of writing, such as editorials or persuasive texts. Through connotative language, writers reveal their biases and opinions while trying to inspire feelings and actions in readers:

A.   Parents won't stop complaining about standardized tests.
B.   Parents continue to raise concerns about standardized tests.

Readers should be able to identify the nuance in meaning between these two sentences. The first one carries a more negative feeling, implying that parents are being bothersome or whiny. Readers of the second sentence, though, might come away with the feeling that parents are concerned and involved in their children's education. Again, the aggregate of even subtle cues can combine to give a specific emotional impression to readers, so from an early age, students should be aware of how language can be used to influence readers' opinions.

Another form of non-literal expression can be found in *figures of speech*. As with connotative language, figures of speech tend to be shared within a cultural group and may be difficult to pick up on for learners outside of that group. In some cases, a figure of speech may be based on the literal denotation of the words it contains, but in other cases, a figure of speech is far removed from its literal meaning. A case in point is *irony*, where what is said is the exact opposite of what is meant:

> The new tax plan is poorly planned, based on faulty economic data, and unable to address the financial struggles of middle class families. Yet legislators remain committed to passing this brilliant proposal.

When the writer refers to the proposal as brilliant, the opposite is implied—the plan is "faulty" and "poorly planned." By using irony, the writer means that the proposal is anything but brilliant by using the word in a non-literal sense.

Another figure of speech is *hyperbole*—extreme exaggeration or overstatement. Statements like, "I love you to the moon and back" or "Let's be friends for a million years" utilize hyperbole to convey a greater depth of emotion, without literally committing oneself to space travel or a life of immortality.

Figures of speech may sometimes use one word in place of another. *Synecdoche*, for example, uses a part of something to refer to its whole. The expression "Don't hurt a hair on her head!" implies protecting more than just an individual hair, but rather her entire body. "The art teacher is training a class of Picassos" uses Picasso, one individual notable artist, to stand in for the entire category of talented artists. Another figure of speech using word replacement is *metonymy*, where a word is replaced with something closely associated to it. For example, news reports may use the word "Washington" to refer to the American government or "the crown" to refer to the British monarch.

Writing Conventions

Verbs, Modifiers, and Pronouns

Verbs

Within the human skeleton, the constant motion of the inner workings of organs may be likened to the verb of a sentence in the grammar skeleton. The verb is the part of speech that describes an action, state of being, or occurrence.

A verb forms the main part of a predicate of a sentence. This means that the verb explains what the noun (person, place, or thing) is doing. A simple example is "Time flies." The verb *flies* explains what the action of the noun, *time*, is doing. This example is a *main* verb.

Auxiliary (helping) verbs are forms of the words *have, do,* and *be,* as well as other auxiliary verbs called *modals*. Modals and semi-modals (modal phrases) express ability, possibility, permission, or obligation. Modals and semi-modal examples are *can/could/be able to, may/might, shall/should, must/have to,* and *will/would.* "I *should* go to the store."

Particles are minor function words like *not, in, out, up,* or *down* that become part of the verb itself. "I might *not*."

Participles are words formed from verbs that are used to modify a noun (or noun phrase), or verb (or verb phrase), or to make a compound verb form. "He is *speaking.*"

Verbs have five basic forms: the *base* form, the *-s* form, the *-ing* form, the *past* form, and the *past participle* form.

The *past* forms are either *regular* (*love/loved*; *hate/hated*) or *irregular* because they don't end by adding the common past tense suffix "-ed" (*go/went*; *fall/fell*; *set/set*). *To be* is an irregular verb with eight forms.

Pronouns

A word used in place of a noun is known as a *pronoun*. Pronouns are words like *I, mine, hers,* and *us*.

Pronouns can be split into different classifications (see below) which make them easier to learn; however, it's not important to memorize the classifications.

- Personal pronouns: refer to people

- First person: we, I, our, mine

- Second person: you, yours

- Third person: he, them

- Possessive pronouns: demonstrate ownership (mine, his, hers, its, ours, theirs, yours)

- Interrogative pronouns: ask questions (what, which, who, whom, whose)

- Relative pronouns: include the five interrogative pronouns and others that are relative (whoever, whomever, that, when, where)

- Demonstrative pronouns: replace something specific (this, that, those, these)

- Reciprocal pronouns: indicate something was done or given in return (each other, one another)

- Indefinite pronouns: have a nonspecific status (anybody, whoever, someone, everybody, somebody)

Indefinite pronouns such as *anybody, whoever, someone, everybody*, and *somebody* command a singular verb form, but others such as *all, none,* and *some* could require a singular or plural verb form.

An *antecedent* is the noun to which a pronoun refers; it needs to be written or spoken before the pronoun is used. For many pronouns, antecedents are imperative for clarity. In particular, many of the personal, possessive, and demonstrative pronouns need antecedents. Otherwise, it would be unclear who or what someone is referring to when they use a pronoun like *he* or *this*.

Pronoun reference means that the pronoun should refer clearly to one, clear, unmistakable noun (the antecedent).

Pronoun-antecedent agreement refers to the need for the antecedent and the corresponding pronoun to agree in gender, person, and number. Here are some examples:

The *kidneys* (plural antecedent) are part of the urinary system. *They* (plural pronoun) serve several roles.

The kidneys are part of the *urinary system* (singular antecedent). *It* (singular pronoun) is also known as the renal system.

The subjective pronouns —*I, you, he/she/it, we, they,* and *who*—are the subjects of the sentence.

Example: *They* have a new house.

The objective pronouns—*me, you* (*singular*)*, him/her, us, them,* and *whom*—are used when something is being done for or given to someone; they are objects of the action.

Example: The teacher has an apple for *us*.

The possessive pronouns—*mine, my, your, yours, his, hers, its, their, theirs, our,* and *ours*—are used to denote that something (or someone) belongs to someone (or something).

Example: It's *their* chocolate cake.

Even Better Example: It's *my* chocolate cake!

One of the greatest challenges and worst abuses of pronouns concerns *who* and *whom*. Just knowing the following rule can eliminate confusion. *Who* is a subjective-case pronoun used only as a subject or subject complement. *Whom* is only objective-case and, therefore, the object of the verb or preposition.

Who is going to the concert?

You are going to the concert with *whom*?

Hint: When using *who* or *whom*, think of whether someone would say *he* or *him*. If the answer is *he*, use *who*. If the answer is *him*, use *whom*. This trick is easy to remember because *he* and *who* both end in vowels, and *him* and *whom* both end in the letter *M*.

Modifier Placement

Modifiers are words or phrases (often adjectives or nouns) that add detail to, explain, or limit the meaning of other parts of a sentence. Look at the following example:

A big pine tree is in the yard.

In the sentence, the words *big* (an adjective) and *pine* (a noun) modify *tree* (the head noun).

All related parts of a sentence must be placed together correctly. *Misplaced* and *dangling modifiers* are common writing mistakes. In fact, they're so common that many people are accustomed to seeing them and can decipher an incorrect sentence without much difficulty.

Since *modifiers* refer to something else in the sentence (*big* and *pine* refer to *tree* in the example above), they need to be placed close to what they modify. If a modifier is so far away that the reader isn't sure what it's describing, it becomes a *misplaced modifier*. For example:

Seat belts almost saved 5,000 lives in 2009.

It's likely that the writer means that the total number of lives saved by seat belts in 2009 is close to 5,000. However, due to the misplaced modifier (*almost*), the sentence actually says there are 5,000 examples when seat belts *almost saved lives*. In this case, the position of the modifier is actually the difference between life and death (at least in the meaning of the sentence). A clearer way to write the sentence is:

Seat belts saved almost 5,000 lives in 2009.

Now that the modifier is close to the 5,000 lives it references, the sentence's meaning is clearer.

Another common example of a misplaced modifier occurs when the writer uses the modifier to begin a sentence. For example:

Having saved 5,000 lives in 2009, Senator Wilson praised the seat belt legislation.

It seems unlikely that Senator Wilson saved 5,000 lives on her own, but that's what the writer is saying in this sentence. To correct this error, the writer should move the modifier closer to the intended object it modifies. Here are two possible solutions:

Having saved 5,000 lives in 2009, the seat belt legislation was praised by Senator Wilson.

Senator Wilson praised the seat belt legislation, which saved 5,000 lives in 2009.

When choosing a solution for a misplaced modifier, look for an option that places the modifier close to the object or idea it describes.

A modifier must have a target word or phrase that it's modifying. Without this, it's a *dangling modifier*. Dangling modifiers are usually found at the beginning of sentences:

After passing the new law, there is sure to be an improvement in highway safety.

This sentence doesn't say anything about who is passing the law. Therefore, "After passing the new law" is a dangling modifier because it doesn't modify anything in the sentence. To correct this type of error, determine what the writer intended the modifier to point to:

After passing the new law, legislators are sure to see an improvement in highway safety.

"After passing the new law" now points to *legislators*, which makes the sentence clearer and eliminates the dangling modifier.

Maintaining Grammatical Agreement

In English writing, certain words connect to other words. People often learn these connections (or *agreements*) as young children and use the correct combinations without a second thought. However, the questions on the test dealing with agreement probably aren't simple ones.

Subject-Verb Agreement

Which of the following sentences is correct?

A large crowd of protesters was on hand.

A large crowd of protesters were on hand.

Many people would say the second sentence is correct, but they'd be wrong. However, they probably wouldn't be alone. Most people just look at two words: *protesters were*. Together they make sense. They sound right. The problem is that the verb *were* doesn't refer to the word *protesters*. Here, the word *protesters* is part of a prepositional phrase that clarifies the actual subject of the sentence (*crowd*). Take the phrase "of protesters" away and re-examine the sentences:

A large crowd was on hand.

A large crowd were on hand.

Without the prepositional phrase to separate the subject and verb, the answer is obvious. The first sentence is correct. On the test, look for confusing prepositional phrases when answering questions about subject-verb agreement. Take the phrase away, and then recheck the sentence.

Noun Agreement

Nouns that refer to other nouns must also match in number. Take the following example:

John and Emily both served as an intern for Senator Wilson.

Two people are involved in this sentence: John and Emily. Therefore, the word *intern* should be plural to match. Here is how the sentence should read:

John and Emily both served as interns for Senator Wilson.

Pronouns are used to replace nouns so sentences don't have a lot of unnecessary repetition. This repetition can make a sentence seem awkward as in the following example:

Seat belts are important because seat belts save lives, but seat belts can't do so unless seat belts are used.

Replacing some of the nouns (*seat belts*) with a pronoun (*they*) improves the flow of the sentence:

> Seat belts are important because they save lives, but they can't do so unless they are used.

A pronoun should agree in number (singular or plural) with the noun that precedes it. Another common writing error is the shift in *noun-pronoun agreement*. Here's an example:

> When people are getting in a car, he should always remember to buckle his seatbelt.

The first half of the sentence talks about a plural (*people*), while the second half refers to a singular person (*he* and *his*). These don't agree, so the sentence should be rewritten as:

> When people are getting in a car, they should always remember to buckle their seatbelt.

Fragments and Run-Ons

A *sentence fragment* is a failed attempt to create a complete sentence because it's missing a required noun or verb. Fragments don't function properly because there isn't enough information to understand the writer's intended meaning. For example:

> Seat belt use corresponds to a lower rate of hospital visits, reducing strain on an already overburdened healthcare system. Insurance claims as well.

Look at the last sentence: *Insurance claims as well*. What does this mean? This is a fragment because it has a noun but no verb, and it leaves the reader guessing what the writer means about insurance claims. Many readers can probably infer what the writer means, but this distracts them from the flow of the writer's argument. Choosing a suitable replacement for a sentence fragment may be one of the questions on the test. The fragment is probably related to the surrounding content, so look at the overall point the writer is trying to make and choose the answer that best fits that idea.

Remember that sometimes a fragment can *look* like a complete sentence or have all the nouns and verbs it needs to make sense. Consider the following two examples:

> Seat belt use corresponds to a lower rate of hospital visits.

> Although seat belt use corresponds to a lower rate of hospital visits.

Both examples above have nouns and verbs, but only the first sentence is correct. The second sentence is a fragment, even though it's actually longer. The key is the writer's use of the word *although*. Starting a sentence with *although* turns that part into a *subordinate clause* (more on that next). Keep in mind that one doesn't have to remember that it's called a subordinate clause on the test. Just be able to recognize that the words form an incomplete thought and identify the problem as a sentence fragment.

A *run-on sentence* is, in some ways, the opposite of a fragment. It contains two or more sentences that have been improperly forced together into one. An example of a run-on sentence looks something like this:

> Seat belt use corresponds to a lower rate of hospital visits it also leads to fewer insurance claims.

Here, there are two separate ideas in one sentence. It's difficult for the reader to follow the writer's thinking because there is no transition from one idea to the next. On the test, choose the best way to

correct the run-on sentence.

Here are two possibilities for the sentence above:

> Seat belt use corresponds to a lower rate of hospital visits. It also leads to fewer insurance claims.

> Seat belt use corresponds to a lower rate of hospital visits, but it also leads to fewer insurance claims.

Both solutions are grammatically correct, so which one is the best choice? That depends on the point that the writer is trying to make. Always read the surrounding text to determine what the writer wants to demonstrate, and choose the option that best supports that thought.

Another type of run-on occurs when writers use inappropriate punctuation:

> This winter has been very cold, some farmers have suffered damage to their crops.

Though a comma has been added, this sentence is still not correct. When a comma alone is used to join two independent clauses, it is known as a *comma splice*. Without an appropriate conjunction, a comma cannot join two independent clauses by itself.

Capitalization, Punctuation, and Spelling

Capitalization

- Capitalize the first word in a sentence and the first word in a quotation:

    > The realtor showed them the house.

    > Robert asked, "When can we get together for dinner again?"

- Capitalize proper nouns and words derived from them:

    > We are visiting Germany in a few weeks.

    > We will stay with our German relatives on our trip.

- Capitalize days of the week, months of the year, and holidays:

    > The book club meets the last Thursday of every month.

    > The baby is due in June.

    > I decided to throw a Halloween party this year.

- Capitalize the main words in titles (referred to as *title case*), but not the articles, conjunctions, or prepositions:

    > *A Raisin in the Sun*

    > *To Kill a Mockingbird*

- Capitalize directional words that are used as names, but not when referencing a direction:

The North won the Civil War.

After making a left, go north on Rt. 476.

She grew up on the West Coast.

The winds came in from the west.

- Capitalize titles that go with names:

    Mrs. McFadden   Sir Alec Guinness   Lt. Madeline Suarez

- Capitalize familial relationships when referring to a *specific* person:

    I worked for my Uncle Steven last summer.

    Did you work for your uncle last summer?

Punctuation

Periods (.) are used to end a sentence that is a statement (*declarative*) or a command (*imperative*). They should not be used in a sentence that asks a question or is an exclamation. Periods are also used in abbreviations, which are shortened versions of words.

- Declarative: The boys refused to go to sleep.
- Imperative: Walk down to the bus stop.
- Abbreviations: Joan Roberts, M.D., Apple Inc., Mrs. Adamson
- If a sentence ends with an abbreviation, it is inappropriate to use two periods. It should end with a single period after the abbreviation.

    The chef gathered the ingredients for the pie, which included apples, flour, sugar, etc.

Question marks (?) are used with direct questions (*interrogative*). An *indirect question* can use a period:

    Interrogative: When does the next bus arrive?

    Indirect Question: I wonder when the next bus arrives.

An *exclamation point (!)* is used to show strong emotion or can be used as an *interjection*. This punctuation should be used sparingly in formal writing situations.

    What an amazing shot!

    Whoa!

In a sentence, *colons* are used before a list, a summary or elaboration, or an explanation related to the preceding information in the sentence:

    There are two ways to reserve tickets for the performance: by phone or in person.

    One thing is clear: students are spending more on tuition than ever before.

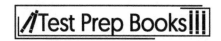

As these examples show, a colon must be preceded by an independent clause. However, the information after the colon may be in the form of an independent clause or in the form of a list.

Semicolons can be used in two different ways—to join ideas or to separate them. In some cases, semicolons can be used to connect what would otherwise be stand-alone sentences. Each part of the sentence joined by a semicolon must be an independent clause. The use of a semicolon indicates that these two independent clauses are closely related to each other:

> The rising cost of childcare is one major stressor for parents; healthcare expenses are another source of anxiety.

> Classes have been canceled due to the snowstorm; check the school website for updates.

Semicolons can also be used to divide elements of a sentence in a more distinct way than simply using a comma. This usage is particularly useful when the items in a list are especially long and complex and contain other internal punctuation.

> Retirees have many modes of income: some survive solely off their retirement checks; others supplement their income through part time jobs, like working in a supermarket or substitute teaching; and others are financially dependent on the support of family members, friends, and spouses.

Dashes are used to set apart groups of words. The em dash (—) is useful because it can separate phrases that would otherwise be in parenthesis, or it can stand in for a colon. The en dash (–) is typically used either to represent a time span or range.

Quotation marks are found wherever there is a direct quote or phrase that needs to be set apart from the other text. They can also be used to indicate the title of a short work.

An *ellipsis* is comprised of three periods which is used to denote something that is missing or has been removed.

Spelling

Both spoken and written words have rhythm that might be defined as *inflection*. This serves to help writers in their choice of words, expression, and correct spelling. When creating original works, do at least one reading aloud. Some inflection is intrinsic to the words, some are added by writers, and some will be inferred when later read. If the written words are not spelled correctly, then what the author intended is not conveyed. Use rhythm as a spelling tool.

Saying and listening to a word serves as the beginning of knowing how to spell it. Keep these subsequent guidelines in mind, remembering there are often exceptions, because the English language is replete with them.

Guideline #1: Syllables must have a vowel

Every syllable in every English word has a vowel. Examples: dog, *haystack*, *answering*, *abstentious* (the longest word that uses the five vowels in order), and *simple*.

In addition to this vowel guideline is a built-in bonus: Guideline #1 helps one see whether the word looks right.

Guideline #2: The silent final -e

The final word example in Guideline #1, *simple*, provides the opportunity to see another guideline with multiple types:

- Because every syllable has a vowel, words like *simple* require the final silent -e.

- In a word that has a vowel-consonant-e combination like the short, simple word at*e*, the silent –e at the end shapes the sound of the earlier vowel. The technical term for this is it "makes the vowel say its name." There are thousands of examples of this guideline; just for starters, look at cut*e*, mat*e*, and tot*e*.

- Let's *dance*...after we leave the *range*! Look what the final silent –e does for the –c and –g: each provides the word's soft sound.

- Other than to *rev* a car's engine, are there other words that ends in a –v? How about a word that ends in a –u? Well some like their cheese ble*u*, there's one, but, while there are more (well, okay, *you*), they are few and far between, and consider words having the ending of the letter –i. Yes, English words generally do not end in –v's, –u's, and –i's, so silent –e to the rescue! Note that it does not change the pronunciation. Examples: believ*e*, lov*e*, and activ*e*; blu*e*, and tru*e*; and two very important –i examples, browni*e* and cooki*e*. (Exceptions to this rule are generally words from other languages.)

Guideline #3: The long and short of it

When the vowel has a short vowel sound as in *mad* or *bed*, only the single vowel is needed. If the word has a long vowel sound, add another vowel, either alongside it or separated by a consonant: bed/*bead*; mad/*made*. When the second vowel is separated by two spaces—*madder*—it does not affect the first vowel's sound.

Guideline #4: What about the –fixes (pre- and suf-)?

A *prefix* is a word, letter, or number that is placed before another. It adjusts or qualifies the root word's meaning. When written alone, prefixes are followed by a dash to indicate that the root word follows. Some of the most common prefixes are the following:

Prefix	Meaning	Example
dis-	not or opposite of	disabled
in-, im-, il-, ir-	not	illiterate
re-	again	return
un-	not	unpredictable
anti-	against	antibacterial
fore-	before	forefront
mis-	wrongly	misunderstand
non-	not	nonsense
over-	more than normal	overabundance
pre-	before	preheat
super-	above	superman

A suffix is a letter or group of letters added at the end of a word to form another word. The word created from the root and suffix is either a different tense of the same root (*help + ed = helped*) or a new word (*help + ful = helpful*). When written alone, suffixes are preceded by a dash to indicate that the root word comes before.

Some of the most common suffixes are the following:

Suffix	Meaning	Example
ed	makes a verb past tense	wash*ed*
ing	makes a verb a present participle verb	wash*ing*
ly	to make characteristic of	love*ly*
s/es	to make more than one	chair*s*, box*es*
able	can be done	deplor*able*
al	having characteristics of	comic*al*
est	comparative	great*est*
ful	full of	wonder*ful*
ism	belief in	commun*ism*
less	without	faith*less*
ment	action or process	accomplish*ment*
ness	state of	happi*ness*
ize, ise	to render, to make	steril*ize*, advert*ise*
cede/ceed/sede	go	con*cede*, pro*ceed*, super*sede*

Here are some helpful tips:

- When adding a suffix that starts with a vowel (for example, -*ed*) to a one-syllable root whose vowel has a short sound and ends in a consonant (for example, *stun*), double the final consonant of the root (*n*).

   stun + ed = stun*n*ed

   Exception: If the past tense verb ends in *x* such as *box*, do not double the *x*.

   box + ed = boxed

- If adding a suffix that starts with a vowel (-*er*) to a multi-syllable word ending in a consonant (*begin*), double the consonant (*n*).

   begin + er = begin*n*er

- If a short vowel is followed by two or more consonants in a word such as *i+t+c+h = itch*, do <u>not</u> double the last consonant.

   itch + ed = itched

- If adding a suffix that starts with a vowel (-*ing*) to a word ending in *e* (for example, *name*), that word's final *e* is generally (but not always) dropped.

  name + ing = naming
  exception: manage + able = manageable

- If adding a suffix that starts with a consonant (-*ness*) to a word ending in *e* (*complete*), the *e* generally (but not always) remains.

  complete + ness = completeness
  exception: judge + ment = judgment

- There is great diversity on handling words that end in *y*. For words ending in a vowel + y, nothing changes in the original word.

  play + ed = played

- For words ending in a consonant + y, change the *y* to *i* when adding any suffix except for –*ing*.

  marry + ed = married
  marry + ing = marrying

Guideline #5: Which came first, the –i or the –e?

"When the letter 'c' you spy, put the 'e' before the 'i.' (Do not be) dec*ei*ved; when the letter 's' you see, put the 'i' before the 'e' (or you might be under) s*ie*ge." This old adage still holds up today regarding words where the "c" and "s" *precede* the "i." Another variation is, "'*i*' before '*e*' except after '*c*' or when sounded as '*a*' as in *neighbor* or *weigh*." Keep in mind that these are only guidelines and that there are always exceptions to every rule.

Guideline #6: Vowels in the right order

A different helpful ditty is, "When two vowels go walking, the first one does the talking." Usually, when two vowels are in a row, the first one often has a long vowel sound and the other is silent. An example is *team*.

When having difficulty spelling words, determine a strategy to help. Work on pronunciations, play word games like Scrabble or Words with Friends, and consider using phonics (sounding words out by slowly and surely stating each syllable). Try using repetition and memorization and picturing the words. Try memory aids like making up silly things. See what works best. For disorders such as dyslexia, know that there are accommodations to help.

Use computer spellcheck; however, do not *rely on* computer spellcheck.

Using Reference Sources

Reference materials are indispensable tools for beginners and experts alike. Becoming a competent English communicator doesn't necessarily mean memorizing every single rule about spelling, grammar, or punctuation—it means knowing where and how to find accurate information about the rules of English usage. Students of English have a wide variety of references materials available to them, and, in an increasingly digitized world, more and more of these materials can be found online or as easily-

accessible phone applications. Educators should introduce students to different types of reference materials as well as when and how to use them.

Dictionary

Dictionaries are readily available in print, digital formats, and as mobile apps. A dictionary offers a wealth of information to users. First, in the absence of spell checking software, a *dictionary* can be used to identify correct spelling and to determine the word's pronunciation—often written using the International Phonetic Alphabet (IPA). Perhaps the best-known feature of a dictionary is its explanation of a word's meanings, as a single word can have multiple definitions. A dictionary organizes these definitions based on their parts of speech and then arranges them from most to least commonly used meanings or from oldest to most modern usage. Many dictionaries also offer information about a word's etymology and usage. With all these functions, then, a dictionary is a basic, essential tool in many situations. Students can turn to a dictionary when they encounter an unfamiliar word or when they see a familiar word used in a new way.

There are many dictionaries to choose from, but perhaps the most highly respected source is the *Oxford English Dictionary* (OED). The OED is a historical dictionary, and as such, all entries include quotes of the word as it has been used throughout history. Users of the OED can get a deeper sense of a word's evolution over time and in different parts of the world. Another standard dictionary in America is *Merriam-Webster*.

Thesaurus

Whereas a dictionary entry lists a word's definitions, a *thesaurus* entry lists a word's *synonyms* and *antonyms*—i.e., words with similar and opposite meanings, respectively. A dictionary can be used to find out what a word means and where it came from, and a thesaurus can be used to understand a word's relationship to other words. A thesaurus can be a powerful vocabulary-building tool. By becoming familiar with synonyms and antonyms, students will be more equipped to use a broad range of vocabulary in their speech and writing. Of course, one thing to be aware of when using a thesaurus is that most words do not have exact synonyms. Rather, there are slight nuances of meaning that can make one word more appropriate than another in a given context. In this case, it is often to the user's advantage to consult a thesaurus side-by-side with a dictionary to confirm any differences in usage between two synonyms. Some digital sources, such as *Dictionary.com*, integrate a dictionary and a thesaurus.

Generally, though, a thesaurus is a useful tool to help writers add variety and precision to their word choice. Consulting a thesaurus can help students elevate their writing to an appropriate academic level by replacing vague or overused words with more expressive or academic ones. Also, word processors often offer a built-in thesaurus, making it easy for writers to look up synonyms and vary word choice as they work.

Glossary

A *glossary* is similar to a dictionary in that it offers an explanation of terms. However, while a dictionary attempts to cover every word in a language, a glossary only focuses on those terms relevant to a specific field. Also, a glossary entry is more likely to offer a longer explanation of a term and its relevance within that field. Glossaries are often found at the back of textbooks or other nonfiction publications in order to explain new or unfamiliar terms to readers. A glossary may also be an entire book on its own that covers all of the essential terms and concepts within a particular profession, field, or other specialized area of knowledge. For learners seeking general definitions of terms from any context, then, a dictionary is an appropriate reference source, but for students of specialized fields, a glossary will usually provide more in-depth information.

Style Manual

Many rules of English usage are standard, but other rules may be more subjective. An example can be seen in the following structures:

A. I went to the store and bought eggs, milk, and bread.
B. I went to the store and bought eggs, milk and bread.

The final comma in a list before *and* or *or* is known as an Oxford comma or serial comma. It is, recommended in some styles, but not in others. To determine the appropriate use of the Oxford comma, writers can consult a style manual.

A *style manual* is a comprehensive collection of guidelines for language use and document formatting. Some fields refer to a common style guide—e.g., the Associated Press or *AP Stylebook*, a standard in American journalism. Individual organizations may rely on their own house style. Regardless, the purpose of a style manual is to ensure uniformity across all documents. Style manuals explain things such as how to format titles, when to write out numbers or use numerals, and how to cite sources. Because there are many different style guides, students should know how and when to consult an appropriate guide. The Chicago Manual of Style is common in the publication of books and academic journals. The Modern Language Association style (MLA) is another commonly used academic style format, while the American Psychological Association style (APA) may be used for scientific publications. Familiarity with using a style guide is particularly important for students who are college bound or pursuing careers in academic or professional writing.

In the examples above, the Oxford comma is recommended by the Chicago Manual of Style, so sentence A would be correct if the writer is using this style. But the comma is not recommended by the *AP Stylebook*, so sentence B would be correct if the writer is using the AP style.

General Grammar and Style References

Any language arts textbook should offer general grammatical and stylistic advice to students, but there are a few well-respected texts that can also be used for reference. *Elements of Style* by William Strunk is regularly assigned to students as a guide on effective written communication, including how to avoid common usage mistakes and how to make the most of parallel structure. *Garner's Modern American Usage* by Bryan Garner is another text that guides students on how to achieve precision and understandability in their writing. Whereas other reference sources discussed above tend to address specific language concerns, these types of texts offer a more holistic approach to cultivating effective language skills.

Electronic Resources

With print texts, it is easy to identify the authors and their credentials, as well as the publisher and their reputation. With electronic resources like websites, though, it can be trickier to assess the reliability of information. Students should be alert when gathering information from the Internet. Understanding the significance of website *domains*—which include identification strings of a site—can help. Website domains ending in *.edu* are educational sites and tend to offer more reliable research in their field. A *.org* ending tends to be used by nonprofit organizations and other community groups, *.com* indicates a privately-owned website, and a *.gov* site is run by the government. Websites affiliated with official

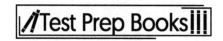

organizations, research groups, or institutes of learning are more likely to offer relevant, fact-checked, and reliable information.

Practice Questions

Read the selection about traveling in an RV and answer Questions 1 – 7.

I have to admit that when my father bought a recreational vehicle (RV), I thought he was making a huge mistake. I didn't really know anything about RVs, but I knew that my dad was as big a "city slicker" as there was. (1) <u>In fact, I even thought he might have gone a little bit crazy.</u> On trips to the beach, he preferred to swim at the pool, and whenever he went hiking, he avoided touching any plants for fear that they might be poison ivy. Why would this man, with an almost irrational fear of the outdoors, want a 40-foot camping behemoth?

(2) <u>The RV</u> was a great purchase for our family and brought us all closer together. Every morning (3) <u>we would wake up, eat breakfast, and broke camp.</u> We laughed at our own comical attempts to back The Beast into spaces that seemed impossibly small. (4) <u>We rejoiced as "hackers."</u> When things inevitably went wrong and we couldn't solve the problems on our own, we discovered the incredible helpfulness and friendliness of the RV community. (5) <u>We even made some new friends in the process.</u>

(6) <u>Above all, it allowed us to share adventures. While traveling across America,</u> which we could not have experienced in cars and hotels. Enjoying a campfire on a chilly summer evening with the mountains of Glacier National Park in the background, or waking up early in the morning to see the sun rising over the distant spires of Arches National Park are memories that will always stay with me and our entire family. (7) <u>Those are also memories that my siblings and me</u> have now shared with our own children.

1. Which of the following would be the best choice for this sentence (reproduced below)?

   In fact, I even thought he might have gone a little bit crazy.

   a. (No change; best as written.)
   b. Move the sentence so that it comes before the preceding sentence.
   c. Move the sentence to the end of the first paragraph.
   d. Omit the sentence.

2. In context, which is the best version of the underlined portion of this sentence (reproduced below)?

   The RV was a great purchase for our family and brought us all closer together.

   a. (No change)
   b. Not surprisingly, the RV
   c. Furthermore, the RV
   d. As it turns out, the RV

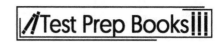

3. Which is the best version of the underlined portion of this sentence (reproduced below)?

Every morning we would wake up, eat breakfast, and broke camp.

a. (No change)
b. we would wake up, eat breakfast, and break camp.
c. would we wake up, eat breakfast, and break camp?
d. we are waking up, eating breakfast, and breaking camp.

4. Which is the best version of the underlined portion of this sentence (reproduced below)?

We rejoiced as "hackers."

a. (No change)
b. To a nagging problem of technology, we rejoiced as "hackers."
c. We rejoiced when we figured out how to "hack" a solution to a nagging technological problem.
d. To "hack" our way to a solution, we had to rejoice.

5. Which is the best version of the underlined portion of this sentence (reproduced below)?

We even made some new friends in the process.

a. (No change)
b. In the process was the friends we were making.
c. We are even making some new friends in the process.
d. We will make new friends in the process.

6. Which is the best version of the underlined portion of this sentence (reproduced below)?

Above all, it allowed us to share adventures. While traveling across America, which we could not have experienced in cars and hotels.

a. (No change)
b. Above all, it allowed us to share adventures while traveling across America
c. Above all, it allowed us to share adventures; while traveling across America
d. Above all, it allowed us to share adventures—while traveling across America

7. Which is the best version of the underlined portion of this sentence (reproduced below)?

Those are also memories that my siblings and me have now shared with our own children.

a. (No change)
b. Those are also memories that me and my siblings
c. Those are also memories that my siblings and I
d. Those are also memories that I and my siblings

Read the following section about Fred Hampton and answer Questions 8 – 20.

Fred Hampton desired to see lasting social change for African American people through nonviolent means and community recognition. (8) <u>In the meantime,</u> he became an African American activist during the American Civil Rights Movement and led the Chicago chapter of the Black Panther Party.

Hampton's Education

Hampton was born and raised (9) <u>in Maywood of Chicago, Illinois in 1948.</u> Gifted academically and a natural athlete, he became a stellar baseball player in high school. (10) <u>After graduating from Proviso East High School in 1966, he later went on to study law at Triton Junior College. While studying at Triton, Hampton joined and became a leader of the National Association for the Advancement of Colored People (NAACP). As a result of his leadership, the NAACP gained more than 500 members.</u> Hampton worked relentlessly to acquire recreational facilities in the neighborhood and improve the educational resources provided to the impoverished black community of Maywood.

The Black Panthers

The Black Panther Party (BPP) (11) <u>was another that</u> formed around the same time as and was similar in function to the NAACP. Hampton was quickly attracted to the (12) <u>Black Panther Party's approach</u> to the fight for equal rights for African Americans. Hampton eventually joined the chapter and relocated to downtown Chicago to be closer to its headquarters.

His charismatic personality, organizational abilities, sheer determination, and rhetorical skills (13) <u>enable him to quickly rise</u> through the chapter's ranks. Hampton soon became the leader of the Chicago chapter of the BPP where he organized rallies, taught political education classes, and established a free medical clinic. (14) <u>He also took part in the community police supervision project. He played an instrumental role</u> in the BPP breakfast program for impoverished African American children.

Hampton's (15) <u>greatest</u> <u>acheivement</u> as the <u>leader</u> of the BPP may be his fight against street gang violence in Chicago. In 1969, (16) <u>Hampton was held by a press conference</u> where he made the gangs agree to a nonaggression pact known as the Rainbow Coalition. As a result of the pact, a multiracial alliance between blacks, Puerto Ricans, and poor youth was developed.

Assassination

(17) <u>As the Black Panther Party's popularity and influence grew, the Federal Bureau of Investigation (FBI) placed the group under constant surveillance.</u> In an attempt to neutralize the party, the FBI launched several harassment campaigns against the BPP, raided its headquarters in Chicago three times, and arrested over one hundred of the group's members. Hampton was shot during such a raid that occurred on the morning of December 4th, 1969.

(18) <u>In 1976; seven years after the event,</u> it was revealed that William O'Neal, Hampton's trusted bodyguard, was an undercover FBI agent. (19) <u>O'Neal will provide</u> the FBI with detailed floor plans of the BPP's headquarters, identifying the exact location of Hampton's bed. It was because of these floor plans that the police were able to target and kill Hampton.

The assassination of Hampton fueled outrage amongst the African American community. It was not until years after the assassination that the police admitted wrongdoing. (20) <u>The Chicago City Council now are commemorating December 4th as Fred Hampton Day.</u>

8. In context, which is the best version of the underlined portion of this sentence (reproduced below)?

<u>In the meantime,</u> he became an African American activist during the American Civil Rights Movement and led the Chicago chapter of the Black Panther Party.

    a. (No change)
    b. Unfortunately,
    c. Finally,
    d. As a result,

9. Which is the best version of the underlined portion of this sentence (reproduced below)?

Hampton was born and raised <u>in Maywood of Chicago, Illinois in 1948.</u>

    a. (No change)
    b. in Maywood, of Chicago, Illinois in 1948.
    c. in Maywood of Chicago, Illinois, in 1948.
    d. in Chicago, Illinois of Maywood in 1948.

10. Which of the following sentences, if any, should begin a new paragraph?

<u>After graduating from Proviso East High School in 1966, he later went on to study law at Triton Junior College. While studying at Triton, Hampton joined and became a leader of the National Association for the Advancement of Colored People (NAACP). As a result of his leadership, the NAACP gained more than 500 members.</u>

    a. (No change; best as written.)
    b. After graduating from Proviso East High School in 1966, he later went on to study law at Triton Junior College.
    c. While studying at Triton, Hampton joined and became a leader of the National Association for the Advancement of Colored People (NAACP).
    d. As a result of his leadership, the NAACP gained more than 500 members.

11. Which of the following facts would be the most relevant to include here?

The Black Panther Party (BPP) <u>was another that</u> formed around the same time as and was similar in function to the NAACP.

    a. (No change; best as written.)
    b. was another activist group that
    c. had a lot of members that
    d. was another school that

12. Which is the best version of the underlined portion of this sentence (reproduced below)?

Hampton was quickly attracted to the <u>Black Panther Party's approach</u> to the fight for equal rights for African Americans.

  a. (No change)
  b. Black Panther Parties approach
  c. Black Panther Partys' approach
  d. Black Panther Parties' approach

13. Which is the best version of the underlined portion of this sentence (reproduced below)?

His charismatic personality, organizational abilities, sheer determination, and rhetorical skills <u>enable him to quickly rise</u> through the chapter's ranks.

  a. (No change)
  b. are enabling him to quickly rise
  c. enabled him to quickly rise
  d. will enable him to quickly rise

14. Which is the best version of the underlined portion of this sentence (reproduced below)?

<u>He also took part in the community police supervision project. He played an instrumental role</u> in the BPP breakfast program for impoverished African American children.

  a. (No change)
  b. He also took part in the community police supervision project but played an instrumental role
  c. He also took part in the community police supervision project, he played an instrumental role
  d. He also took part in the community police supervision project and played an instrumental role

15. Which of these, if any, is misspelled?

Hampton's (15) <u>greatest</u> <u>acheivement</u> as the <u>leader</u> of the BPP may be his fight against street gang violence in Chicago.

  a. (No change; best as written.)
  b. greatest
  c. acheivement
  d. leader

16. Which is the best version of the underlined portion of this sentence (reproduced below)?

In 1969, <u>Hampton was held by a press conference</u> where he made the gangs agree to a nonaggression pact known as the Rainbow Coalition.

  a. (No change)
  b. Hampton held a press conference
  c. Hampton, holding a press conference
  d. Hampton to hold a press conference

17. Which is the best version of the underlined portion of this sentence (reproduced below)?

As the Black Panther Party's popularity and influence grew, the Federal Bureau of Investigation (FBI) placed the group under constant surveillance.

a. (No change)
b. The Federal Bureau of Investigation (FBI) placed the group under constant surveillance as the Black Panther Party's popularity and influence grew.
c. Placing the group under constant surveillance, the Black Panther Party's popularity and influence grew.
d. As their influence and popularity grew, the FBI placed the group under constant surveillance.

18. Which is the best version of the underlined portion of this sentence (reproduced below)?

In 1976; seven years after the event, it was revealed that William O'Neal, Hampton's trusted bodyguard, was an undercover FBI agent.

a. (No change)
b. In 1976, seven years after the event,
c. In 1976 seven years after the event,
d. In 1976. Seven years after the event,

19. Which is the best version of the underlined portion of this sentence (reproduced below)?

O'Neal will provide the FBI with detailed floor plans of the BPP's headquarters, identifying the exact location of Hampton's bed.

a. (No change)
b. O'Neal provides
c. O'Neal provided
d. O'Neal, providing

20. Which is the best version of the underlined portion of this sentence (reproduced below)?

The Chicago City Council now are commemorating December 4ᵗʰ as Fred Hampton Day.

a. (No change)
b. Fred Hampton Day by the Chicago City Council, December 4, is now commemorated.
c. Now commemorated December 4ᵗʰ is Fred Hampton Day.
d. The Chicago City Council now commemorates December 4ᵗʰ as Fred Hampton Day.

Read the essay entitled "Education is Essential to Civilization" and answer Questions 21 – 35.

Early in my career, (21) a master's teacher shared this thought with me "Education is the last bastion of civility." While I did not completely understand the scope of those words at the time, I have since come to realize the depth, breadth, truth, and significance of what he said. (22) Education provides society with a vehicle for (23) raising it's children to be civil, decent, human beings with something valuable to contribute to the world. It is really what makes us human and what (24) distinguishes us as civilised creatures.

Being "civilized" humans means being "whole" humans. Education must address the mind, body, and soul of students. (25) It would be detrimental to society, only meeting the needs of the mind, if our schools were myopic in their focus. As humans, we are multi-dimensional, multi-faceted beings who need more than head knowledge to survive. (26) The human heart and psyche have to be fed in order for the mind to develop properly, and the body must be maintained and exercised to help fuel the working of the brain. Education is a basic human right, and it allows us to sustain a democratic society in which participation is fundamental to its success. It should inspire students to seek better solutions to world problems and to dream of a more equitable society. Education should never discriminate on any basis, and it should create individuals who are self-sufficient, patriotic, and tolerant of (27) others' ideas.

(28) All children can learn. Although not all children learn in the same manner. All children learn best, however, when their basic physical needs are met and they feel safe, secure, and loved. Students are much more responsive to a teacher who values them and shows them respect as individual people. Teachers must model at all times the way they expect students to treat them and their peers. If teachers set high expectations for (29) there students, the students will rise to that high level. Teachers must make the well-being of students their primary focus and must not be afraid to let students learn from their own mistakes.

In the modern age of technology, a teacher's focus is no longer the "what" of the content, (30) but more importantly, the 'why.' Students are bombarded with information and have access to ANY information they need right at their fingertips. Teachers have to work harder than ever before to help students identify salient information (31) so to think critically about the information they encounter. Students have to (32) read between the lines, identify bias, and determine who they can trust in the milieu of ads, data, and texts presented to them.

Schools must work in consort with families in this important mission. While children spend most of their time in school, they are dramatically and indelibly shaped (33) with the influences of their family and culture. Teachers must not only respect this fact, (34) but must strive to include parents in the education of their children and must work to keep parents informed of progress and problems. Communication between classroom and home is essential for a child's success.

Humans have always aspired to be more, do more, and to better ourselves and our communities. This is where education lies, right at the heart of humanity's desire to be all that we can be. Education helps us strive for higher goals and better treatment of ourselves and others. I shudder to think what would become of us if education ceased to be the "last bastion of civility." (35) We must be unapologetic about expecting excellence from our students? Our very existence depends upon it.

21. Which is the best version of the underlined portion of this sentence (reproduced below)?

Early in my career, a master's teacher shared this thought with me "Education is the last bastion of civility."

a. (No change)
b. a master's teacher shared this thought with me: "Education is the last bastion of civility."
c. a master's teacher shared this thought with me: "Education is the last bastion of civility".
d. a master's teacher shared this thought with me. "Education is the last bastion of civility."

22. Which is the best version of the underlined portion of this sentence (reproduced below)?

    Education provides society with a vehicle

    a. (No change)
    b. Education provide
    c. Education will provide
    d. Education providing

23. Which is the best version of the underlined portion of this sentence (reproduced below)?

    for _raising it's children to be_ civil, decent, human beings with something valuable to contribute to the world.

    a. (No change)
    b. raises its children to be
    c. raising its' children to be
    d. raising its children to be

24. Which of these, if any, is misspelled?

    It is really what makes us human and what _distinguishes_ us as _civilised_ _creatures._

    a. (No change; best as written.)
    b. distinguishes
    c. civilised
    d. creatures

25. Which is the best version of the underlined portion of this sentence (reproduced below)?

    It would be detrimental to society, only meeting the needs of the mind, if our schools were myopic in their focus.

    a. (No change)
    b. It would be detrimental to society if our schools were myopic in their focus, only meeting the needs of the mind.
    c. Only meeting the needs of our mind, our schools were myopic in their focus, detrimental to society.
    d. Myopic is the focus of our schools, being detrimental to society for only meeting the needs of the mind.

26. Which of these sentences, if any, should begin a new paragraph?

The human heart and psyche have to be fed in order for the mind to develop properly, and the body must be maintained and exercised to help fuel the working of the brain. Education is a basic human right, and it allows us to sustain a democratic society in which participation is fundamental to its success. It should inspire students to seek better solutions to world problems and to dream of a more equitable society.

a. (No change; best as written.)
b. The human heart and psyche have to be fed in order for the mind to develop properly, and the body must be maintained and exercised to help fuel the working of the brain.
c. Education is a basic human right, and it allows us to sustain a democratic society in which participation is fundamental to its success.
d. It should inspire students to seek better solutions to world problems and to dream of a more equitable society.

27. Which is the best version of the underlined portion of this sentence (reproduced below)?

Education should never discriminate on any basis, and it should create individuals who are self-sufficient, patriotic, and tolerant of <u>others' ideas.</u>

a. (No change)
b. other's ideas
c. others ideas
d. others's ideas

28. Which is the best version of the underlined portion of this sentence (reproduced below)?

<u>All children can learn. Although not all children learn in the same manner.</u>

a. (No change)
b. All children can learn although not all children learn in the same manner.
c. All children can learn although, not all children learn in the same manner.
d. All children can learn, although not all children learn in the same manner.

29. Which is the best version of the underlined portion of this sentence (reproduced below)?

If teachers set high expectations for <u>there students</u>, the students will rise to that high level.

a. (No change)
b. they're students
c. their students
d. thare students

30. Which is the best version of the underlined portion of this sentence (reproduced below)?

In the modern age of technology, a teacher's focus is no longer the "what" of the content, <u>but more importantly, the 'why.'</u>

a. (No change)
b. but more importantly, the "why."
c. but more importantly, the 'why'.
d. but more importantly, the "why".

31. Which is the best version of the underlined portion of this sentence (reproduced below)?

Teachers have to work harder than ever before to help students identify salient information <u>so to think critically</u> about the information they encounter.

a. (No change)
b. and to think critically
c. but to think critically
d. nor to think critically

32. Which is the best version of the underlined portion of this sentence (reproduced below)?

Students have to <u>read between the lines, identify bias, and determine</u> who they can trust in the milieu of ads, data, and texts presented to them.

a. (No change)
b. read between the lines, identify bias, and determining
c. read between the lines, identifying bias, and determining
d. reads between the lines, identifies bias, and determines

33. Which is the best version of the underlined portion of this sentence (reproduced below)?

While children spend most of their time in school, they are dramatically and indelibly shaped <u>with the influences</u> of their family and culture.

a. (No change)
b. for the influences
c. to the influences
d. by the influences

34. Which is the best version of the underlined portion of this sentence (reproduced below)?

Teachers must not only respect this fact, <u>but must strive</u> to include parents in the education of their children and must work to keep parents informed of progress and problems.

a. (No change)
b. but to strive
c. but striving
d. but strived

35. Which is the best version of the underlined portion of this sentence (reproduced below)?

We must be unapologetic about expecting excellence from our students? Our very existence depends upon it.

a. (No change)
b. We must be unapologetic about expecting excellence from our students, our very existence depends upon it.
c. We must be unapologetic about expecting excellence from our students—our very existence depends upon it.
d. We must be unapologetic about expecting excellence from our students our very existence depends upon it.

Read the following passage and answer Questions 36 – 40.

Although many Missourians know that Harry S. Truman and Walt Disney hailed from their great state, probably far fewer know that it was also home to the remarkable George Washington Carver. (36) <u>As a child, George was driven to learn, and he loved painting.</u> At the end of the Civil War, Moses Carver, the slave owner who owned George's parents, decided to keep George and his brother and raise them on his farm.

He even went on to study art while in college but was encouraged to pursue botany instead. He spent much of his life helping others (37) <u>by showing them better ways to farm, his ideas improved agricultural productivity</u> in many countries. One of his most notable contributions to the newly emerging class of Black farmers was to teach them the negative effects of agricultural monoculture, i.e. (38) <u>growing the same crops in the same fields year after year, depleting the soil of much needed nutrients and results in a lesser yielding crop.</u>

Carver was an innovator, always thinking of new and better ways to do things, and is most famous for his over three hundred uses for the peanut. Toward the end of his career, (39) <u>Carver returns</u> to his first love of art. Through his artwork, he hoped to inspire people to see the beauty around them and to do great things themselves. (40) <u>Because Carver died,</u> he left his money to help fund ongoing agricultural research. Today, people still visit and study at the George Washington Carver Foundation at Tuskegee Institute.

36. Which of the following would be the best choice for this sentence (reproduced below)?

As a child, George was driven to learn, and he loved painting.

a. (No change)
b. Move to the end of the first paragraph.
c. Move to the beginning of the first paragraph.
d. Move to the end of the second paragraph.

37. Which is the best version of the underlined portion of this sentence (reproduced below)?

He spent much of his life helping others <u>by showing them better ways to farm, his ideas improved agricultural productivity</u> in many countries.

a. (No change)
b. by showing them better ways to farm his ideas improved agricultural productivity
c. by showing them better ways to farm . . . his ideas improved agricultural productivity
d. by showing them better ways to farm; his ideas improved agricultural productivity

38. Which is the best version of the underlined portion of this sentence (reproduced below)?

One of his most notable contributions to the newly emerging class of Black farmers *was to teach them the negative effects of agricultural monoculture, i.e. <u>growing the same crops in the same fields year after year, depleting the soil of much needed nutrients and results in a lesser yielding crop.</u>*

a. (No change)
b. growing the same crops in the same fields year after year, depleting the soil of much needed nutrients and resulting in a lesser yielding crop.
c. growing the same crops in the same fields year after year, depletes the soil of much needed nutrients and resulting in a lesser yielding crop.
d. grows the same crops in the same fields year after year, depletes the soil of much needed nutrients and resulting in a lesser yielding crop.

39. Which is the best version of the underlined portion of this sentence (reproduced below)?

Toward the end of his career, <u>Carver returns</u> to his first love of art.

a. (No change)
b. Carver is returning
c. Carver returned
d. Carver was returning

40. Which is the best version of the underlined portion of this sentence (reproduced below)?

<u>Because Carver died,</u> he left his money to help fund ongoing agricultural research.

a. (No change)
b. Although Carver died,
c. When Carver died,
d. Finally Carver died,

Read the following passage and answer Questions 41 – 50.

(41) <u>Christopher Columbus is often credited for discovering America. This is incorrect.</u> First, it is impossible to "discover" something where people already live; however, Christopher Columbus did explore places in the New World that were previously untouched by Europe, (42) <u>so the ships set sail from Palos, Spain.</u> Another correction must be made, as well: Christopher Columbus was not the first European explorer to reach the present day Americas! (43)

Nevertheless, it was Leif Erikson who first came to the New World and contacted the natives, nearly five hundred years before Christopher Columbus.

Leif Erikson, the son of Erik the Red (a famous Viking outlaw and explorer in his own right), was born in either (44) 970 or 980. Depending on which historian you seek. (45) His own family, though, did not raise Leif, which was a Viking tradition. Instead, one of Erik's prisoners taught Leif reading and writing, languages, sailing, and weaponry. At age 12, Leif was considered a man and returned to his family. He killed a man during a dispute shortly after his return, and the council banished the Erikson clan to Greenland.

In 999, Leif left Greenland and traveled to Norway where he would serve as a guard to King Olaf Tryggvason. It was there that he became a convert to Christianity. (46) Later trying to return home, Leif with the intention of taking supplies and spreading Christianity to Greenland, however his ship was blown off course and he arrived in a strange new land: present day Newfoundland, Canada.

When he finally returned to his adopted homeland, Greenland, (47) Leif consults with a merchant who had also seen the shores of this previously unknown land we now know as Canada. The son of the legendary Viking explorer then gathered a crew of 35 men and set sail. Leif became the first European to touch foot in the New World as he explored present-day Baffin Island and Labrador, Canada. His crew called the land Vinland since it was plentiful with grapes.

During their time in present-day Newfoundland, Leif's expedition made contact with the natives whom they referred to as Skraelings (48) (which translates to 'wretched ones' in Norse). There are several secondhand accounts of their meetings. Some contemporaries described trade between the peoples. (49) Other accounts describes clashes where the Skraelings defeated the Viking explorers with long spears, while still others claim the Vikings dominated the natives. Regardless of the circumstances, it seems that the Vikings made contact of some kind. This happened around 1000, nearly five hundred years before Columbus famously sailed the ocean blue.

Eventually, in 1003, Leif set sail for home and arrived at Greenland with a ship full of timber. (50) In 1020, seventeen years later. The legendary Viking died. Many believe that Leif Erikson should receive more credit for his contributions in exploring the New World.

41. Which is the best version of the underlined portion of this sentence (reproduced below)?

Christopher Columbus is often credited for discovering America. This is incorrect.

a. (No change)
b. Christopher Columbus is often credited for discovering America this is incorrect.
c. Christopher Columbus is often credited for discovering America, this is incorrect.
d. Christopher Columbus is often credited for discovering America: this is incorrect.

42. Which of the following facts would be the most relevant to include here?

however, Christopher Columbus did explore places in the New World that were previously untouched by Europe, <u>so the ships set sail from Palos, Spain.</u>

a. (No change; best as written.)
b. so Columbus discovered Watling Island in the Bahamas.
c. so the ships were named them the Santa María, the Pinta, and the Niña.
d. so the term "explorer" would be more accurate.

43. Which is the best version of the underlined portion of this sentence (reproduced below)?

<u>Nevertheless,</u> it was Leif Erikson who first came to the New World and contacted the natives, nearly five hundred years before Christopher Columbus.

a. (No change)
b. Rather,
c. Finally,
d. Suddenly,

44. Which is the best version of the underlined portion of this sentence (reproduced below)?

Leif Erikson, the son of Erik the Red (a famous Viking outlaw and explorer in his own right), was born in either <u>970 or 980. Depending on which historian you seek.</u>

a. (No change)
b. 970 or 980! depending on which historian you seek.
c. 970 or 980, depending on which historian you seek.
d. 970 or 980; depending on which historian you seek.

45. Which of the following would be the best choice for this sentence?

<u>His own family, though, did not raise Leif, which was a Viking tradition.</u>

a. (No change; best as written.)
b. Move to the end of the second paragraph.
c. Move to the beginning of the second paragraph.
d. Switch with the following sentence.

46. Which is the best version of the underlined portion of this sentence (reproduced below)?

<u>Later trying to return home, Leif</u> with the intention of taking supplies and spreading Christianity to Greenland, however his ship was blown off course and he arrived in a strange new land: present day Newfoundland, Canada.

a. (No change)
b. To return home later, Leif
c. Leif later tried to return home
d. Leif to return home tried later

47. Which is the best version of the underlined portion of this sentence (reproduced below)?

When he finally returned to his adopted homeland, Greenland, Leif consults with a merchant who had also seen the shores of this previously unknown land we now know as Canada.

a. (No change)
b. Leif consulted
c. Leif consulting
d. Leif was consulted

48. Which is the best version of the underlined portion of this sentence (reproduced below)?

During their time in present-day Newfoundland, Leif's expedition made contact with the natives whom they referred to as Skraelings (which translates to 'wretched ones' in Norse).

a. (No change)
b. (which translates to "wretched ones" in Norse.)
c. (which translates to 'wretched ones' in Norse.)
d. (which translates to "wretched ones" in Norse).

49. Which is the best version of the underlined portion of this sentence (reproduced below)?

Other accounts describes clashes where the Skraelings defeated the Viking explorers with long spears, while still others claim the Vikings dominated the natives.

a. (No change)
b. Other account's describe
c. Other accounts describe
d. Others account's describes

50. Which is the best version of the underlined portion of this sentence (reproduced below)?

In 1020, seventeen years later. The legendary Viking died

a. (No change)
b. In 1020, seventeen years later; the legendary Viking died.
c. In 1020 seventeen years later the legendary Viking died.
d. In 1020, seventeen years later, the legendary Viking died.

51. What is the structure of the following sentence?
   The restaurant is unconventional because it serves both Chicago style pizza and New York style pizza.

a. Simple
b. Compound
c. Complex
d. Compound-complex

52. The following sentence contains what kind of error?

> This summer, I'm planning to travel to Italy, take a Mediterranean cruise, going to Pompeii, and eat a lot of Italian food.

a. Parallelism
b. Sentence fragment
c. Misplaced modifier
d. Subject-verb agreement

53. The following sentence contains what kind of error?

> Forgetting that he was supposed to meet his girlfriend for dinner, Anita was mad when Fred showed up late.

a. Parallelism
b. Run-on sentence
c. Misplaced modifier
d. Subject-verb agreement

54. The following sentence contains what kind of error?

> Some workers use all their sick leave, other workers cash out their leave.

a. Parallelism
b. Comma splice
c. Sentence fragment
d. Subject-verb agreement

55. A student writes the following in an essay:

> Protestors filled the streets of the city. Because they were dissatisfied with the government's leadership.

Which of the following is an appropriately-punctuated correction for this sentence?
a. Protestors filled the streets of the city, because they were dissatisfied with the government's leadership.
b. Protesters, filled the streets of the city, because they were dissatisfied with the government's leadership.
c. Because they were dissatisfied with the government's leadership protestors filled the streets of the city.
d. Protestors filled the streets of the city because they were dissatisfied with the government's leadership.

56. While studying vocabulary, a student notices that the words *circumference*, *circumnavigate*, and *circumstance* all begin with the prefix *circum–*. The student uses her knowledge of affixes to infer that all of these words share what related meaning?
a. Around, surrounding
b. Travel, transport
c. Size, measurement
d. Area, location

57. A local newspaper is looking for writers for a student column. A student would like to submit his article to the newspaper, but he isn't sure how to format his article according to journalistic standards. What resource should he use?
    a. A thesaurus
    b. A dictionary
    c. A style guide
    d. A grammar book

58. A student encounters the word *aficionado* and wants to learn more about it. It doesn't sound like other English words he knows, so the student is curious to identify the word's origin. What resource should he consult?
    a. A thesaurus
    b. A dictionary
    c. A style guide
    d. A grammar book

59. Which of the following refers to what an author wants to express about a given subject?
    a. Primary purpose
    b. Plot
    c. Main idea
    d. Characterization

60. Which organizational style is used in the following passage?
    There are several reasons why the new student café has not been as successful as expected. One factor is that prices are higher than originally advertised, so many students cannot afford to buy food and beverages there. Also, the café closes rather early; as a result, students go out in town to other late-night gathering places rather than meeting friends at the café on campus.

    a. Cause and effect order
    b. Compare and contrast order
    c. Spatial order
    d. Time order

Essay Prompt

Directions: The HiSET writing portion of the exam will allow you 45 minutes to write an essay. This essay is a test of your writing skills. Please do the following in your essay:

- Develop a position through explaining the supporting reasons and examples from the two passages and from personal experience.

- Organize ideas clearly, using an introduction, conclusion, body paragraphs, and effective transitions.

- Use appropriate word choice, different sentence construction, and a consistent style.

- Use proper grammar and writing conventions.

The two passages below are written to disagree with one another on the same issue of importance. Please read both passages carefully and determine the strengths and weaknesses of each argument. Then write an essay explaining your own opinion on the issue.

A school administration has asked the teachers to research cell phone use in the classroom. Then it asked them to state their opinions in an essay. The excerpts below were taken from two different papers.

Passage 1

In the modern classroom, cell phones have become indispensable. Cell phones, which are essentially handheld computers, allow students to take notes, connect to the web, perform complex computations, teleconference, and participate in surveys.

Additionally, due to their mobility and excellent reception, cell phones are necessary in emergencies. According to a 2005 study conducted by Dr. Havish and Dr. Braum, 85% of students said that they felt safer having access to their cell phones in class. For them, it was about having contact to the "outside world" if anything were to happen inside the classroom. Also, they were able to have direct communication with their families if one of them became sick or injured in some way.

Unlike tablets, laptops, or computers, cell phones are a readily available and free resource. Most school district budgets are already strained to begin with. According to University of Texas' technological journal *Bot*, since today's student is already strongly rooted in technology, "when teachers incorporate cell phones, they're 'speaking' the student's language," (Dr. Branson, 2010) which increases the chance of higher engagement.

Passage 2

As with most forms of technology, there is an appropriate time and place for the use of cell phones. Students are comfortable with cell phones, so it makes sense when teachers allow cell phone use at their discretion. Allowing cell phone use can prove advantageous if done correctly.

Unfortunately, if that's not the case—and often it isn't—then a sizable percentage of students pretend to pay attention while *surreptitiously* playing on their phones. It is a well-known fact that a large percentage of teachers across America disagree with the use of cell phones in the classroom, because students end up ignoring their lectures and instead play on their phones. With this information in mind, it can be said that cell phones are actually *hindering* our education as a country.

This type of disrespectful behavior is often justified by the argument that cell phones are not only a privilege but also a right (*Journal of Florida Technology,* 2012, p. 184). Under this logic, confiscating phones is akin to rummaging through students' backpacks. This is in stark contrast to several decades ago when teachers regulated where and when students accessed information.

Write an essay explaining your own position on the issue of whether or not to allow cell phone use in the classroom.

Make sure to use evidence from the passages provided along with reasons and examples from your own experience to support your position. Your essay should acknowledge opposing ideas. Please review your essay once you have finished for correct punctuation, grammar, and spelling.

Answer Explanations

1. B: Move the sentence so that it comes before the preceding sentence. For this question, place the underlined sentence in each prospective choice's position. To keep it as-is is incorrect because the father "going crazy" doesn't logically follow the fact that he was a "city slicker." Choice *C* is incorrect because the sentence in question is not a concluding sentence and does not transition smoothly into the second paragraph. Choice *D* is incorrect because the sentence doesn't necessarily need to be omitted since it logically follows the very first sentence in the passage.

2. D: Choice *D* is correct because "As it turns out" indicates a contrast from the previous sentiment, that the RV was a great purchase. Choice *A* is incorrect because the sentence needs an effective transition from the paragraph before. Choice *B* is incorrect because the text indicates it *is* surprising that the RV was a great purchase because the author was skeptical beforehand. Choice *C* is incorrect because the transition "Furthermore" does not indicate a contrast.

3. B: This sentence calls for parallel structure. Choice *B* is correct because the verbs "wake," "eat," and "break" are consistent in tense and parts of speech. Choice *A* is incorrect because the words "wake" and "eat" are present tense while the word "broke" is in past tense. Choice *C* is incorrect because this turns the sentence into a question, which doesn't make sense within the context. Choice *D* is incorrect because it breaks tense with the rest of the passage. "Waking," "eating," and "breaking" are all present participles, and the context around the sentence is in past tense.

4. C: Choice *C* is correct because it is clear and fits within the context of the passage. Choice *A* is incorrect because "We rejoiced as 'hackers'" does not give a reason why hacking was rejoiced. Choice *B* is incorrect because it does not mention a solution being found and is therefore not specific enough. Choice *D* is incorrect because the meaning is eschewed by the helping verb "had to rejoice," and the sentence suggests that rejoicing was necessary to "hack" a solution.

5. A: The original sentence is correct because the verb tense as well as the meaning aligns with the rest of the passage. Choice *B* is incorrect because the order of the words makes the sentence more confusing than it otherwise would be. Choice *C* is incorrect because "We are even making" is in present tense. Choice *D* is incorrect because "We will make" is future tense. The surrounding text of the sentence is in past tense.

6. B: Choice *B* is correct because there is no punctuation needed if a dependent clause ("while traveling across America") is located behind the independent clause ("it allowed us to share adventures"). Choice *A* is incorrect because there are two dependent clauses connected and no independent clause, and a complete sentence requires at least one independent clause. Choice *C* is incorrect because of the same reason as Choice *A*. Semicolons have the same function as periods: there must be an independent clause on either side of the semicolon. Choice *D* is incorrect because the dash simply interrupts the complete sentence.

7. C: The rules for "me" and "I" is that one should use "I" when it is the subject pronoun of a sentence, and "me" when it is the object pronoun of the sentence. Break the sentence up to see if "I" or "me" should be used. To say "Those are memories that I have now shared" is correct, rather than "Those are memories that me have now shared." Choice *D* is incorrect because "my siblings" should come before "I."

8. D: Choice *D* is correct because Fred Hampton becoming an activist was a direct result of him wanting to see lasting social change for Black people. Choice *A* doesn't make sense because "In the meantime" denotes something happening at the same time as another thing. Choice *B* is incorrect because the text's tone does not indicate that becoming a civil rights activist is an unfortunate path. Choice *C* is incorrect because "Finally" indicates something that comes last in a series of events, and the word in question is at the beginning of the introductory paragraph.

9. C: Choice *C* is correct because there should be a comma between the city and state, as well as after the word "Illinois." Commas should be used to separate all geographical items within a sentence. Choice *A* is incorrect because it does not include the comma after "Illinois." Choice *B* is incorrect because the comma after "Maywood" interrupts the phrase, "Maywood of Chicago." Finally, Choice *D* is incorrect because the order of the sentence designates that Chicago, Illinois is in Maywood, which is incorrect.

10. C: This is a difficult question. The paragraph is incorrect as-is because it is too long and thus loses the reader halfway through. Choice *C* is correct because if the new paragraph began with "While studying at Triton," we would see a smooth transition from one paragraph to the next. We can also see how the two paragraphs are logically split in two. The first half of the paragraph talks about where he studied. The second half of the paragraph talks about the NAACP and the result of his leadership in the association. If we look at the passage as a whole, we can see that there are two main topics that should be broken into two separate paragraphs.

11. B: The BPP "was another activist group that . . ." We can figure out this answer by looking at context clues. We know that the BPP is "similar in function" to the NAACP. To find out what the NAACP's function is, we must look at the previous sentences. We know from above that the NAACP is an activist group, so we can assume that the BPP is also an activist group.

12. A: Choice *A* is correct because the Black Panther Party is one entity; therefore, the possession should show the "Party's approach" with the apostrophe between the "y" and the "s." Choice *B* is incorrect because the word "Parties" should not be plural. Choice *C* is incorrect because the apostrophe indicates that the word "Partys" is plural. The plural of "party" is "parties." Choice *D* is incorrect because, again, the word "parties" should not be plural; instead, it is one unified party.

13. C: Choice *C* is correct because the passage is told in past tense, and "enabled" is a past tense verb. Choice *A*, "enable," is present tense. Choice *B*, "are enabling," is a present participle, which suggests a continuing action. Choice *D*, "will enable," is future tense.

14. D: Choice *D* is correct because the conjunction "and" is the best way to combine the two independent clauses. Choice *A* is incorrect because the word "he" becomes repetitive since the two clauses can be joined together. Choice *B* is incorrect because the conjunction "but" indicates a contrast, and there is no contrast between the two clauses. Choice *C* is incorrect because the introduction of the comma after "project" with no conjunction creates a comma splice.

15. C: The word "acheivement" is misspelled. Remember the rules for "*i* before *e* except after *c*." Choices *B* and *D*, "greatest" and "leader," are both spelled correctly.

16. B: Choice *B* is correct because it provides the correct verb tense and verb form. Choice *A* is incorrect; Hampton was not "held by a press conference"—rather, he held a press conference. The passage indicates that he "made the gangs agree to a nonaggression pact," implying that it was Hampton who was doing the speaking for this conference. Choice *C* is incorrect because, with this use of the sentence,

it would create a fragment because the verb "holding" has no helping verb in front of it. Choice D is incorrect because it adds an infinitive ("to hold") where a past tense form of a verb should be.

17. A: Choice A is correct because it provides the most clarity. Choice B is incorrect because it doesn't name the group until the end, so the phrase "the group" is vague. Choice C is incorrect because it indicates that the BPP's popularity grew as a result of placing the group under constant surveillance, which is incorrect. Choice D is incorrect because there is a misplaced modifier; this sentence actually says that the FBI's influence and popularity grew, which is incorrect.

18. B: Choice B is correct. Choice A is incorrect because there should be an independent clause on either side of a semicolon, and the phrase "In 1976" is not an independent clause. Choice C is incorrect because there should be a comma after introductory phrases in general, such as "In 1976," and Choice C omits a comma. Choice D is incorrect because the sentence "In 1976." is a fragment.

19. C: Choice C is correct because the past tense verb "provided" fits in with the rest of the verb tense throughout the passage. Choice A, "will provide," is future tense. Choice B, "provides," is present tense. Choice D, "providing," is a present participle, which means the action is continuous.

20. D: The correct answer is Choice D because this statement provides the most clarity. Choice A is incorrect because the noun "Chicago City Council" acts as one, so the verb "are" should be singular, not plural. Choice B is incorrect because it is perhaps the most confusingly worded out of all the answer choices; the phrase "December 4" interrupts the sentence without any indication of purpose. Choice C is incorrect because it is too vague and leaves out *who* does the commemorating.

21. B: Choice B is correct. Here, a colon is used to introduce an explanation. Colons either introduce explanations or lists. Additionally, the quote ends with the punctuation inside the quotes, unlike Choice C.

22. A: The verb tense in this passage is predominantly in the present tense, so Choice A is the correct answer. Choice B is incorrect because the subject and verb do not agree. It should be "Education provides," not "Education provide." Choice C is incorrect because the passage is in present tense, and "Education will provide" is future tense. Choice D doesn't make sense when placed in the sentence.

23. D: The possessive form of the word "it" is "its." The contraction "it's" denotes "it is." Thus, Choice A is wrong. The word "raises" in Choice B makes the sentence grammatically incorrect. Choice C adds an apostrophe at the end of "its." While adding an apostrophe to most words would indicate possession, adding 's to the word "it" indicates a contraction.

24. C: The word *civilised* should be spelled *civilized.* The words "distinguishes" and "creatures" are both spelled correctly.

25. B: Choice B is correct because it provides clarity by describing what "myopic" means right after the word itself. Choice A is incorrect because the explanation of "myopic" comes before the word; thus, the meaning is skewed. It's possible that Choice C makes sense within context. However, it's not the best way to say this because the commas create too many unnecessary phrases. Choice D is confusingly worded. Using "myopic focus" is not detrimental to society; however, the way D is worded makes it seem that way.

26. C: Again, we see where the second paragraph can be divided into two parts due to separate topics. The paragraph's first main focus is education addressing the mind, body, and soul. This first section, then, could end with the concluding sentence, "The human heart and psyche . . ." The next sentence to start a new paragraph would be "Education is a basic human right." The rest of this paragraph talks about what education is and some of its characteristics.

27. A: Choice *A* is correct because the phrase "others' ideas" is both plural and indicates possession. Choice *B* is incorrect because "other's" indicates only one "other" that's in possession of "ideas," which is incorrect. Choice *C* is incorrect because no possession is indicated. Choice *D* is incorrect because the word "other" does not end in *s*. Others's is not a correct form of the plural possessive word.

28. D: This sentence must have a comma before "although" because the word "although" is connecting two independent clauses. Thus, Choices *B* and *C* are incorrect. Choice *A* is incorrect because the second sentence in the underlined section is a fragment.

29. C: Choice *C* is the correct choice because the word "their" indicates possession, and the text is talking about "their students," or the students of someone. Choice *A*, "there," means at a certain place and is incorrect. Choice *B*, "they're," is a contraction and means "they are." Choice *D* is not a word.

30. B: Choice *B* uses all punctuation correctly in this sentence. In American English, single quotes should only be used if they are quotes within a quote, making Choices *A* and *C* incorrect. Additionally, punctuation should go inside quotation marks with a few exceptions, making Choice *D* incorrect.

31. B: Choice *B* is correct because the conjunction "and" is used to connect phrases that are to be used jointly, such as teachers working hard to help students "identify salient information" and to "think critically." The conjunctions *so*, *but*, and *nor* are incorrect in the context of this sentence.

32. A: Choice *A* has consistent parallel structure with the verbs "read," "identify," and "determine." Choices *B* and *C* have faulty parallel structure with the words "determining" and "identifying." Choice *D* has incorrect subject/verb agreement. The sentence should read, "Students have to read . . . identify . . . and determine."

33. D: The correct choice for this sentence is that "they are . . . shaped by the influences." The prepositions "for," "to," and "with" do not make sense in this context. People are *shaped by*, not *shaped for, shaped to,* or *shaped with*.

34. A: To see which answer is correct, it might help to place the subject, "Teachers," near the verb. Choice *A* is correct: "Teachers . . . must strive" makes grammatical sense here. Choice *B* is incorrect because "Teachers . . . to strive" does not make grammatical sense. Choice *C* is incorrect because "Teachers must not only respect . . . but striving" eschews parallel structure. Choice *D* is incorrect because it is in past tense, and this passage is in present tense.

35. C: Choice *C* is correct because it uses an em-dash. Em-dashes are versatile. They can separate phrases that would otherwise be in parenthesis, or they can stand in for a colon. In this case, a colon would be another decent choice for this punctuation mark because the second sentence expands upon the first sentence. Choice *A* is incorrect because the statement is not a question. Choice *B* is incorrect because adding a comma here would create a comma splice. Choice *D* is incorrect because this creates a run-on sentence since the two sentences are independent clauses.

36. B: The best place for this sentence given all the answer choices is at the end of the first paragraph. Choice *A* is incorrect; the passage is told in chronological order, and leaving the sentence as-is defies that order, since we haven't been introduced to who raised George. Choice *C* is incorrect because this

sentence is not an introductory sentence. It does not provide the main topic of the paragraph. Choice *D* is incorrect because again, it defies chronological order. By the end of paragraph two we have already gotten to George as an adult, so this sentence would not make sense here.

37. D: Out of these choices, a semicolon would be the best fit because there is an independent clause on either side of the semicolon, and the two sentences closely relate to each other. Choice *A* is incorrect because putting a comma between two independent clauses (i.e. complete sentences) creates a comma splice. Choice *B* is incorrect; omitting punctuation here creates a run-on sentence. Choice *C* is incorrect because an ellipsis (. . .) is used to designate an omission in the text.

38. B: This is another example of parallel structure. Choice *A* is incorrect because the verbs in the original sentence are "growing," "depleting," and "results," the last of which has a different form than the first two. Choices *C* and *D* add "depletes" and "grows," both of which abandon the "-ing" verbs.

39. C: Choice *C* is correct because it keeps with the verb tense in the rest of the passage: past tense. Choice *A* is in present tense, which is incorrect. Choice *B* is present progressive, which means there is a continual action, which is also incorrect. Choice *D* is incorrect because "was returning" is past progressive tense, which means that something was happening continuously at some point in the past.

40. C: The correct choice is the subordinating conjunction, "When." We should look at the clues around the phrase to see what fits best. Carver left his money "when he died." Choice *A*, "Because," could perhaps be correct, but "When" is the more appropriate word to use here. Choice *B* is incorrect; "Although" denotes a contrast, and there is no contrast here. Choice *D* is incorrect because "Finally" indicates something at the very end of a list or series, and there is no series at this point in the text.

41. A: There should be no change here. Both underlined sentences are complete and do not need changing. Choice B is incorrect because there is no punctuation between the two independent clauses, it is considered a run-on. Choice *C* is incorrect because placing a comma between two independent clauses creates a comma splice. Choice *D* is incorrect. The underlined portion could *possibly* act with a colon. However, it's not the best choice, so omit Choice *D*.

42. D: Choice *D* is correct. The text before this underlined phrase talks about the difference between "discovery" and "exploration," so making a decision on what term to label Columbus would be the best choice. The other three choices may be true to the historical narrative of Columbus; however, they do not fit within the surrounding text.

43. B: This question seeks to determine the best introductory word for the main point of the following sentence. Choice *B* is correct; the word "Rather" indicates something unexpected. "Rather" fits in this sentence because it is "unexpected" that Leif Erikson first came to the New World and not Columbus. Choice *A* is incorrect; "Nevertheless" means "all the same," and does not fit with the sentiment of this sentence. Choice *C* is incorrect because "Finally" is used to indicate the last point in a series, and we do not have a listed series here. The word "Suddenly" is used to indicate something that has happened quickly or unexpectedly. Thus, Choice *D* is incorrect.

44. C: Choice *C* is correct; the underlined phrase consists of part of an independent clause and a dependent clause ("Depending on which historian you seek.") The dependent clause cannot stand by itself. Thus, the best choice is to connect the two clauses with a comma. Choices *A* and *D* do not work because you must have two independent clauses on either side of a period as well as a semicolon. Choice *B* is incorrect because an exclamation point is used to show excitement and does not fit the tone here.

45. A: There should be no change. The sentence fits perfectly before the current one because in question is who raised Leif. Choice *B* is incorrect because this narrative is in chronological order, and by the end of the second paragraph, Leif is already an adult. Choice *C* wouldn't work because the sentence is not an introductory sentence. Rather, it shares the details of Leif's childhood. Finally, Choice *D* is incorrect because there is already a transition, "Instead," to lead into the next sentence.

46. C: To find out the best answer, try out each answer choice. Choice *A* is incorrect; it might make sense that Leif is "later trying to return home." However, the next sentence says "Leif with the intention of taking supplies," and is not grammatically correct. Choice *B* is also incorrect because we would have the same problem with "Leif with the intention of taking supplies." Choice *D* is not a good answer choice because it inverts words that are otherwise clear with Choice *C*, "Leif later tried to return home with the intention of taking supplies."

47. B: The most appropriate verb for this sentence is Choice *B*, "Leif consulted." Choice *A* is in present tense and therefore does not fit with the rest of the passage. Choice *C* is incorrect because "consulting" is present progressive tense and also does not fit with the consistent past tense of the passage. Choice *D*, "Leif was consulted with a merchant," doesn't make sense. Leif can consult with a merchant or be consulted by a merchant.

48. D: Choice *D* uses the correct punctuation. American English uses double quotes unless placing quotes within a quote (which would then require single quotes). Thus, Choices *A* and *C* are incorrect. Choice *B* is incorrect because the period should go outside of the parenthesis, not inside.

49. C: Choice *C* is correct. The subject and verb agree with each other (accounts describe), and there is no apostrophe because no possession is being shown. Choices *B* and *D* are incorrect because there is no possession—"accounts" is simply plural. Choice *A* is incorrect because the subject and verb do not agree with each other (accounts describes).

50. D: Choice *D* is correct because the interrupting phrase, "seventeen years later," is separated by commas. Choice *A* is incorrect because putting a period between "later" and "The" causes the first sentence to become a fragment. Choice *B* is incorrect because of the same reason: the semicolon should have an independent clause on either side of it, and the first half of the sentence is not an independent clause. Choice *C* needs commas to separate the interrupting phrase or else the words become mashed together, causing confusion.

51. C: A complex sentence joins an independent or main clause with a dependent or subordinate clause. In this case, the main clause is "The restaurant is unconventional." This is a clause with one subject-verb combination that can stand alone as a grammatically-complete sentence. The dependent clause is "because it serves both Chicago style pizza and New York style pizza." This clause begins with the subordinating conjunction *because* and also consists of only one subject-verb combination. *A* is incorrect because a simple sentence consists of only one verb-subject combination—one independent clause. *B* is incorrect because a compound sentence contains two independent clauses connected by a conjunction. *D* is incorrect because a complex-compound sentence consists of two or more independent clauses and one or more dependent clauses.

52. A: Parallelism refers to consistent use of sentence structure or word form. In this case, the list within the sentence does not utilize parallelism; three of the verbs appear in their base form—*travel, take,* and *eat*—but one appears as a gerund—*going*. A parallel version of this sentence would be "This summer, I'm planning to travel to Italy, take a Mediterranean cruise, go to Pompeii, and eat a lot of Italian food." *B* is incorrect because this description is a complete sentence. *C* is incorrect as a misplaced modifier is a modifier that is not located appropriately in relation to the word or words they modify. *D* is incorrect because subject-verb agreement refers to the appropriate conjugation of a verb in relation to its subject.

53. C: In this sentence, the modifier is the phrase "Forgetting that he was supposed to meet his girlfriend for dinner." This phrase offers information about Fred's actions, but the noun that immediately follows it is Anita, creating some confusion about the "do-er" of the phrase. A more appropriate sentence arrangement would be "Forgetting that he was supposed to meet his girlfriend for dinner, Fred made Anita mad when he showed up late." *A* is incorrect as parallelism refers to the consistent use of sentence structure and verb tense, and this sentence is appropriately consistent. *B* is incorrect as a run-on sentence does not contain appropriate punctuation for the number of independent clauses presented, which is not true of this description. *D* is incorrect because subject-verb agreement refers to the appropriate conjugation of a verb relative to the subject, and all verbs have been properly conjugated.

54. B: A comma splice occurs when a comma is used to join two independent clauses together without the additional use of an appropriate conjunction. One way to remedy this problem is to replace the comma with a semicolon. Another solution is to add a conjunction: "Some workers use all their sick leave, but other workers cash out their leave." *A* is incorrect as parallelism refers to the consistent use of sentence structure and verb tense; all tenses and structures in this sentence are consistent. *C* is incorrect because a sentence fragment is a phrase or clause that cannot stand alone—this sentence contains two independent clauses. *D* is incorrect because subject-verb agreement refers to the proper conjugation of a verb relative to the subject, and all verbs have been properly conjugated.

55. D: The problem in the original passage is that the second sentence is a dependent clause that cannot stand alone as a sentence; it must be attached to the main clause found in the first sentence. Because the main clause comes first, it does not need to be separated by a comma. However, if the dependent clause came first, then a comma would be necessary, which is why Choice *C* is incorrect. *A* and *B* also insert unnecessary commas into the sentence.

56. A: The affix *circum–* originates from Latin and means *around or surrounding*. It is also related to other round words, such as circle and circus. The rest of the choices do not relate to the affix *circum–* and are therefore incorrect.

57. C: A style guide offers advice about proper formatting, punctuation, and usage when writing for a specific field, such as journalism or scientific research. The other resources would not offer similar information. A dictionary is useful for looking up definitions; a thesaurus is useful for looking up synonyms and antonyms. A grammar book is useful for looking up specific grammar topics. Thus, Choices *A, C,* and *D* are incorrect.

58. B: A word's origin is also known as its *etymology*. In addition to offering a detailed list of a word's various meanings, a dictionary also provides information about a word's history, such as when it first came into use, what language it originated from, and how its meaning may have changed over time. A thesaurus is for identifying synonyms and antonyms, so *A* is incorrect. A style guide provides formatting, punctuation, and syntactical advice for a specific field, and a grammar book is related to the appropriate placement of words and punctuation, which does not provide any insight into a word's meaning. Therefore, Choices *A*, *C*, and *D* are incorrect.

59. C: The main idea of a piece is its central theme or subject and what the author wants readers to know or understand after they read. Choice *A* is incorrect because the primary purpose is the reason that a piece was written, and while the main idea is an important part of the primary purpose, the above elements are not developed with that intent. Choice *B* is incorrect because while the plot refers to the events that occur in a narrative, organization, tone, and supporting details are not used only to develop plot. Choice *D* is incorrect because characterization is the description of a person.

60. A: The passage describes a situation and then explains the causes that led to it. Also, it utilizes cause and effect signal words, such as *causes, factors, so,* and *as a result*. *B* is incorrect because a compare and contrast order considers the similarities and differences of two or more things. *C* is incorrect because spatial order describes where things are located in relation to each other. Finally, *D* is incorrect because time order describes when things occurred chronologically.

Mathematics

Numbers and Operations on Numbers

Properties of Operations with Real Numbers, Including Rational and Irrational Numbers

The mathematical number system is made up of two general types of numbers: real and complex. *Real numbers* are those that are used in normal settings, while *complex numbers* are those composed of both a real number and an imaginary one. Imaginary numbers are the result of taking the square root of -1, and $\sqrt{-1} = i$.

The real number system is often explained using a Venn diagram similar to the one below. After a number has been labeled as a real number, further classification occurs when considering the other groups in this diagram. If a number is a never-ending, non-repeating decimal, it falls in the irrational category. Otherwise, it is rational. More information on these types of numbers is provided in the previous section. Furthermore, if a number does not have a fractional part, it is classified as an integer, such as -2, 75, or zero. Whole numbers are an even smaller group that only includes positive integers and zero. The last group of natural numbers is made up of only positive integers, such as 2, 56, or 12.

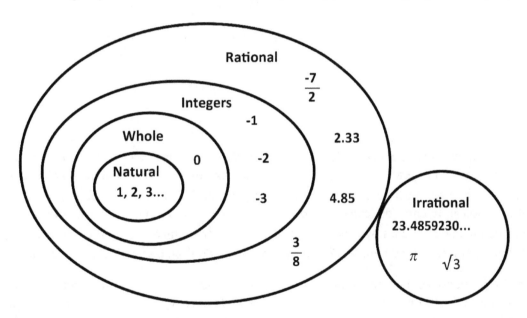

Real numbers can be compared and ordered using the number line. If a number falls to the left on the real number line, it is less than a number on the right. For example, $-2 < 5$ because -2 falls to the left of zero, and 5 falls to the right. Numbers to the left of zero are negative while those to the right are positive.

Complex numbers are made up of the sum of a real number and an imaginary number. Some examples of complex numbers include $6 + 2i$, $5 - 7i$, and $-3 + 12i$. Adding and subtracting complex numbers is similar to collecting like terms. The real numbers are added together, and the imaginary numbers are added together. For example, if the problem asks to simplify the expression $6 + 2i - 3 + 7i$, the 6 and

(-3) are combined to make 3, and the $2i$ and $7i$ combine to make $9i$. Multiplying and dividing complex numbers is similar to working with exponents.

One rule to remember when multiplying is that:

$$i \times i = -1$$

For example, if a problem asks to simplify the expression $4i(3 + 7i)$, the $4i$ should be distributed throughout the 3 and the $7i$. This leaves the final expression $12i - 28$. The 28 is negative because $i \times i$ results in a negative number. The last type of operation to consider with complex numbers is the conjugate. The *conjugate* of a complex number is a technique used to change the complex number into a real number. For example, the conjugate of $4 - 3i$ is $4 + 3i$. Multiplying $(4 - 3i)(4 + 3i)$ results in $16 + 12i - 12i + 9$, which has a final answer of $16 + 9 = 25$.

The order of operations—PEMDAS—simplifies longer expressions with real or imaginary numbers. Each operation is listed in the order of how they should be completed in a problem containing more than one operation. Parenthesis can also mean grouping symbols, such as brackets and absolute value. Then, exponents are calculated. Multiplication and division should be completed from left to right, and addition and subtraction should be completed from left to right. The following shows step-by-step how an expression is simplified using the order of operations:

$$25 \div (8 - 3)^2 - 1$$

$$25 \div (5)^2 - 1$$

$$25 \div 25 - 1$$

$$1 - 1$$

$$0$$

Simplification of another type of expression occurs when radicals are involved. As explained previously, root is another word for radical. For example, the following expression is a radical that can be simplified: $\sqrt{24x^2}$. First, the number must be factored out to the highest perfect square. Any perfect square can be taken out of a radical. Twenty-four can be factored into 4 and 6, and 4 can be taken out of the radical. $\sqrt{4} = 2$ can be taken out, and 6 stays underneath. If $x > 0$, x can be taken out of the radical because it is a perfect square. The simplified radical is $2x\sqrt{6}$. An approximation can be found using a calculator.

There are also properties of numbers that are true for certain operations. The *commutative* property allows the order of the terms in an expression to change while keeping the same final answer. Both addition and multiplication can be completed in any order and still obtain the same result. However, order does matter in subtraction and division. The *associative* property allows any terms to be "associated" by parenthesis and retain the same final answer. For example, $(4 + 3) + 5 = 4 + (3 + 5)$. Both addition and multiplication are associative; however, subtraction and division do not hold this property. The *distributive* property states that $a(b + c) = ab + ac$. It is a property that involves both addition and multiplication, and the a is distributed onto each term inside the parentheses.

Integers can be factored into prime numbers. To *factor* is to express as a product. For example, $6 = 3 \times 2$, and $6 = 6 \times 1$. Both are factorizations, but the expression involving the factors of 3 and 2 is known as a *prime factorization* because it is factored into a product of two *prime numbers*—integers which do not have any factors other than themselves and 1. A *composite number* is a positive integer

that can be divided into at least one other integer other than itself and 1, such as 6. Integers that have a factor of 2 are even, and if they are not divisible by 2, they are odd. Finally, a *multiple* of a number is the product of that number and a counting number—also known as a *natural number*. For example, some multiples of 4 are 4, 8, 12, 16, etc.

Properties of Rational and Irrational Numbers

All real numbers can be separated into two groups: rational and irrational numbers. *Rational numbers* are any numbers that can be written as a fraction, such as $\frac{1}{3}, \frac{7}{4}$, and -25. Alternatively, *irrational numbers* are those that cannot be written as a fraction, such as numbers with never-ending, non-repeating decimal values. Many irrational numbers result from taking roots, such as $\sqrt{2}$ or $\sqrt{3}$. An irrational number may be written as:

$$34.5684952\ldots$$

The ellipsis (…) represents the line of numbers after the decimal that does not repeat and is never-ending.

When rational and irrational numbers interact, there are different types of number outcomes. For example, when adding or multiplying two rational numbers, the result is a rational number. No matter what two fractions are added or multiplied together, the result can always be written as a fraction. The following expression shows two rational numbers multiplied together:

$$\frac{3}{8} \times \frac{4}{7} = \frac{12}{56}$$

The product of these two fractions is another fraction that can be simplified to $\frac{3}{14}$.

As another interaction, rational numbers added to irrational numbers will always result in irrational numbers. No part of any fraction can be added to a never-ending, non-repeating decimal to make a rational number. The same result is true when multiplying a rational and irrational number. Taking a fractional part of a never-ending, non-repeating decimal will always result in another never-ending, non-repeating decimal. An example of the product of rational and irrational numbers is shown in the following expression: $2 \times \sqrt{7}$.

The last type of interaction concerns two irrational numbers, where the sum or product may be rational or irrational depending on the numbers being used. The following expression shows a rational sum from two irrational numbers:

$$\sqrt{3} + \left(6 - \sqrt{3}\right) = 6$$

The product of two irrational numbers can be rational or irrational. A rational result can be seen in the following expression:

$$\sqrt{2} \times \sqrt{8} = \sqrt{2 \times 8} = \sqrt{16} = 4$$

An irrational result can be seen in the following:

$$\sqrt{3} \times \sqrt{2} = \sqrt{6}$$

Rewriting Expressions Involving Radicals and Rational Exponents

Exponents are used in mathematics to express a number or variable multiplied by itself a certain number of times. For example, x^3 means x is multiplied by itself three times. In this expression, x is called the *base*, and 3 is the *exponent*. Exponents can be used in more complex problems when they contain fractions and negative numbers.

Fractional exponents can be explained by looking first at the inverse of exponents, which are *roots*. Given the expression x^2, the square root can be taken, $\sqrt{x^2}$, cancelling out the 2 and leaving x by itself, if x is positive. Cancellation occurs because $\sqrt{x}$ can be written with exponents, instead of roots, as $x^{\frac{1}{2}}$. The numerator of 1 is the exponent, and the denominator of 2 is called the root (which is why it's referred to as *square root*). Taking the square root of x^2 is the same as raising it to the $\frac{1}{2}$ power. Written out in mathematical form, it takes the following progression:

$$\sqrt{x^2} = (x^2)^{\frac{1}{2}} = x$$

From properties of exponents, $2 \times \frac{1}{2} = 1$ is the actual exponent of x. Another example can be seen with $x^{\frac{4}{7}}$. The variable x, raised to four-sevenths, is equal to the seventh root of x to the fourth power: $\sqrt[7]{x^4}$. In general,

$$x^{\frac{1}{n}} = \sqrt[n]{x}$$

and

$$x^{\frac{m}{n}} = \sqrt[n]{x^m}$$

Negative exponents also involve fractions. Whereas y^3 can also be rewritten as $\frac{y^3}{1}$, y^{-3} can be rewritten as $\frac{1}{y^3}$. A negative exponent means the exponential expression must be moved to the opposite spot in a fraction to make the exponent positive. If the negative appears in the numerator, it moves to the denominator. If the negative appears in the denominator, it is moved to the numerator. In general, $a^{-n} = \frac{1}{a^n}$, and a^{-n} and a^n are reciprocals.

Take, for example, the following expression:

$$\frac{a^{-4}b^2}{c^{-5}}$$

Since a is raised to the negative fourth power, it can be moved to the denominator. Since c is raised to the negative fifth power, it can be moved to the numerator. The b variable is raised to the positive second power, so it does not move.

The simplified expression is as follows:

$$\frac{b^2c^5}{a^4}$$

In mathematical expressions containing exponents and other operations, the order of operations must be followed. *PEMDAS* states that exponents are calculated after any parenthesis and grouping symbols but before any multiplication, division, addition, and subtraction.

Scientific Notation

Scientific Notation is used to represent numbers that are either very small or very large. For example, the distance to the sun is approximately 150,000,000,000 meters. Instead of writing this number with so many zeros, it can be written in scientific notation as 1.5×10^{11} meters. The same is true for very small numbers, but the exponent becomes negative. If the mass of a human cell is 0.000000000001 kilograms, that measurement can be easily represented by 1.0×10^{-12} kilograms. In both situations, scientific notation makes the measurement easier to read and understand. Each number is translated to an expression with one digit in the tens place times an expression corresponding to the zeros.

When two measurements are given and both involve scientific notation, it is important to know how these interact with each other:

- In addition and subtraction, the exponent on the ten must be the same before any operations are performed on the numbers. For example, $(1.3 \times 10^4) + (3.0 \times 10^3)$ cannot be added until one of the exponents on the ten is changed. The 3.0×10^3 can be changed to 0.3×10^4, then the 1.3 and 0.3 can be added. The answer comes out to be 1.6×10^4.

- For multiplication, the first numbers can be multiplied and then the exponents on the tens can be added. Once an answer is formed, it may have to be converted into scientific notation again depending on the change that occurred.

- The following is an example of multiplication with scientific notation:

$$(4.5 \times 10^3) \times (3.0 \times 10^{-5}) = 13.5 \times 10^{-2}$$

- Since this answer is not in scientific notation, the decimal is moved over to the left one unit, and 1 is added to the ten's exponent. This results in the final answer: 1.35×10^{-1}.

- For division, the first numbers are divided, and the exponents on the tens are subtracted. Again, the answer may need to be converted into scientific notation form, depending on the type of changes that occurred during the problem.

- *Order of magnitude* relates to scientific notation and is the total count of powers of 10 in a number. For example, there are 6 orders of magnitude in 1,000,000. If a number is raised by an order of magnitude, it is multiplied times 10. Order of magnitude can be helpful in estimating results using very large or small numbers. An answer should make sense in terms of its order of magnitude.

- For example, if area is calculated using two dimensions with 6 orders of magnitude, because area involves multiplication, the answer should have around 12 orders of magnitude. Also, answers can be estimated by rounding to the largest place value in each number. For example, 5,493,302×2,523,100 can be estimated by 5×3 = 15 with 6 orders of magnitude.

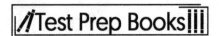

Reasoning Quantitatively and Using Units to Solve Problems

It is important to be able to reason quantitatively when working with mathematical problems. In some ways, mathematics can be thought of as a foreign language. As one gains fluency, they should develop the ability to correctly represent a given problem, work through the procedures to find a solution, and take pause throughout the process to evaluate the logic behind the steps and intermediate answers found.

The meaning behind the numbers involved, including their units and magnitude, should be considered as the values are manipulated. For example, if a problem is investigating the speed at which a car traveled, students should be mindful that the units should be in miles per hour or kilometers per hour. Also, they should consider logical driving speeds. It is important to not only memorize formulas and procedures, but to understand the meaning and purpose behind them so that they are correctly applied to various situations. Whenever possible, the calculated solution should be verified for accuracy before moving onto the next problem. To accomplish this, the inverse operation or procedure can sometimes be applied.

As a simple example, if asked to calculate the product of nine and four ($9 \times 4 = 36$), the answer can be double-checked by using the inverse operation: ($\frac{36}{9} = 4$). When it is not possible to rely on this method of double-checking answers, one should consider if the answer makes sense logically. This ability showcases a deeper understanding of the mathematical principles at play. As another example, if a problem is asking for the total length of fencing needed to enclose a small vegetable garden, it is reasonable that the answer will be a certain amount of feet. Any calculated solutions that include units of measurement that differ from this (inches, miles, square feet, etc.) can be immediately disregarded because they would not be logical. Solutions to quantitative problems should be verified as reasonable to prevent careless mistakes and unnecessary errors and to ensure that the proper procedures and calculations are carried out.

Dimensional analysis is the process of converting between different units using equivalent measurement statements. For example, running a 5K is the same as running approximately 3.1 miles. This conversion can be found by knowing that 1 kilometer is equal to approximately 0.62 miles.

The following calculation shows how to convert kilometers into miles. The original units need to be opposite one another in each of the two fractions: one in the original amount and one in the denominator of the conversion factor. This specific example consists of 5 km being multiplied times the conversion factor .62 mi/km. By design, quantities in kilometers are opposite one another and therefore cancel, leaving 3.11 miles as the converted result.

$$5km \times \left(\frac{0.62 miles}{1 km}\right) = 3.11 \; miles$$

Units are also important throughout formulas in calculating quantities such as volume and area. To find the volume of a pyramid, the following formula is used: $V = \frac{1}{3}Bh$. B is the area of the base, and h is the height. In the example shown below, two of the same type of dimension are composed of two different units. All dimensions must be converted to the same units before plugging values into the formula for volume.

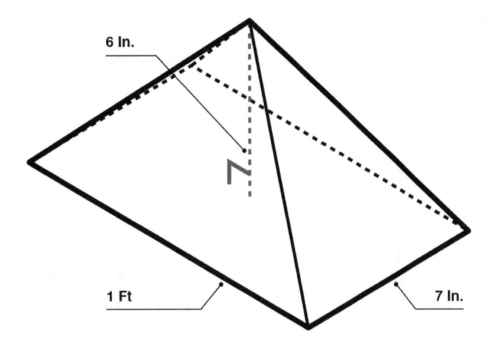

In this case, all lengths will be converted to inches. To find the area of the base, it's necessary to convert 1 ft. to 12 inches. Then, the area of the base can be calculated as $B = 12\ in \times 7\ in = 84\ in^2$. B can then be substituted into the volume formula as follows: $V = \frac{1}{3}(84in^2)(6in) = 168\ in^3$.

Formulas are a common situation in which units need to be interpreted and used. However, graphs can also carry meaning through units. The following graph is an example. It represents a graph of the position of an object over time. The *m* axis represents the number of meters the object is from the starting point at time *s*, in seconds. Interpreting this graph, the origin shows that at time zero seconds, the object is zero meters away from the starting point. As the time increases to one second, the position increases to five meters away. This trend continues until 6 seconds, where the object is 30 meters away from the starting position. After this point in time—since the graph remains horizontal from 6 to 10 seconds—the object must have stopped moving.

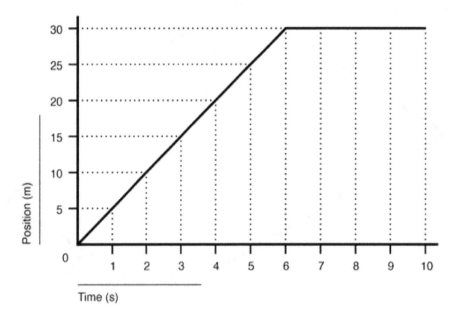

When solving problems with units, it's important to consider the reasonableness of the answer. If conversions are used, it's helpful to have an estimated value to compare the final answer to. This way, if the final answer is too distant from the estimate, it will be obvious that a mistake was made.

Choosing a Level of Accuracy Appropriate to Limitations on Measurement

Precision and accuracy are used to describe groups of measurements. *Precision* describes a group of measures that are very close together, regardless of whether the measures are close to the true value. *Accuracy* describes how close the measures are to the true value. The following graphic illustrates the different combinations that may occur with different groups of measures:

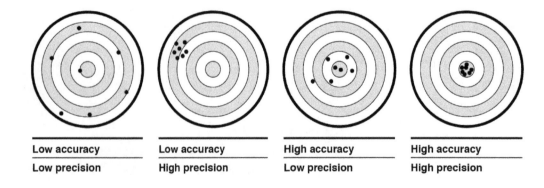

Low accuracy
Low precision

Low accuracy
High precision

High accuracy
Low precision

High accuracy
High precision

Since accuracy refers to the closeness of a value to the true measurement, the level of accuracy depends on the object measured and the instrument used to measure it. This will vary depending on the situation. If measuring the mass of a set of dictionaries, kilograms may be used as the units. In this case, it is not vitally important to have a high level of accuracy. If the measurement is a few grams away from the true value, the discrepancy might not make a big difference in the problem.

In a different situation, the level of accuracy may be more significant. Pharmacists need to be sure they are very accurate in their measurements of medicines that they give to patients. In this case, the level of accuracy is vitally important and not something to be estimated. In the dictionary situation, the measurements were given as whole numbers in kilograms. In the pharmacist's situation, the measurements for medicine must be taken to the milligram and sometimes further, depending on the type of medicine.

When considering the accuracy of measurements, the error in each measurement can be shown as absolute and relative. *Absolute error* tells the actual difference between the measured value and the true value. The *relative error* tells how large the error is in relation to the true value. There may be two problems where the absolute error of the measurements is 10 grams. For one problem, this may mean the relative error is very small because the measured value is 14,990 grams, and the true value is 15,000 grams. Ten grams in relation to the true value of 15,000 is small: 0.06%. For the other problem, the measured value is 290 grams, and the true value is 300 grams. In this case, the 10-gram absolute error means a high relative error because the true value is smaller. The relative error is 10/300 = 0.03, or 3%.

Solving Multistep Real-World and Mathematical Problems Involving Rational Numbers in Any Form

*Ratio*s are used to show the relationship between two quantities. The ratio of oranges to apples in the grocery store may be 3 to 2. That means that for every 3 oranges, there are 2 apples. This comparison can be expanded to represent the actual number of oranges and apples, such as 36 oranges to 24 apples. Another example may be the number of boys to girls in a math class. If the ratio of boys to girls is given as 2 to 5, that means there are 2 boys to every 5 girls in the class. Ratios can also be compared if the units in each ratio are the same. The ratio of boys to girls in the math class can be compared to the ratio of boys to girls in a science class by stating which ratio is higher and which is lower.

Rates are used to compare two quantities with different units. *Unit rates* are the simplest form of rate. With unit rates, the denominator in the comparison of two units is one. For example, if someone can type at a rate of 1000 words in 5 minutes, then his or her unit rate for typing is $\frac{1000}{5} = 200$ words in one minute or 200 words per minute. Any rate can be converted into a unit rate by dividing to make the denominator one. 1000 words in 5 minutes has been converted into the unit rate of 200 words per minute.

Ratios and rates can be used together to convert rates into different units. For example, if someone is driving 50 kilometers per hour, that rate can be converted into miles per hour by using a ratio known as the *conversion factor*. Since the given value contains kilometers and the final answer needs to be in miles, the ratio relating miles to kilometers needs to be used. There are 0.62 miles in 1 kilometer. This, written as a ratio and in fraction form, is

$$\frac{0.62 \ miles}{1 \ km}$$

To convert 50km/hour into miles per hour, the following conversion needs to be set up:

$$\frac{50\ km}{hour} \times \frac{0.62\ miles}{1\ km} = 31\ miles\ per\ hour$$

The ratio between two similar geometric figures is called the *scale factor*. In the following example, there are two similar triangles. The scale factor from figure A to figure B is 2 because the length of the corresponding side of the larger triangle, 14, is twice the corresponding side on the smaller triangle, 7.

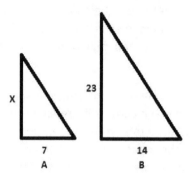

This scale factor can also be used to find the value of X. Since the scale factor from small to large is 2, the larger number, 23, can be divided by 2 to find the missing side: X=11.5. The scale factor can also be represented in the equation $2A = B$ because two times the lengths of A gives the corresponding lengths of B. This is the idea behind similar triangles.

Much like a scale factor can be written using an equation like $2A = B$, a *proportional relationship* is represented by the equation $Y = kX$. X and Y are proportional because as values in X increase, the values in Y also increase. A relationship that is inversely proportional can be represented by the equation $Y = \frac{k}{X}$, where the value of Y decreases as the value of X increases and vice versa. The following graph represents these two types of relationships between x and y. The grey line represents a proportional relationship because the y-values increase as the x-values increase. The black line represents an inversely-proportional relationship because the y-values decrease as the x-values increase.

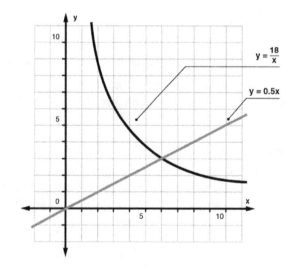

Proportional reasoning can be used to solve problems involving ratios, percentages, and averages. Ratios can be used in setting up proportions and solving them to find unknowns. For example, if someone averages 10 pages of math homework completed in 3 nights, how long would it take him or her to complete 22 pages? Both ratios can be written as fractions. The second ratio would contain the unknown. The following proportion represents this problem where x is the unknown number of nights:

$$\frac{10\ pages}{3\ nights} = \frac{22\ pages}{x\ nights}$$

Solving this proportion entails cross-multiplying and results in the following equation: $10x = 22 \times 3$. Simplifying and solving for x results in the exact solution: $x = 6.6\ nights$. The result would be rounded up to 7 because the homework would actually be completed on the 7th night.

The following problem uses ratios involving percentages:

If 20% of the class is girls and 30 students are in the class, how many girls are in the class?

To set up this problem, it is helpful to use the common proportion: $\frac{\%}{100} = \frac{is}{of}$. Within the proportion, % is the percentage of girls, 100 is the total percentage of the class, *is* is the number of girls, and *of* is the total number of students in the class. Most percentage problems can be written using this language. To solve this problem, the proportion should be set up as $\frac{20}{100} = \frac{x}{30}$, then solved for x. Cross-multiplying results in the equation $20 \times 30 = 100x$, which results in the solution $x = 6$. There are 6 girls in the class.

Problems involving volume, length, and other units can also be solved using ratios. If the following graphic of a cone is given, the problem may ask for the volume to be found.

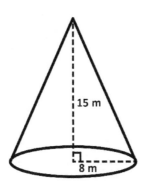

15 m

8 m

Referring to the formulas provided on the test, the volume of a cone is given as: $V = \pi r^2 \frac{h}{3}$, where r is the radius, and h is the height. Plugging $r = 8$ and $h = 15$ from the graphic into the formula, the following is obtained: $V = \pi(8^2)\frac{15}{3}$. Therefore, volume of the cone is found to be 1005.3m³. Sometimes, answers in different units are sought. If this problem wanted the answer in liters, 1005.3m³ would need to be converted. Using the equivalence statement 1m³ = 1000L, the following ratio would be used to solve for liters: $1005.3m^3 \times \frac{1000L}{1m^3}$. Cubic meters in the numerator and denominator cancel each other out, and the answer is converted to 1,005,300 liters, or 1.0053×10^6 L.

Other conversions can also be made between different given and final units. If the temperature in a pool is 30°C, what is the temperature of the pool in degrees Fahrenheit? To convert these units, an equation

is used relating Celsius to Fahrenheit. The following equation is used: $T_{°F} = 1.8T_{°C} + 32$. Plugging in the given temperature and solving the equation for T yields the result: $T_{°F} = 1.8(30) + 32 = 86°F$. Both units in the metric system and U.S. customary system are widely used.

Measurement/Geometry

Using Transformations to Show Congruence or Similarity

Transformations in the Plane

A *transformation* occurs when a shape is altered in the plane where it exists. There are three major types of transformation: translations, reflections, and rotations. A *translation* consists of shifting a shape in one direction. A *reflection* results when a shape is transformed over a line to its mirror image. Finally, a *rotation* occurs when a shape moves in a circular motion around a specified point. The object can be turned clockwise or counterclockwise and, if rotated 360 degrees, returns to its original location.

Distance and Angle Measure

The three major types of transformations preserve distance and angle measurement. The shapes stay the same, but they are moved to another place in the plane. Therefore, the distance between any two points on the shape doesn't change. Also, any original angle measure between two line segments doesn't change. However, there are transformations that don't preserve distance and angle measurements, including those that don't preserve the original shape. For example, transformations that involve stretching and shrinking shapes don't preserve distance and angle measures. In these cases, the input variables are multiplied by either a number greater than one (*stretch*) or less than one (*shrink*).

Rigid Motion

A *rigid motion* is a transformation that preserves distance and length. Every line segment in the resulting image is congruent to the corresponding line segment in the pre-image. Congruence between two figures means a series of transformations (or a rigid motion) can be defined that maps one of the figures onto the other. Basically, two figures are congruent if they have the same shape and size.

Dilation

A shape is dilated, or a *dilation* occurs, when each side of the original image is multiplied by a given scale factor. If the scale factor is less than 1 and greater than 0, the dilation contracts the shape, and the resulting shape is smaller. If the scale factor equals 1, the resulting shape is the same size, and the dilation is a rigid motion. Finally, if the scale factor is greater than 1, the resulting shape is larger and the dilation expands the shape. The *center of dilation* is the point where the distance from it to any point on the new shape equals the scale factor times the distance from the center to the corresponding point in the pre-image. Dilation isn't an isometric transformation because distance isn't preserved. However, angle measure, parallel lines, and points on a line all remain unchanged. The following figure is an example of translation, rotation, dilation, and reflection:

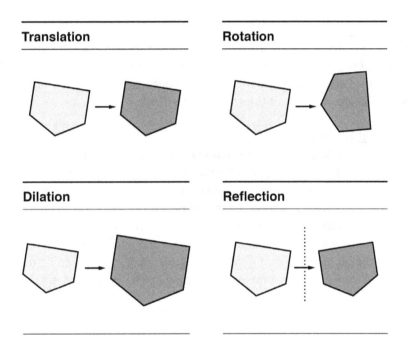

Determining Congruence

Two figures are congruent if there is a rigid motion that can map one figure onto the other. Therefore, all pairs of sides and angles within the image and pre-image must be congruent. For example, in triangles, each pair of the three sides and three angles must be congruent. Similarly, in two four-sided figures, each pair of the four sides and four angles must be congruent.

To prove theorems about triangles, basic definitions involving triangles (e.g., equilateral, isosceles, etc.) need to be known. Proven theorems concerning lines and angles can be applied to prove theorems about triangles. Common theorems to be proved include: the sum of all angles in a triangle equals 180 degrees; the sum of the lengths of two sides of a triangle is greater than the length of the third side; the base angles of an isosceles triangle are congruent; the line segment connecting the midpoint of two sides of a triangle is parallel to the third side and its length is half the length of the third side; and the medians of a triangle all meet at a single point.

Triangle Congruence

There are five theorems to show that triangles are congruent when it's unknown whether each pair of angles and sides are congruent. Each theorem is a shortcut that involves different combinations of sides and angles that must be true for the two triangles to be congruent. For example, *side-side-side (SSS)* states that if all sides are equal, the triangles are congruent. *Side-angle-side (SAS)* states that if two pairs of sides are equal and the included angles are congruent, then the triangles are congruent. Similarly, *angle-side-angle (ASA)* states that if two pairs of angles are congruent and the included side lengths are equal, the triangles are similar. *Angle-angle-side (AAS)* states that two triangles are congruent if they have two pairs of congruent angles and a pair of corresponding equal side lengths that aren't included. Finally, *hypotenuse-leg (HL)* states that if two right triangles have equal hypotenuses and an equal pair of shorter sides, then the triangles are congruent. An important item to note is that angle-angle-angle *(AAA)* is not enough information to have congruence. It's important to understand why these rules work by using rigid motions to show congruence between the triangles with the given properties. For example, three reflections are needed to show why *SAS* follows from the definition of congruence.

Similarity for Two Triangles

If two angles of one triangle are congruent with two angles of a second triangle, the triangles are similar. This is because, within any triangle, the sum of the angle measurements is 180 degrees. Therefore, if two are congruent, the third angle must also be congruent because their measurements are equal. Three congruent pairs of angles mean that the triangles are similar.

Proving Congruence and Similarity

The criteria needed to prove triangles are congruent involves both angle and side congruence. Both pairs of related angles and sides need to be of the same measurement to use congruence in a proof. The criteria to prove similarity in triangles involves proportionality of side lengths. Angles must be congruent in similar triangles; however, corresponding side lengths only need to be a constant multiple of each other. Once similarity is established, it can be used in proofs as well. Relationships in geometric figures other than triangles can be proven using triangle congruence and similarity. If a similar or congruent triangle can be found within another type of geometric figure, their criteria can be used to prove a relationship about a given formula. For example, a rectangle can be broken up into two congruent triangles.

Properties of Polygons and Circles

A polygon is a closed two-dimensional figure consisting of three or more sides. Polygons can be either convex or concave. A polygon that has interior angles all measuring less than 180° is convex. A concave polygon has one or more interior angles measuring greater than 180°. Examples are shown below.

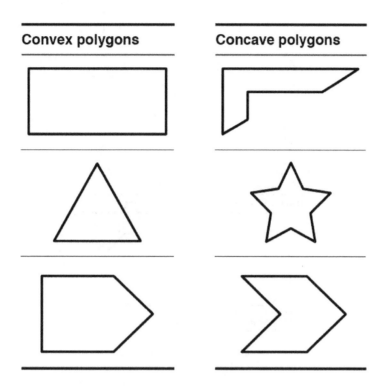

Polygons can be classified by the number of sides (also equal to the number of angles) they have. The following are the names of polygons with a given number of sides or angles:

# of sides	3	4	5	6	7	8	9	10
Name of polygon	Triangle	Quadrilateral	Pentagon	Hexagon	Septagon (or heptagon)	Octagon	Nonagon	Decagon

Equiangular polygons are polygons in which the measure of every interior angle is the same. The sides of equilateral polygons are always the same length. If a polygon is both equiangular and equilateral, the polygon is defined as a regular polygon. Examples are shown below.

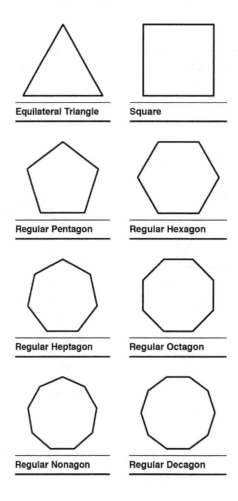

Equilateral Triangle   Square

Regular Pentagon   Regular Hexagon

Regular Heptagon   Regular Octagon

Regular Nonagon   Regular Decagon

Triangles can be further classified by their sides and angles. A triangle with its largest angle measuring 90° is a right triangle.

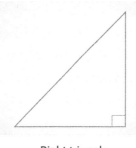

Right triangle

A triangle with the largest angle less than 90° is an acute triangle. A triangle with the largest angle greater than 90° is an obtuse triangle.

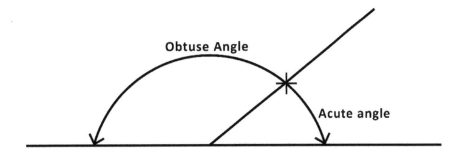

A triangle consisting of two equal sides and two equal angles is an isosceles triangle. A triangle with three equal sides and three equal angles is an equilateral triangle. A triangle with no equal sides or angles is a scalene triangle.

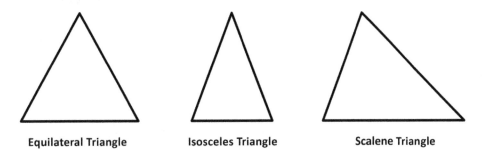

Quadrilaterals can be further classified according to their sides and angles. A quadrilateral with exactly one pair of parallel sides is called a trapezoid. A quadrilateral that shows both pairs of opposite sides parallel is a parallelogram. Parallelograms include rhombuses, rectangles, and squares. A rhombus has four equal sides. A rectangle has four equal angles (90° each). A square has four 90° angles and four equal sides. Therefore, a square is both a rhombus and a rectangle.

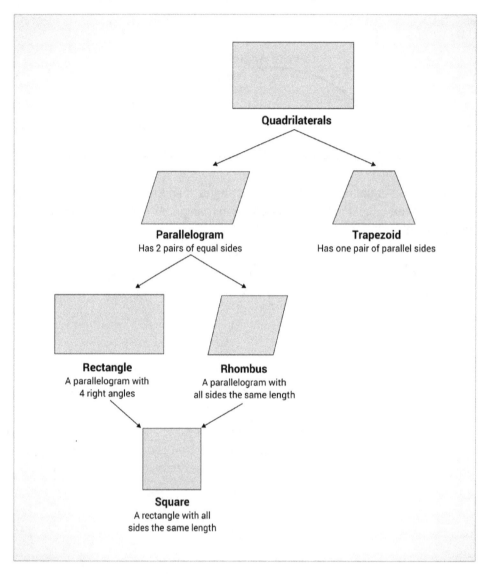

There are many key facts related to geometry that are applicable. The sum of the measures of the angles of a triangle are 180°, and for a quadrilateral, the sum is 360°. Rectangles and squares each have four right angles. A *right angle* has a measure of 90°.

Perimeter

The *perimeter* is the distance around a figure or the sum of all sides of a polygon.

The *formula for the perimeter of a square* is four times the length of a side. For example, the following square has side lengths of 5 meters:

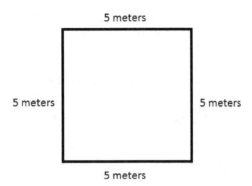

The perimeter is 20 meters because 4 times 5 is 20.

The *formula for a perimeter of a rectangle* is the sum of twice the length and twice the width. For example, if the length of a rectangle is 10 inches and the width 8 inches, then the perimeter is 36 inches because:

$$P = 2l + 2w$$

$$2(10) + 2(8)$$

$$20 + 16 = 36 \text{ inches}$$

Area

The area is the amount of space inside of a figure, and there are formulas associated with area.

The area of a triangle is the product of ½ the base and height. For example, if the base of the triangle is 2 feet and the height is 4 feet, then the area is 4 square feet. The following equation shows the formula used to calculate the area of the triangle:

$$A = \frac{1}{2}bh = \frac{1}{2}(2)(4) = 4 \text{ square feet}$$

The area of a square is the length of a side squared. For example, if a side of a square is 7 centimeters, then the area is 49 square centimeters. The formula for this example is $A = s^2 = 7^2 = 49$ square centimeters. An example is if the rectangle has a length of 6 inches and a width of 7 inches, then the area is 42 square inches:

$$A = lw = 6(7) = 42 \text{ square inches}$$

The area of a trapezoid is ½ the height times the sum of the bases. For example, if the length of the bases are 2.5 and 3 feet and the height 3.5 feet, then the area is 9.625 square feet. The following formula shows how the area is calculated:

$$A = \frac{1}{2}h(b_1 + b_2)$$

$$\frac{1}{2}(3.5)(2.5 + 3)$$

$$\frac{1}{2}(3.5)(5.5) = 9.625 \text{ square feet}$$

The perimeter of a figure is measured in single units, while the area is measured in square units.

If a quadrilateral is inscribed in a circle, the sum of its opposite angles is 180 degrees. Consider the quadrilateral ABCD centered at the point O:

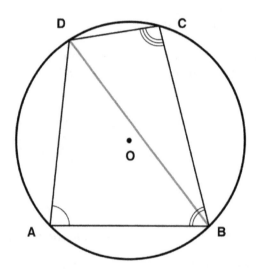

Each of the four line segments within the quadrilateral is a chord of the circle. Consider the diagonal DB. Angle DAB is an inscribed angle leaning on the arc DCB. Therefore, angle DAB is half the measure of the arc DCB. Conversely, angle DCB is an inscribed angle leaning on the arc DAB. Therefore, angle DCB is half the measure of the arc DAB. The sum of arcs DCB and DAB is 360 degrees because they make up the entire circle. Therefore, the sum of angles DAB and DCB equals half of 360 degrees, which is 180 degrees.

Degrees are used to express the size of an angle. A complete circle is represented by 360°, and a half circle is represented by 180°. In addition, a right angle fills one quarter of a circle and is represented by 90°.

The equation used to find the area of a circle is $A = \pi r^2$. For example, if a circle has a radius of 5 centimeters, the area is computed by substituting 5 for the radius: $(5)^2$. Using this reasoning, to find half of the area of a circle, the formula is $A = .5\pi r^2$. Similarly, to find the quarter of an area of a circle,

the formula is $A = .25\pi r^2$. To find any fractional area of a circle, a student can use the formula $A = \frac{C}{360}\pi r^2$, where C is the number of degrees of the central angle of the sector.

Other related concepts for circles include the diameter and circumference. *Circumference* is the distance around a circle. The formula for circumference is $C = 2\pi r$. The *diameter* of a circle is the distance across a circle through its center point. The formula for circumference can also be thought of as $C = dr$ where d is the circle's diameter, since the diameter of a circle is *2r*.

A *circle* can be defined as the set of all points that are the same distance (known as the radius, r) from a single point C (known as the center of the circle). The center has coordinates (h, k), and any point on the circle can be labelled with coordinates (x, y).

As shown below, a *right triangle* is formed with these two points:

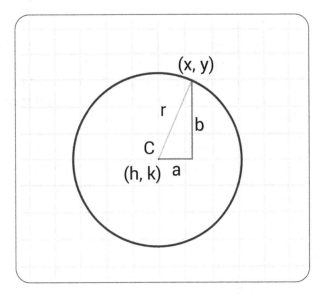

The Pythagorean theorem states that $a^2 + b^2 = r^2$. However, a can be replaced by $|x - h|$ and b can be replaced by $|y - k|$ by using the *distance formula* which is:

$$d = \sqrt{(x_2 - x_1)^2 + (y_2 - y_1)^2}$$

That substitution results in:

$$(x - h)^2 + (y - k)^2 = r^2$$

This is the formula for finding the equation of any circle with a center (h, k) and a radius r. Note that sometimes C is used instead of r.

The Pythagorean Theorem

The *Pythagorean theorem* is an important relationship between the three sides of a right triangle. It states that the square of the side opposite the right triangle, known as the *hypotenuse* (denoted as c^2), is equal to the sum of the squares of the other two sides ($a^2 + b^2$). Thus, $a^2 + b^2 = c^2$.

The theorem can be seen in the following image:

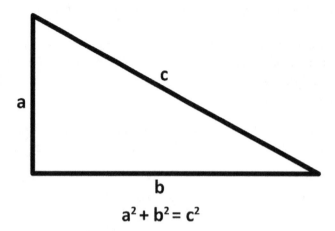

$$a^2 + b^2 = c^2$$

Both the trigonometric functions and the Pythagorean theorem can be used in problems that involve finding either a missing side or a missing angle of a right triangle. To do so, one must look to see what sides and angles are given and select the correct relationship that will help find the missing value. These relationships can also be used to solve application problems involving right triangles. Often, it's helpful to draw a figure to represent the problem to see what's missing.

As an example of the theorem, suppose that Shirley has a rectangular field that is 5 feet wide and 12 feet long, and she wants to split it in half using a fence that goes from one corner to the opposite corner. How long will this fence need to be? To figure this out, note that this makes the field into two right triangles, whose hypotenuse will be the fence dividing it in half. Therefore, the fence length is given by $\sqrt{5^2 + 12^2} = \sqrt{169} = 13$ feet long.

Similar and Congruent Triangles

Suppose that Lara is 5 feet tall and is standing 30 feet from the base of a light pole, and her shadow is 6 feet long. How high is the light on the pole? To figure this out, it helps to make a sketch of the situation:

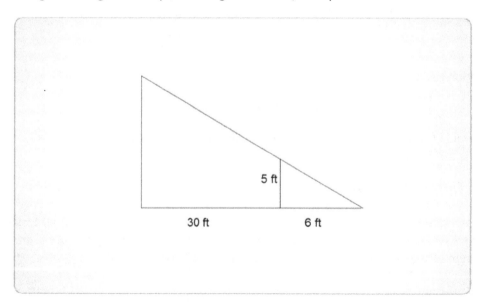

The light pole is the left side of the triangle. Lara is the 5-foot vertical line. Test takers should notice that there are two right triangles here, and that they have all the same angles as one another. Therefore, they form similar triangles. So, the ratio of proportionality between them must be found.

The bases of these triangles are known. The small triangle, formed by Lara and her shadow, has a base of 6 feet. The large triangle formed by the light pole along with the line from the base of the pole out to the end of Lara's shadow is $30 + 6 = 36$ feet long. So, the ratio of the big triangle to the little triangle is $\frac{36}{6} = 6$. The height of the little triangle is 5 feet. Therefore, the height of the big triangle will be $6 \cdot 5 = 30$ feet, meaning that the light is 30 feet up the pole.

Using Volume and Surface Area Formulas

Surface area and volume are two- and three-dimensional measurements. Surface area measures the total surface space of an object, like the six sides of a cube. Questions about surface area will ask how much of something is needed to cover a three-dimensional object, like wrapping a present. **Volume** is the measurement of how much space an object occupies, like how much space is in the cube. Volume questions will ask how much of something is needed to completely fill the object. The most common surface area and volume questions deal with spheres, cubes, and rectangular prisms.

The formula for a cube's surface area is $SA = 6 \times s^2$, where s is the length of a side. A cube has 6 equal sides, so the formula expresses the area of all the sides. Volume is simply measured by taking the cube of the length, so the formula is $V = s^3$.

The surface area formula for a rectangular prism or a general box is $SA = 2(lw + lh + wh)$, where l is the length, h is the height, and w is the width. The volume formula is $V = l \times w \times h$, which is the cube's volume formula adjusted for the unequal lengths of a box's sides.

The formula for a sphere's surface area is $SA = 4\pi r^2$, where r is the sphere's radius. The surface area formula is the area for a circle multiplied by four. To measure volume, the formula is $V = \frac{4}{3}\pi r^3$.

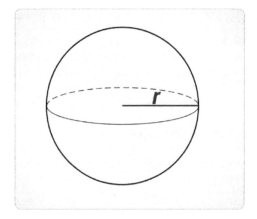

A **rectangular pyramid** is a figure with a rectangular base and four triangular sides that meet at a single vertex. If the rectangle has sides of lengths x and y, then the volume will be given by $V = \frac{1}{3}xyh$.

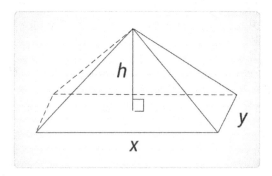

To find the surface area, the dimensions of each triangle must be known. However, these dimensions can differ depending on the problem in question. Therefore, there is no general formula for calculating total surface area.

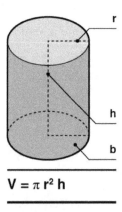

$$V = \pi r^2 h$$

The formula to find the volume of a cylinder is $\pi r^2 h$. This formula contains the formula for the area of a circle (πr^2) because the base of a cylinder is a circle. To calculate the volume of a cylinder, the slices of

circles needed to build the entire height of the cylinder are added together. For example, if the radius is 5 feet and the height of the cylinder is 10 feet, the cylinder's volume is calculated by using the following equation: $\pi 5^2 \times 10$. Substituting 3.14 for π, the volume is 785 ft³.

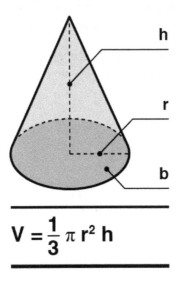

$$V = \frac{1}{3} \pi r^2 h$$

The formula used to calculate the volume of a cone is $\frac{1}{3}\pi r^2 h$. Essentially, the area of the base of the cone is multiplied by the cone's height. In a real-life example where the radius of a cone is 2 meters and the height of a cone is 5 meters, the volume of the cone is calculated by utilizing the formula $\frac{1}{3}\pi 2^2 \times 5$. After substituting 3.14 for π, the volume is 20.9 m³.

Concepts of Density

The *density* of a substance is the ratio of mass to area or volume. It's a relationship between the mass and how much space the object actually takes up. Knowing which units to use in each situation is crucial. Population density is an example of a real-life situation that's modeled by using density concepts. It involves calculating the ratio of the number of people to the number of square miles. The amount of material needed per a specific unit of area or volume is another application. For example, estimating the number of BTUs per cubic foot of a home is a measurement that relates to heating or cooling the house based on the desired temperature and the house's size.

Solving Problems Involving Angles

In geometry, a *line* connects two points, has no thickness, and extends indefinitely in both directions beyond each point. If the length is finite, it's known as a *line segment* and has two *endpoints*. A *ray* is the straight portion of a line that has one endpoint and extends indefinitely in the other direction. An *angle* is formed when two rays begin at the same endpoint and extend indefinitely. The endpoint of an angle is called a *vertex*. *Adjacent angles* are two side-by-side angles formed from the same ray that have the same endpoint. Angles are measured in *degrees* or *radians*, which is a measure of *rotation*. A *full rotation* equals 360 degrees or 2π radians, which represents a circle. Half a rotation equals 180 degrees or π radians and represents a half-circle. Subsequently, 90 degrees ($\frac{\pi}{2}$ radians) represents a quarter of a circle, which is known as a *right angle*. Any angle less than 90 degrees is an *acute angle*, and any angle greater than 90 degrees is an *obtuse angle*. Angle measurement is additive. When an angle is broken into two non-overlapping angles, the total measure of the larger angle equals the sum of the two smaller angles. Lines are *coplanar* if they're located in the same plane. Two lines are *parallel* if they are coplanar, extend in the same direction, and never cross. If lines do cross, they're labeled as *intersecting lines* because they "intersect" at one point. If they intersect at more than one point, they're the same line. *Perpendicular lines* are coplanar lines that form a right angle at their point of intersection.

Supplementary angles add up to 180 degrees. *Vertical angles* are two nonadjacent angles formed by two intersecting lines. For example, in the following picture, angles 4 and 2 are vertical angles and so are angles 1 and 3:

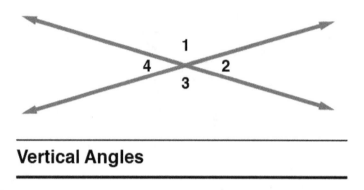

Vertical Angles

Corresponding angles are two angles in the same position whenever a straight line (known as a *transversal*) crosses two others. If the two lines are parallel, the corresponding angles are equal. In the following diagram, angles 1 and 3 are corresponding angles but aren't equal to each other:

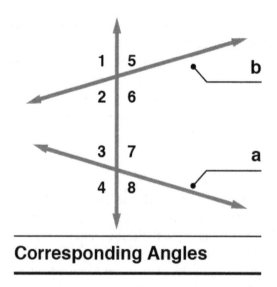

Corresponding Angles

Alternate interior angles are also a pair of angles formed when two lines are crossed by a transversal. They are opposite angles that exist inside of the two lines. In the corresponding angles diagram above,

angles 2 and 7 are alternate interior angles, as well as angles 6 and 3. *Alternate exterior angles* are opposite angles formed by a transversal but, in contrast to interior angles, exterior angles exist outside the two original lines. Therefore, angles 1 and 8 are alternate exterior angles and so are angles 5 and 4. Finally, *consecutive interior angles* are pairs of angles formed by a transversal. These angles are located on the same side of the transversal and inside the two original lines. Therefore, angles 2 and 3 are a pair of consecutive interior angles, and so are angles 6 and 7. These definitions are instrumental in solving many problems that involve determining relationships between angles.

Data Analysis/Probability/Statistics

Summarizing and Interpreting Data, Making Predictions, and Solving Problems

Summarizing Data

Most statistics involve collecting a large amount of data, analyzing it, and then making decisions based on previously known information. These decisions also can be measured through additional data collection and then analyzed. Therefore, the cycle can repeat itself over and over. Representing the data visually is a large part of the process, and many plots on the real number line exist that allow this to be done. For example, a *dot plot* uses dots to represent data points above the number line. Also, a *histogram* represents a data set as a collection of rectangles, which illustrate the frequency distribution of the data. Finally, a *box plot* (also known as a *box and whisker plot*) plots a data set on the number line by segmenting the distribution into four quartiles that are divided equally in half by the median. Here's an example of a box plot, a histogram, and a dot plot for the same data set:

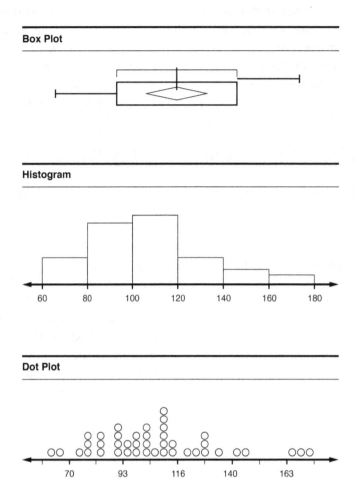

Interpreting Data

Comparing data sets within statistics can mean many things. The first way to compare data sets is by looking at the center and spread of each set. The center of a data set can mean two things: median or mean. The *median* is the value that's halfway into each data set, and it splits the data into two intervals. The *mean* is the average value of the data within a set. It's calculated by adding up all of the data in the set and dividing the total by the number of data points. Outliers can significantly impact the mean. Additionally, two completely different data sets can have the same mean. For example, a data set with values ranging from zero to 100 and a data set with values ranging from 44 to 56 can both have means of 50. The first data set has a much wider range, which is known as the *spread* of the data. This measures how varied the data is within each set. Spread can be defined further as either interquartile range or standard deviation. The *interquartile range (IQR)* is the range of the middle 50 percent of the data set. This range can be seen in the large rectangle on a box plot. The *standard deviation* quantifies the amount of variation with respect to the mean. A lower standard deviation shows that the data set doesn't differ greatly from the mean. A larger standard deviation shows that the data set is spread out farther from the mean.

Given a data set X consisting of data points $(x_1, x_2, x_3, \ldots x_n)$, the *variance* of X is defined to be:

$$\frac{\sum_{i=1}^{n}(x_i - \bar{X})^2}{n}$$

This means that the variance of X is the average of the squares of the differences between each data point and the mean of X.

Given a data set X consisting of data points $(x_1, x_2, x_3, \ldots x_n)$, the **standard deviation** of X is defined to be:

$$s_x = \sqrt{\frac{\sum_{i=1}^{n}(x_i - \bar{X})^2}{n}}$$

x is each value in the data set, $\bar{x}$ is the mean, and n is the total number of data points in the set.

In other words, the standard deviation is the square root of the variance. Both the variance and the standard deviation are measures of how much the data tend to be spread out. When the standard deviation is low, the data points are mostly clustered around the mean. When the standard deviation is high, it generally indicates that the data are quite spread out, or else that there are a few substantial outliers.

As a simple example, compute the standard deviation for the data set (1, 3, 3, 5). The first step is to compute the mean, which is $\frac{1+3+3+5}{4} = \frac{12}{4} = 3$. Next, the variance of X is found with the formula:

$$\sum_{i=1}^{4}(x_i - \bar{X})^2 = (1-3)^2 + (3-3)^2 + (3-3)^2 + (5-3)^2$$

$$-2^2 + 0^2 + 0^2 + 2^2 = 8$$

Therefore, the variance is $\frac{8}{4} = 2$. Taking the square root, the standard deviation is found to be $\sqrt{2}$.

The shape of a data set is another way to compare two or more sets of data. If a data set isn't symmetric around its mean, it's said to be *skewed*. If the tail to the left of the mean is longer, it's said to be *skewed to the left*. In this case, the mean is less than the median. Conversely, if the tail to the right of the mean is longer, it's said to be *skewed to the right* and the mean is greater than the median. When classifying a data set according to its shape, its overall *skewness* is being discussed. If the mean and median are equal, the data set isn't *skewed*; it is *symmetric*, and is considered normally distributed.

An outlier is a data point that lies a great distance away from the majority of the data set. It also can be labeled as an extreme value. Technically, an outlier is any value that falls 1.5 times the IQR above the upper quartile or 1.5 times the IQR below the lower quartile. The effect of outliers in the data set is seen visually because they affect the mean. If there's a large difference between the mean and mode, outliers are the cause. The mean shows bias towards the outlying values. However, the median won't be affected as greatly by outliers.

Representing Data

Chart is a broad term that refers to a variety of ways to represent data.

To graph relations, the *Cartesian plane* is used. This means to think of the plane as being given a grid of squares, with one direction being the *x*-axis and the other direction the *y*-axis. Generally, the independent variable is placed along the horizontal axis, and the dependent variable is placed along the vertical axis. Any point on the plane can be specified by saying how far to go along the *x*-axis and how far along the *y*-axis with a pair of numbers (x, y). Specific values for these pairs can be given names such as $C = (-1, 3)$. Negative values mean to move left or down; positive values mean to move right or up. The point where the axes cross one another is called the *origin*. The origin has coordinates $(0, 0)$ and is usually called *O* when given a specific label. An illustration of the Cartesian plane, along with the plotted points $(2, 1)$ and $(-1, -1)$, is below.

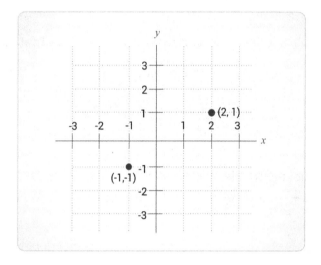

A *line plot* is a diagram that shows quantity of data along a number line. It is a quick way to record data in a structure similar to a bar graph without needing to do the required shading of a bar graph. Here is an example of a line plot:

A *tally chart* is a diagram in which tally marks are utilized to represent data. Tally marks are a means of showing a quantity of objects within a specific classification. Here is an example of a tally chart:

Number of days with rain	Number of weeks
0	||
1	⊮ŤŤ
2	⊮ŤŤ |
3	⊮ŤŤ ||||
4	⊮ŤŤ ⊮ŤŤ ⊮ŤŤ
5	⊮ŤŤ
6	⊮ŤŤ |
7	||||

Data is often recorded using fractions, such as half a mile, and understanding fractions is critical because of their popular use in real-world applications. Also, it is extremely important to label values with their units when using data. For example, regarding length, the number 2 is meaningless unless it is attached to a unit. Writing 2 cm shows that the number refers to the length of an object.

A *picture graph* is a diagram that shows pictorial representation of data being discussed. The symbols used can represent a certain number of objects. Notice how each fruit symbol in the following graph represents a count of two fruits. One drawback of picture graphs is that they can be less accurate if each symbol represents a large number. For example, if each banana symbol represented ten bananas, and students consumed 22 bananas, it may be challenging to draw and interpret two and one-fifth bananas as a frequency count of 22.

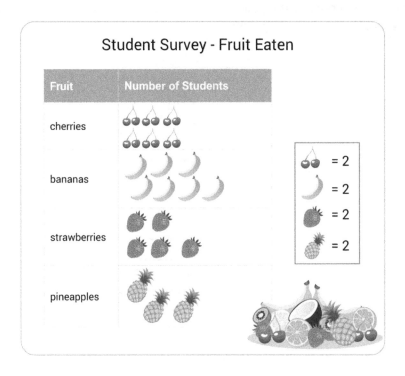

A circle graph, also called a pie chart, shows categorical data with each category representing a percentage of the whole data set. To make a circle graph, the percent of the data set for each category must be determined. To do so, the frequency of the category is divided by the total number of data points and converted to a percent. For example, if 80 people were asked what their favorite sport is and 20 responded basketball, basketball makes up 25% of the data ($\frac{20}{80} = .25 = 25\%$). Each category in a data set is represented by a *slice* of the circle proportionate to its percentage of the whole.

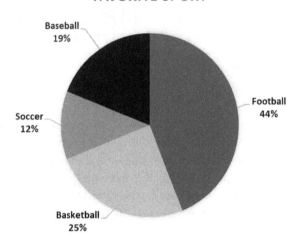

A scatter plot displays the relationship between two variables. Values for the independent variable, typically denoted by *x*, are paired with values for the dependent variable, typically denoted by *y*. Each set of corresponding values are written as an ordered pair (*x, y*). To construct the graph, a coordinate grid is labeled with the *x*-axis representing the independent variable and the *y*-axis representing the dependent variable. Each ordered pair is graphed.

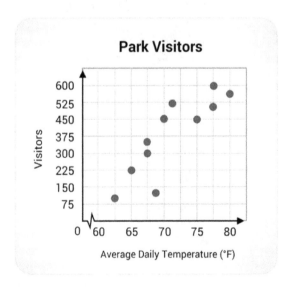

Like a scatter plot, a line graph compares two variables that change continuously, typically over time. Paired data values (ordered pair) are plotted on a coordinate grid with the *x*- and *y*-axis representing the two variables. A line is drawn from each point to the next, going from left to right. A double line graph simply displays two sets of data that contain values for the same two variables. The double line graph below displays the profit for given years (two variables) for Company A and Company B (two data sets).

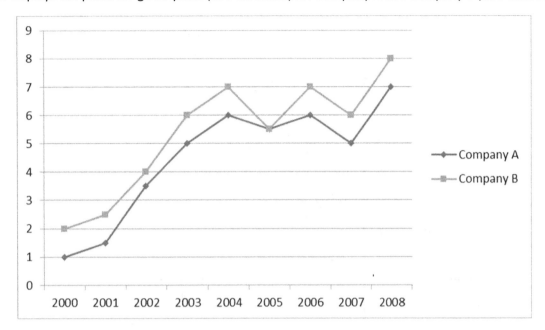

Choosing the appropriate graph to display a data set depends on what type of data is included in the set and what information must be shown.

Scatter plots and line graphs can be used to display data consisting of two variables. Examples include height and weight, or distance and time. A correlation between the variables is determined by examining the points on the graph. Line graphs are used if each value for one variable pairs with a distinct value for the other variable. Line graphs show relationships between variables.

Identifying Line of Best Fit

Regression lines are a way to calculate a relationship between the independent variable and the dependent variable. A straight line means that there's a linear trend in the data. Technology can be used to find the equation of this line (e.g., a graphing calculator or Microsoft Excel®). In either case, all of the data points are entered, and a line is "fit" that best represents the shape of the data. Other functions used to model data sets include quadratic and exponential models. Here's an example of a data set and its regression line:

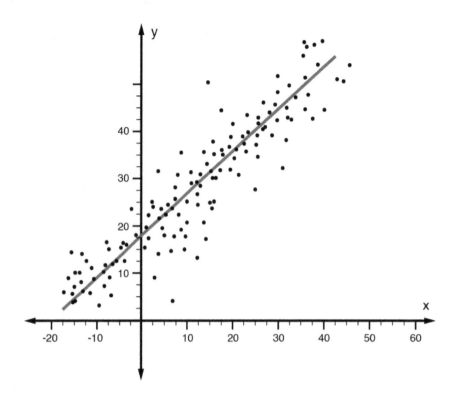

Estimating Data Points

Regression lines can be used to estimate data points not already given. For example, if an equation of a line is found that fit the temperature and beach visitor data set, its input is the average daily temperature and its output is the projected number of visitors. Thus, the number of beach visitors on a 100-degree day can be estimated. The output is a data point on the regression line, and the number of daily visitors is expected to be greater than on a 96-degree day because the regression line has a positive slope.

Interpreting the Regression Line

The formula for a regression line is $y = mx + b$, where m is the slope and b is the y-intercept. Both the slope and y-intercept are found in the *Method of Least Squares*, which is the process of finding the equation of the line through minimizing residuals. The slope represents the rate of change in y as x gets larger. Therefore, because y is the dependent variable, the slope actually provides the predicted values given the independent variable. The y-intercept is the predicted value for when the independent variable equals zero. In the temperature example, the y-intercept is the expected number of beach visitors for a very cold average daily temperature of zero degrees.

Probabilities of Single and Compound Events

A *simple event* consists of only one outcome. The most popular simple event is flipping a coin, which results in either heads or tails. A *compound event* results in more than one outcome and consists of more than one simple event. An example of a compound event is flipping a coin while tossing a die. The

result is either heads or tails on the coin and a number from one to six on the die. The probability of a simple event is calculated by dividing the number of possible outcomes by the total number of outcomes. Therefore, the probability of obtaining heads on a coin is $\frac{1}{2}$, and the probability of rolling a 6 on a die is $\frac{1}{6}$. The probability of compound events is calculated using the basic idea of the probability of simple events. If the two events are independent, the probability of one outcome is equal to the product of the probabilities of each simple event. For example, the probability of obtaining heads on a coin and rolling a 6 is equal to $\frac{1}{2} \times \frac{1}{6} = \frac{1}{12}$. The probability of either A or B occurring is equal to the sum of the probabilities minus the probability that both A and B will occur. Therefore, the probability of obtaining either heads on a coin or rolling a 6 on a die is:

$$\frac{1}{2} + \frac{1}{6} - \frac{1}{12} = \frac{7}{12}$$

The two events aren't mutually exclusive because they can happen at the same time. If two events are mutually exclusive, and the probability of both events occurring at the same time is zero, the probability of event A or B occurring equals the sum of both probabilities. An example of calculating the probability of two mutually exclusive events is determining the probability of pulling a king or a queen from a deck of cards. The two events cannot occur at the same time.

Uniform and Non-Uniform Probability Models

A *uniform probability model* is one where each outcome has an equal chance of occurring, such as the probabilities of rolling each side of a die. A *non-uniform probability model* is one where each outcome has an unequal chance of occurring. In a uniform probability model, the conditional probability formulas for $P(B|A)$ and $P(A|B)$ can be multiplied by their respective denominators to obtain two formulas for $P(A \text{ and } B)$. Therefore, the multiplication rule is derived as:

$$P(A \text{ and } B) = P(A)P(B|A) = P(B)P(A|B)$$

In a model, if the probability of either individual event is known and the corresponding conditional probability is known, the multiplication rule allows the probability of the joint occurrence of A and B to be calculated.

Measuring Probabilities with Two-Way Frequency Tables

When measuring event probabilities, two-way frequency tables can be used to report the raw data and then used to calculate probabilities. If the frequency tables are translated into relative frequency tables, the probabilities presented in the table can be plugged directly into the formulas for conditional probabilities. By plugging in the correct frequencies, the data from the table can be used to determine if events are independent or dependent.

Differing Probabilities

The probability that event A occurs differs from the probability that event A occurs given B. When working within a given model, it's important to note the difference. $P(A|B)$ is determined using the formula $P(A|B) = \frac{P(A \text{ and } B)}{P(B)}$ and represents the total number of A's outcomes left that could occur after B occurs. $P(A)$ can be calculated without any regard for B. For example, the probability of a student finding a parking spot on a busy campus is different once class is in session.

The Addition Rule

The probability of event A or B occurring isn't equal to the sum of each individual probability. The probability that both events can occur at the same time must be subtracted from this total. This idea is shown in the *addition rule*:

$$P(A \text{ or } B) = P(A) + P(B) - P(A \text{ and } B)$$

The addition rule is another way to determine the probability of compound events that aren't mutually exclusive. If the events are mutually exclusive, the probability of both A and B occurring at the same time is 0.

Approximating the Probability of a Chance Event

Probability is a measure of how likely an event is to occur. Probability is written as a fraction between zero and one. If an event has a probability of zero, the event will never occur. If an event has a probability of one, the event will definitely occur. If the probability of an event is closer to zero, the event is unlikely to occur. If the probability of an event is closer to one, the event is more likely to occur. For example, a probability of $\frac{1}{2}$ means that the event is equally as likely to occur as it is not to occur. An example of this is tossing a coin. To calculate the probability of an event, the number of favorable outcomes is divided by the number of total outcomes. For example, suppose you have 2 raffle tickets out of 20 total tickets sold. The probability that you win the raffle is calculated:

$$\frac{number \ of \ favorable \ outcomes}{total \ number of \ outcomes} = \frac{2}{20}$$

$$\frac{2}{20} = \frac{1}{10} \text{ (always reduce fractions)}$$

Therefore, the probability of winning the raffle is $\frac{1}{10}$ or 0.1.

Chance is the measure of how likely an event is to occur, written as a percent. If an event will never occur, the event has a 0% chance. If an event will certainly occur, the event has a 100% chance. If an event will sometimes occur, the event has a chance somewhere between 0% and 100%. To calculate chance, probability is calculated, and the fraction is converted to a percent.

The probability of multiple events occurring can be determined by multiplying the probability of each event. For example, suppose you flip a coin with heads and tails, and roll a six-sided die numbered one through six. To find the probability that you will flip heads AND roll a two, the probability of each event is determined, and those fractions are multiplied. The probability of flipping heads is $\frac{1}{2} \left(\frac{1 \ side \ with \ heads}{2 \ sides \ total} \right)$, and the probability of rolling a two is $\frac{1}{6} \left(\frac{1 \ side \ with \ a \ 2}{6 \ total \ sides} \right)$. The probability of flipping heads AND rolling a 2 is: $\frac{1}{2} \times \frac{1}{6} = \frac{1}{12}$.

The above scenario with flipping a coin and rolling a die is an example of independent events. Independent events are circumstances in which the outcome of one event does not affect the outcome of the other event. Conversely, dependent events are ones in which the outcome of one event affects the outcome of the second event. Consider the following scenario: a bag contains 5 black marbles and 5 white marbles. What is the probability of picking 2 black marbles without replacing the marble after the first pick?

The probability of picking a black marble on the first pick is:

$$\frac{5}{10}\left(\frac{5 \text{ black marbles}}{10 \text{ total marbles}}\right)$$

Assuming that a black marble was picked, there are now 4 black marbles and 5 white marbles for the second pick. Therefore, the probability of picking a black marble on the second pick is:

$$\frac{4}{9}\left(\frac{4 \text{ black marbles}}{9 \text{ total marbles}}\right)$$

To find the probability of picking two black marbles, the probability of each is multiplied:

$$\frac{5}{10} \times \frac{4}{9} = \frac{20}{90} = \frac{2}{9}$$

Using Measures of Center to Draw Inferences About Populations

The center of a set of data (statistical values) can be represented by its mean, median, or mode. These are sometimes referred to as measures of central tendency.

Mean

The first property that can be defined for this set of data is the *mean*. This is the same as the average. To find the mean, add up all the data points, then divide by the total number of data points. For example, suppose that in a class of 10 students, the scores on a test were 50, 60, 65, 65, 75, 80, 85, 85, 90, 100. Therefore, the average test score will be:

$$\frac{50 + 60 + 65 + 65 + 75 + 80 + 85 + 85 + 90 + 100}{10} = 75.5$$

The mean is a useful number if the distribution of data is normal (more on this later), which roughly means that the frequency of different outcomes has a single peak and is roughly equally distributed on both sides of that peak. However, it is less useful in some cases where the data might be split or where there are some *outliers*. Outliers are data points that are far from the rest of the data. For example, suppose there are 10 executives and 90 employees at a company. The executives make $1000 per hour, and the employees make $10 per hour.

Therefore, the average pay rate will be:

$$\frac{\$1000 \times 10 + \$10 \times 90}{100} = \$109 \text{ per hour}$$

In this case, this average is not very descriptive since it's not close to the actual pay of the executives or the employees.

Median

Another useful measurement is the *median*. In a data set, the median is the point in the middle. The middle refers to the point where half the data comes before it and half comes after, when the data is recorded in numerical order. For instance, these are the speeds of the fastball of a pitcher during the last inning that he pitched (in order from least to greatest):

90, 92, 93, 93, 95, 96, 97, 97, 97

There are nine total numbers, so the middle or *median* number is the 5th one, which is 95.

In cases where the number of data points is an even number, then the average of the two middle points is taken. In the previous example of test scores, the two middle points are 75 and 80. Since there is no single point, the average of these two scores needs to be found. The average is:

$$\frac{75 + 80}{2} = 77.5$$

The median is generally a good value to use if there are a few outliers in the data. It prevents those outliers from affecting the "middle" value as much as when using the mean.

Since an outlier is a data point that is far from most of the other data points in a data set, this means an outlier also is any point that is far from the median of the data set. The outliers can have a substantial effect on the mean of a data set, but they usually do not change the median or mode, or do not change them by a large quantity. For example, consider the data set (3, 5, 6, 6, 6, 8). This has a median of 6 and a mode of 6, with a mean of $\frac{34}{6} \approx 5.67$. Now, suppose a new data point of 1000 is added so that the data set is now (3, 5, 6, 6, 6, 8, 1000). The median and mode, which are both still 6, remain unchanged. However, the average is now $\frac{1034}{7}$, which is approximately 147.7. In this case, the median and mode will be better descriptions for most of the data points.

Outliers in a given data set are sometimes the result of an error by the experimenter, but oftentimes, they are perfectly valid data points that must be taken into consideration.

Mode
One additional measure to define for X is the *mode*. This is the data point that appears most frequently. If two or more data points all tie for the most frequent appearance, then each of them is considered a mode. In the case of the test scores, where the numbers were 50, 60, 65, 65, 75, 80, 85, 85, 90, 100, there are two modes: 65 and 85.

Using Statistics to Gain Information About a Population

Statistics involves making decisions and predictions about larger data sets based on smaller data sets. Basically, the information from one part or subset can help predict what happens in the entire data set or population at large. The entire process involves guessing, and the predictions and decisions may not be 100 percent correct all of the time; however, there is some truth to these predictions, and the decisions do have mathematical support. The smaller data set is called a *sample* and the larger data set (in which the decision is being made) is called a *population*. A *random sample* is used as the sample, which is an unbiased collection of data points that represents the population as well as it can. There are many methods of forming a random sample, and all adhere to the fact that every potential data point has a predetermined probability of being chosen. Statistical inference, based in probability theory, makes calculated assumptions about an entire population based on data from a sample set from that population.

A population is the entire set of people or things of interest. Suppose a study is intended to determine the number of hours of sleep per night for college females in the U.S. The population would consist of EVERY college female in the country. A sample is a subset of the population that may be used for the study. It would not be practical to survey every female college student, so a sample might consist of 100 students per school from 20 different colleges in the country. From the results of the survey, a sample

statistic can be calculated. A sample statistic is a numerical characteristic of the sample data, including mean and variance. A sample statistic can be used to estimate a corresponding population parameter. A population parameter is a numerical characteristic of the entire population. Suppose the sample data had a mean (average) of 5.5. This sample statistic can be used as an estimate of the population parameter (average hours of sleep for every college female in the U.S.).

Confidence Intervals

A population parameter is usually unknown and therefore is estimated using a sample statistic. This estimate may be highly accurate or relatively inaccurate based on errors in sampling. A confidence interval indicates a range of values likely to include the true population parameter. These are constructed at a given confidence level, such as 95%. This means that if the same population is sampled repeatedly, the true population parameter would occur within the interval for 95% of the samples.

The accuracy of a population parameter based on a sample statistic may also be affected by measurement error, which is the difference between a quantity's true value and its measured value. Measurement error can be divided into random error and systematic error. An example of random error for the previous scenario would be a student reporting 8 hours of sleep when she actually sleeps 7 hours per night. Systematic errors are those attributed to the measurement system. Suppose the sleep survey gave response options of 2, 4, 6, 8, or 10 hours. This would lead to systematic measurement error.

Algebraic Concepts

Interpreting Parts of an Expression

A *term* is either a number, a variable, or the product of the two. The *coefficient* is the number part of a term that is made up of a constant and a variable. Algebraic expressions are built out of monomials. A *monomial* is a variable raised to some power multiplied by a constant: ax^n, where a is any constant and n is a whole number. A constant is also a monomial.

A *polynomial* is a sum of monomials. Examples of polynomials include $3x^4 + 2x^2 - x - 3$ and $\frac{4}{5}x^3$. The latter is also a monomial. If the highest power of x is 1, the polynomial is called *linear*. If the highest power of x is 2, it is called *quadratic*.

Performing Arithmetic Operations on Polynomials and Rational Expressions

Addition and subtraction operations can be performed on polynomials with like terms. *Like terms* refers to terms that have the same variable and exponent. The two following polynomials can be added together by collecting like terms:

$$(x^2 + 3x - 4) + (4x^2 - 7x + 8)$$

The x^2 terms can be added as $x^2 + 4x^2 = 5x^2$. The x terms can be added as $3x + -7x = -4x$, and the constants can be added as $-4 + 8 = 4$. The following expression is the result of the addition:

$$5x^2 - 4x + 4$$

When subtracting polynomials, the same steps are followed, only subtracting like terms together.

Multiplication of polynomials can also be performed. Given the two polynomials, $(y^3 - 4)$ and $(x^2 + 8x - 7)$, each term in the first polynomial must be multiplied by each term in the second polynomial. The steps to multiply each term in the given example are as follows:

$$(y^3 \times x^2) + (y^3 \times 8x) + (y^3 \times -7) + (-4 \times x^2) + (-4 \times 8x) + (-4 \times -7)$$

Simplifying each multiplied part, yields:

$$x^2y^3 + 8xy^3 - 7y^3 - 4x^2 - 32x + 28$$

None of the terms can be combined because there are no like terms in the final expression. Any polynomials can be multiplied by each other by following the same set of steps, then collecting like terms at the end.

Polynomial Identities

Difference of squares refers to a binomial composed of the difference of two squares. For example, $a^2 - b^2$ is a difference of squares. It can be written $(a)^2 - (b)^2$, and it can be factored into $(a - b)(a + b)$. Recognizing the difference of squares allows the expression to be rewritten easily because of the form it takes. For some expressions, factoring consists of more than one step. When factoring, it's important to always check to make sure that the result cannot be factored further. If it can, then the expression should be split further. If it cannot be, the factoring step is complete, and the expression is completely factored.

A sum and difference of cubes is another way to factor a polynomial expression. When the polynomial takes the form of addition or subtraction of two terms that can be written as a cube, a formula is given. The following graphic shows the factorization of a difference of cubes:

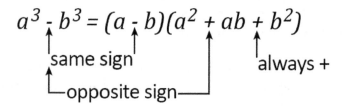

This form of factoring can be useful in finding the zeros of a function of degree 3. For example, when solving $x^3 - 27 = 0$, this rule needs to be used. $x^3 - 27$ is first written as the difference two cubes, $(x)^3 - (3)^3$ and then factored into $(x - 3)(x^2 + 3x + 9)$. This expression may not be factored any further. Each factor is then set equal to zero. Therefore, one solution is found to be $x = 3$, and the other two solutions must be found using the quadratic formula. A sum of squares would have a similar process. The formula for factoring a sum of cubes is:

$$a^3 + b^3 = (a + b)(a^2 - ab + b^2)$$

The opposite of factoring is multiplying. Multiplying a square of a binomial involves the following rules:

$$(a + b)^2 = a^2 + 2ab + b^2$$

$$(a - b)^2 = a^2 - 2ab + b^2$$

The binomial theorem for expansion can be used when the exponent on a binomial is larger than 2, and the multiplication would take a long time. The binomial theorem is given as:

$$(a + b)^n = \sum_{k=0}^{n} \binom{n}{k} a^{n-k} b^k$$

$$\text{where } \binom{n}{k} = \frac{n!}{k!(n-k)!}$$

The *Remainder Theorem* can be helpful when evaluating polynomial functions $P(x)$ for a given value of x. A polynomial can be divided by $(x - a)$, if there is a remainder of 0. This also means that $P(a) = 0$ and $(x - a)$ is a factor of $P(x)$. In a similar sense, if P is evaluated at any other number b, $P(b)$ is equal to the remainder of dividing $P(x)$ by $(x - b)$.

For example, consider:

$$P(x) = x^3 - 7x - 6$$

$$P(4) = 30 \text{ because}$$

$$
\begin{array}{r}
x^2 + 4x + 9 \\
x - 4 \overline{\smash{\big)}\ x^3 + 0x^2 - 7x - 6} \\
\underline{x^3 + 4x^2} \\
4x^2 - 7x - 6 \\
\underline{4x^2 + 16x} \\
9x - 6 \\
\underline{9x + 36} \\
30
\end{array}
$$

Rational Expressions

A fraction, or ratio, wherein each part is a polynomial, defines *rational expressions*. Some examples include $\frac{2x+6}{x}$, $\frac{1}{x^2-4x+8}$, and $\frac{z^2}{x+5}$. Exponents on the variables are restricted to whole numbers, which means roots and negative exponents are not included in rational expressions.

Rational expressions can be transformed by factoring. For example, the expression $\frac{x^2-5x+6}{(x-3)}$ can be rewritten by factoring the numerator to obtain $\frac{(x-3)(x-2)}{(x-3)}$. Therefore, the common binomial $(x-3)$ can cancel so that the simplified expression is $\frac{(x-2)}{1} = (x-2)$.

Additionally, other rational expressions can be rewritten to take on different forms. Some may be factorable in themselves, while others can be transformed through arithmetic operations. Rational expressions are closed under addition, subtraction, multiplication, and division by a nonzero expression. *Closed* means that if any one of these operations is performed on a rational expression, the result will still be a rational expression. The set of all real numbers is another example of a set closed under all four operations.

Adding and subtracting rational expressions is based on the same concepts as adding and subtracting simple fractions. For both concepts, the denominators must be the same for the operation to take place. For example, here are two rational expressions:

$$\frac{x^3-4}{(x-3)} + \frac{x+8}{(x-3)}$$

Since the denominators are both $(x-3)$, the numerators can be combined by collecting like terms to form:

$$\frac{x^3+x+4}{(x-3)}$$

If the denominators are different, they need to be made common (the same) by using the Least Common Denominator (LCD). Each denominator needs to be factored, and the LCD contains each factor that appears in any one denominator the greatest number of times it appears in any denominator. The original expressions need to be multiplied times a form of 1, which will turn each denominator into the LCD. This process is like adding fractions with unlike denominators. It is also important when working with rational expressions to define what value of the variable makes the denominator zero. For this particular value, the expression is undefined.

Multiplication of rational expressions is performed like multiplication of fractions. The numerators are multiplied; then, the denominators are multiplied. The final fraction is then simplified. The expressions are simplified by factoring and cancelling out common terms. In the following example, the numerator of the second expression can be factored first to simplify the expression before multiplying:

$$\frac{x^2}{(x-4)} \times \frac{x^2-x-12}{2}$$

$$\frac{x^2}{(x-4)} \times \frac{(x-4)(x+3)}{2}$$

The $(x-4)$ on the top and bottom cancel out:

$$\frac{x^2}{1} \times \frac{(x+3)}{2}$$

Then multiplication is performed, resulting in:

$$\frac{x^3 + 3x^2}{2}$$

Dividing rational expressions is similar to the division of fractions, where division turns into multiplying by a reciprocal. The following expression can be rewritten as a multiplication problem:

$$\frac{x^2 - 3x + 7}{x - 4} \div \frac{x^2 - 5x + 3}{x - 4}$$

$$\frac{x^2 - 3x + 7}{x - 4} \times \frac{x - 4}{x^2 - 5x + 3}$$

The $x - 4$ cancels out, leaving:

$$\frac{x^2 - 3x + 7}{x^2 - 5x + 3}$$

The final answers should always be completely simplified. If a function is composed of a rational expression, the zeros of the graph can be found from setting the polynomial in the numerator as equal to zero and solving. The values that make the denominator equal to zero will either exist on the graph as a hole or a vertical asymptote.

Writing Expressions in Equivalent Forms

Algebraic expressions are made up of numbers, variables, and combinations of the two, using mathematical operations. Expressions can be rewritten based on their factors. For example, the expression $6x + 4$ can be rewritten as $2(3x + 2)$ because 2 is a factor of both $6x$ and 4. More complex expressions can also be rewritten based on their factors. The expression $x^4 - 16$ can be rewritten as $(x^2 - 4)(x^2 + 4)$. This is a different type of factoring, where a difference of squares is factored into a sum and difference of the same two terms. With some expressions, the factoring process is simple and only leads to a different way to represent the expression. With others, factoring and rewriting the expression leads to more information about the given problem.

In the following quadratic equation, factoring the binomial leads to finding the zeros of the function:

$$x^2 - 5x + 6 = y$$

This equations factors into $(x - 3)(x - 2) = y$, where 2 and 3 are found to be the zeros of the function when y is set equal to zero. The zeros of any function are the x-values where the graph of the function on the coordinate plane crosses the x-axis.

Factoring an equation is a simple way to rewrite the equation and find the zeros, but factoring is not possible for every quadratic. Completing the square is one way to find zeros when factoring is not an option. The following equation cannot be factored: $x^2 + 10x - 9 = 0$. The first step in this method is to move the constant to the right side of the equation, making it $x^2 + 10x = 9$. Then, the coefficient of x is divided by 2 and squared. This number is then added to both sides of the equation, to make the equation still true. For this example, $\left(\frac{10}{2}\right)^2 = 25$ is added to both sides of the equation to obtain:

$$x^2 + 10x + 25 = 9 + 25$$

This expression simplifies to $x^2 + 10x + 25 = 34$, which can then be factored into $(x + 5)^2 = 34$. Solving for x then involves taking the square root of both sides and subtracting 5. This leads to two zeros of the function:

$$x = \pm\sqrt{34} - 5$$

Depending on the type of answer the question seeks, a calculator may be used to find exact numbers.

Given a quadratic equation in standard form— $ax^2 + bx + c = 0$ —the sign of a tells whether the function has a minimum value or a maximum value. If $a > 0$, the graph opens up and has a minimum value. If $a < 0$, the graph opens down and has a maximum value. Depending on the way the quadratic equation is written, multiplication may need to occur before a max/min value is determined.

Exponential expressions can also be rewritten, just as quadratic equations. Properties of exponents must be understood. Multiplying two exponential expressions with the same base involves adding the exponents:

$$a^m a^n = a^{m+n}$$

Dividing two exponential expressions with the same base involves subtracting the exponents:

$$\frac{a^m}{a^n} = a^{m-n}$$

Raising an exponential expression to another exponent includes multiplying the exponents:

$$(a^m)^n = a^{mn}$$

The zero power always gives a value of 1: $a^0 = 1$. Raising either a product or a fraction to a power involves distributing that power:

$$(ab)^m = a^m b^m \text{ and } \left(\frac{a}{b}\right)^m = \frac{a^m}{b^m}$$

Finally, raising a number to a negative exponent is equivalent to the reciprocal including the positive exponent:

$$a^{-m} = \frac{1}{a^m}$$

Finding the Zeros of a Function

The zeros of a function are the points where its graph crosses the x-axis. At these points, $y = 0$. One way to find the zeros is to analyze the graph. If given the graph, the x-coordinates can be found where the line crosses the x-axis. Another way to find the zeros is to set $y = 0$ in the equation and solve for x. Depending on the type of equation, this could be done by using opposite operations, by factoring the equation, by completing the square, or by using the quadratic formula. If a graph does not cross the x-axis, then the function may have complex roots.

Solving Linear Equations and Inequalities in One Variable

The sum of a number and 5 is equal to -8 times the number. To find this unknown number, a simple equation can be written to represent the problem. Key words such as difference, equal, and times are

used to form the following equation with one variable: $n + 5 = -8n$. When solving for n, opposite operations are used. First, n is subtracted from $-8n$ across the equals sign, resulting in $5 = -9n$. Then, -9 is divided on both sides, leaving $n = -\frac{5}{9}$. This solution can be graphed on the number line with a dot as shown below:

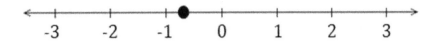

If the problem were changed to say, "The sum of a number and 5 is greater than -8 times the number," then an inequality would be used instead of an equation. Using key words again, *greater than* is represented by the symbol >. The inequality $n + 5 > -8n$ can be solved using the same techniques, resulting in $n < -\frac{5}{9}$. The only time solving an inequality differs from solving an equation is when a negative number is either multiplied times or divided by each side of the inequality. The sign must be switched in this case. For this example, the graph of the solution changes to the following graph because the solution represents all real numbers less than $-\frac{5}{9}$. Not included in this solution is $-\frac{5}{9}$ because it is a *less than* symbol, not *equal to*.

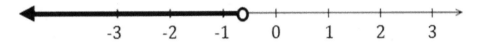

Equations and inequalities in two variables represent a relationship. Jim owns a car wash and charges $40 per car. The rent for the facility is $350 per month. An equation can be written to relate the number of cars Jim cleans to the money he makes per month. Let x represent the number of cars and y represent the profit Jim makes each month from the car wash. The equation $y = 40x - 350$ can be used to show Jim's profit or loss. Since this equation has two variables, the coordinate plane can be used to show the relationship and predict profit or loss for Jim. The following graph shows that Jim must wash at least nine cars to pay the rent, where $x = 9$. Anything nine cars and above yield a profit shown in the value on the y-axis.

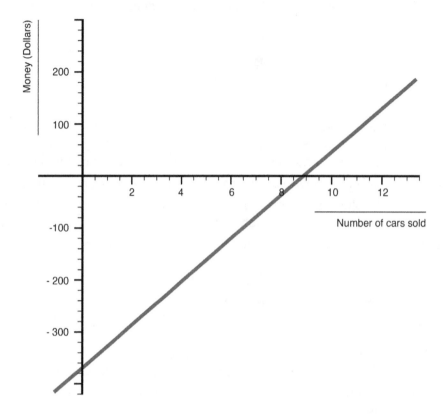

With a single equation in two variables, the solutions are limited only by the situation the equation represents. When two equations or inequalities are used, more constraints are added. For example, in a system of linear equations, there is often—although not always—only one answer. The point of intersection of two lines is the solution. For a system of inequalities, there are infinitely many answers.

The intersection of two solution sets gives the solution set of the system of inequalities. In the following graph, the darker shaded region is where two inequalities overlap. Any set of x and y found in that region satisfies both inequalities. The line with the positive slope is solid, meaning the values on that line are included in the solution.

The line with the negative slope is dotted, so the coordinates on that line are not included.

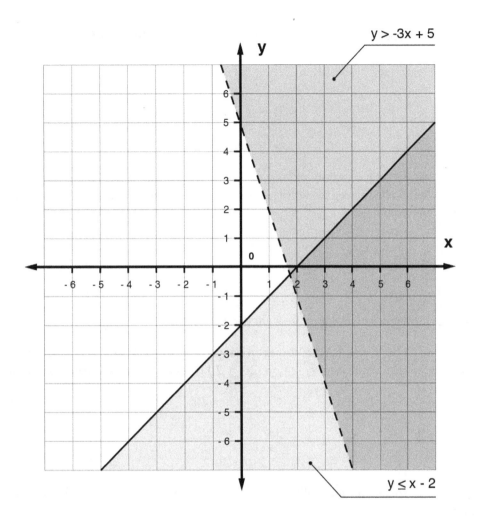

$y > -3x + 5$

$y \leq x - 2$

Formulas with two variables are equations used to represent a specific relationship. For example, the formula $d = rt$ represents the relationship between distance, rate, and time. If Bob travels at a rate of 35 miles per hour on his road trip from Westminster to Seneca, the formula $d = 35t$ can be used to represent his distance traveled in a specific length of time. Formulas can also be used to show different roles of the variables, transformed without any given numbers. Solving for r, the formula becomes $\frac{d}{t} = r$. The t is moved over by division so that *rate* is a function of distance and time.

The letters in an equation are variables as they stand for unknown quantities that you are trying to solve for. The numbers attached to the variables by multiplication are called coefficients. X is commonly used as a variable, though any letter can be used. For example, in $3x - 7 = 20$, the variable is $3x$, and it needs to be isolated. The numbers (also called constants) are -7 and 20. That means $3x$ needs to be on one side of the equals sign (either side is fine), and all the numbers need to be on the other side of the equals sign.

To accomplish this, the equation must be manipulated by performing opposite operations of what already exists. Remember that addition and subtraction are opposites and that multiplication and division are opposites. Any action taken to one side of the equation must be taken on the other side to maintain equality.

Therefore, since the 7 is being subtracted, it can be moved to the right side of the equation by adding seven to both sides:

$$3x - 7 = 20$$

$$3x - 7 + 7 = 20 + 7$$

$$3x = 27$$

Now that the variable $3x$ is on one side and the constants (now combined into one constant) are on the other side, the 3 needs to be moved to the right side. 3 and x are being multiplied together, so 3 needs to be divided from each side.

$$\frac{3x}{3} = \frac{27}{3}$$

$$x = 9$$

Now x has been completely isolated, and thus we know its value.

The solution is found to be $x = 9$. This solution can be checked for accuracy by plugging $x = 9$ in the original equation. After simplifying the equation, $20 = 20$ is found, which is a true statement:

$$3 \times 9 - 7 = 20$$

$$27 - 7 = 20$$

$$20 = 20$$

Equations that require solving for a variable (*algebraic equations*) come in many forms. Here are some more examples:

No coefficient attached to the variable:

$$x + 8 = 20$$

$$x + 8 - 8 = 20 - 8$$

$$x = 12$$

A fractional coefficient:

$$\frac{1}{2}z + 24 = 36$$

$$\frac{1}{2}z + 24 - 24 = 36 - 24$$

$$\frac{1}{2}z = 12$$

Now we multiply the fraction by its inverse:

$$\frac{2}{1} \times \frac{1}{2}z = 12 \times \frac{2}{1}$$

$$z = 24$$

Multiple examples of x:

$$14x + x - 4 = 3x + 2$$

All examples of x can be combined.

$$15x - 4 = 3x + 2$$

$$15x - 4 + 4 = 3x + 2 + 4$$

$$15x = 3x + 6$$

$$15x - 3x = 3x + 6 - 3x$$

$$12x = 6$$

$$\frac{12x}{12} = \frac{6}{12}$$

$$x = \frac{1}{2}$$

Solving Quadratic Equations

Equations with one variable can be solved using the addition principle and multiplication principle. If $a = b$, then $a + c = b + c$, and $ac = bc$. Given the equation $2x - 3 = 5x + 7$, the first step is to combine the variable terms and the constant terms. Using the principles, expressions can be added and subtracted onto and off both sides of the equals sign, so the equation turns into $-10 = 3x$. Dividing by 3 on both sides through the multiplication principle with $c = \frac{1}{3}$ results in the final answer of $x = \frac{-10}{3}$.

Some equations have a higher degree and are not solved by simply using opposite operations. When an equation has a degree of 2, completing the square is an option. For example, the quadratic equation $x^2 - 6x + 2 = 0$ can be rewritten by completing the square. The goal of completing the square is to get the equation into the form $(x - p)^2 = q$. Using the example, the constant term 2 first needs to be moved over to the opposite side by subtracting. Then, the square can be completed by adding 9 to both sides, which is the square of half of the coefficient of the middle term $-6x$. The current equation is $x^2 - 6x + 9 = 7$. The left side can be factored into a square of a binomial, resulting in $(x - 3)^2 = 7$. To solve for x, the square root of both sides should be taken, resulting in:

$$(x - 3) = \pm\sqrt{7}$$

$$x = 3 \pm \sqrt{7}$$

Other ways of solving quadratic equations include graphing, factoring, and using the quadratic formula. The equation $y = x^2 - 4x + 3$ can be graphed on the coordinate plane, and the solutions can be observed where it crosses the x-axis. The graph will be a parabola that opens up with two solutions at 1 and 3.

The equation can also be factored to find the solutions. The original equation, $y = x^2 - 4x + 3$ can be factored into $y = (x - 1)(x - 3)$. Setting this equal to zero, the x-values are found to be 1 and 3, just as on the graph. Solving by factoring and graphing are not always possible.

The method of completing the square can be used in finding another method, the quadratic formula. It can be used to solve any quadratic equation. This formula may be the longest method for solving quadratic equations and is commonly used as a last resort after other methods are ruled out.

It can be helpful in memorizing the formula to see where it comes from, so here are the steps involved.

The most general form for a quadratic equation is $ax^2 + bx + c = 0$.

First, dividing both sides by a leaves us with $x^2 + \frac{b}{a}x + \frac{c}{a} = 0$.

To complete the square on the left-hand side, $\frac{c}{a}$ can be subtracted on both sides to get:

$$x^2 + \frac{b}{a}x = -\frac{c}{a}$$

$(\frac{b}{2a})^2$ is then added to both sides.

This gives:

$$x^2 + \frac{b}{a}x + (\frac{b}{2a})^2 = (\frac{b}{2a})^2 - \frac{c}{a}$$

The left can now be factored and the right-hand side simplified to give:

$$(x + \frac{b}{2a})^2 = \frac{b^2 - 4ac}{4a}$$

Taking the square roots gives:

$$x + \frac{b}{2a} = \pm\frac{\sqrt{b^2 - 4ac}}{2a}$$

Solving for x yields the quadratic formula:

$$x = \frac{-b \pm \sqrt{b^2 - 4ac}}{2a}$$

Where a, b, and c are the coefficients in the original equation in standard form $y = ax^2 + bx + c$. For the above example,

$$x = \frac{4 \pm \sqrt{(-4)^2 - 4(1)(3)}}{2(1)}$$

$$\frac{4 \pm \sqrt{16 - 12}}{2} = \frac{4 \pm 2}{2} = 1, 3$$

The expression underneath the radical is called the *discriminant*. Without working out the entire formula, the value of the discriminant can reveal the nature of the solutions. If the value of the discriminant $b^2 - 4ac$ is positive, then there will be two real solutions. If the value is zero, there will be one real solution. If the value is negative, the two solutions will be imaginary or complex. If the solutions are complex, it means that the parabola never touches the x-axis. An example of a complex solution can be found by solving the following quadratic: $y = x^2 - 4x + 8$. By using the quadratic formula, the solutions are found to be:

$$x = \frac{4 \pm \sqrt{(-4)^2 - 4(1)(8)}}{2(1)}$$

$$\frac{4 \pm \sqrt{16 - 32}}{2}$$

$$\frac{4 \pm \sqrt{-16}}{2} = 2 \pm 2i$$

The solutions both have a real part, 2, and an imaginary part, $2i$.

Solving Simple Rational and Radical Equations in One Variable

A *rational expression* is an expression that has the form $\frac{p(x)}{q(x)}$, where $p(x)$ and $q(x)$ are both polynomials. To solve equations or inequalities involving rational expressions, one typically rewrites the expression to get rid of the denominator; as a result, the problem becomes an equation or inequality involving polynomials. One can then apply the techniques mentioned above to complete the solution.

For example, consider the problem $\frac{3x+2}{x-4} = 2$. One can start by multiplying both sides of the equation by $x - 4$. This results in the equation $3x + 2 = 2x - 8$. Now this equation can be solved like any other linear equation. Subtracting $2x$ from both sides and subtracting 2 from both sides gives the solution $x = -10$.

When an equation or an inequality involves radicals, all the radicals must be moved to one side. Then, one can raise both sides to the appropriate power to get rid of the radicals. Remember that the quantity inside a square root must be non-negative. When dealing with inequalities, remember that multiplying both sides by a negative quantity reverses the direction of the inequality.

For example, $\sqrt{x + 1} - 2 = 2$. The first step is to isolate the radical, so add 2 to both sides. This addition results in $\sqrt{x + 1} = 4$. Square both sides, and the result is $x + 1 = 16$, or $x = 15$.

When dealing with multiple radicals, proceed by first isolating one radical, squaring both sides to remove it, and then repeating this process to remove the remaining radicals. Consider this equation:

$$\sqrt{3x - 1} + 1 = \sqrt{x + 1} + 2$$

Start by subtracting 1 from both sides, isolating the radical on the left, which results in:

$$\sqrt{3x - 1} = \sqrt{x + 1} + 1$$

Now square both sides:

$$3x - 1 = \left(\sqrt{x + 1} + 1\right)^2 = x + 1 + 2\sqrt{x + 1} + 1$$

or

$$3x - 1 = x + 2\sqrt{x + 1} + 2$$

Isolate the radical on the right: $2x - 3 = 2\sqrt{x + 1}$. Now square both sides, which results in:

$$4x^2 - 12x + 9 = 4x + 4$$

This problem can now be solved by using the quadratic formula.

Solving Systems of Equations

A *system of equations* is a group of equations that have the same variables or unknowns. These equations can be linear, but they are not always so. Finding a solution to a system of equations means finding the values of the variables that satisfy each equation. For a linear system of two equations and two variables, there could be a single solution, no solution, or infinitely many solutions.

A single solution occurs when there is one value for x and y that satisfies the system. This is shown on the graph where the lines cross at exactly one point. When there is no solution, the lines are parallel and do not ever cross. With infinitely many solutions, the equations may look different, but they are the same line. One equation will be a multiple of the other, and on the graph, they lie on top of each other. These three types of systems of linear equations are shown below:

Parallel Lines	Intersecting Lines	Coincident Lines
Inconsistent	Independent	Dependent

The process of elimination can be used to solve a system of equations. For example, the following equations make up a system: $x + 3y = 10$ and $2x - 5y = 9$. Immediately adding these equations does not eliminate a variable, but it is possible to change the first equation by multiplying the whole equation by -2. This changes the first equation to $-2x - 6y = -20$. The equations can be then added to obtain $-11y = -11$. Solving for y yields $y = 1$. To find the rest of the solution, 1 can be substituted in for y in either original equation to find the value of $x = 7$. The solution to the system is (7, 1) because it makes both equations true, and it is the point in which the lines intersect. If the system is *dependent*—having infinitely many solutions—then both variables will cancel out when the elimination method is used, resulting in an equation that is true for many values of x and y. Since the system is dependent, both equations can be simplified to the same equation, or line.

A system can also be solved using *substitution*. This involves solving one equation for a variable and then plugging that solved equation into the other equation in the system. For example, $x - y = -2$ and $3x + 2y = 9$ can be solved using substitution. The first equation can be solved for x, where $x = -2 + y$. Then it can be plugged into the other equation, $3(-2 + y) + 2y = 9$. Solving for y yields $-6 + 3y + 2y = 9$, where $y = 3$. If $y = 3$, then $x = 1$. This solution can be checked by plugging in these values for the variables in each equation to see if it makes a true statement.

Finally, a solution to a system of equations can be found graphically. The solution to a linear system is the point or points where the lines cross. The values of x and y represent the coordinates (x, y) where the lines intersect. Using the same system of equations as above, they can be solved for y to put them in slope-intercept form, $y = mx + b$. These equations become $y = x + 2$ and $y = -\frac{3}{2}x + 4.5$. The slope is the coefficient of x, and the y-intercept is the constant value. This system with the solution is shown below:

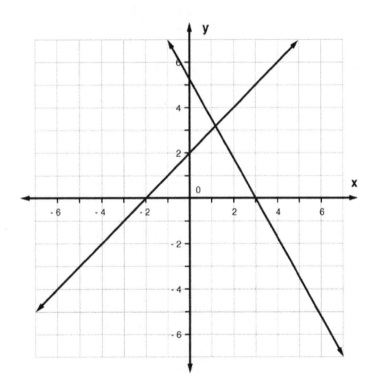

A system of equations may also be made up of a linear and a quadratic equation. These systems may have one solution, two solutions, or no solutions. The graph of these systems involves one straight line and one parabola. Algebraically, these systems can be solved by solving the linear equation for one variable and plugging that answer in to the quadratic equation. If possible, the equation can then be solved to find part of the answer. The graphing method is commonly used for these types of systems. On a graph, these two lines can be found to intersect at one point, at two points across the parabola, or at no points.

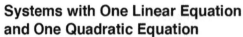

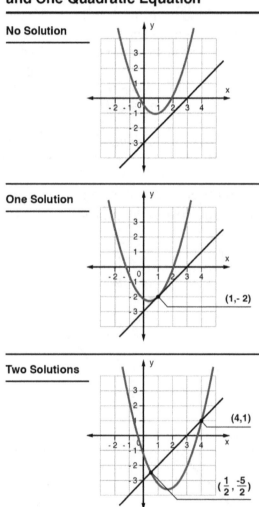

Matrices can also be used to solve systems of linear equations. Specifically, for systems, the coefficients of the linear equations in standard form are the entries in the matrix. Using the same system of linear equations as above, $x - y = -2$ and $3x + 2y = 9$, the matrix to represent the system is:

$$\begin{bmatrix} 1 & -1 \\ 3 & 2 \end{bmatrix} \begin{bmatrix} x \\ y \end{bmatrix} = \begin{bmatrix} -2 \\ 9 \end{bmatrix}$$

To solve this system using matrices, the inverse matrix must be found.

For a general 2x2 matrix, $\begin{bmatrix} a & b \\ c & d \end{bmatrix}$, the inverse matrix is found by the expression:

$$\frac{1}{ad - bc} \begin{bmatrix} d & -b \\ -c & a \end{bmatrix}$$

The inverse matrix for the given system above is:

$$\frac{1}{2 - -3} \begin{bmatrix} 2 & 1 \\ -3 & 1 \end{bmatrix} = \frac{1}{5} \begin{bmatrix} 2 & 1 \\ -3 & 1 \end{bmatrix}$$

The next step in solving is to multiply this identity matrix times the system matrix above. This is given by the following equation:

$$\frac{1}{5} \begin{bmatrix} 2 & 1 \\ -3 & 1 \end{bmatrix} \begin{bmatrix} 1 & -1 \\ 3 & 2 \end{bmatrix} \begin{bmatrix} x \\ y \end{bmatrix}$$

$$\begin{bmatrix} -2 \\ 9 \end{bmatrix} \begin{bmatrix} 2 & 1 \\ -3 & 1 \end{bmatrix} \frac{1}{5}$$

which simplifies to:

$$\frac{1}{5} \begin{bmatrix} 5 & 0 \\ 0 & 5 \end{bmatrix} \begin{bmatrix} x \\ y \end{bmatrix} = \frac{1}{5} \begin{bmatrix} 5 \\ 15 \end{bmatrix}$$

Solving for the solution matrix, the answer is:

$$\begin{bmatrix} 1 & 0 \\ 0 & 1 \end{bmatrix} \begin{bmatrix} x \\ y \end{bmatrix} = \begin{bmatrix} 1 \\ 3 \end{bmatrix}$$

Since the first matrix is the identity matrix, the solution is $x = 1$ and $y = 3$.

Finding solutions to systems of equations is essentially finding what values of the variables make both equations true. It is finding the input value that yields the same output value in both equations. For functions $g(x)$ and $f(x)$, the equation $g(x) = f(x)$ means the output values are being set equal. Solving for the value of x means finding the x-coordinate that gives the same output to both functions. For example, $f(x) = x + 2$ and $g(x) = -3x + 10$ is a system of equations. Setting $f(x) = g(x)$ yields the equation $x + 2 = -3x + 10$. Solving for x gives the x-coordinate $x = 2$ where the two lines cross. This value can also be found by using a table or a graph. On a table, both equations could be given the same inputs, and the outputs could be recorded to find the point(s) where the lines crossed. Any method of solving finds the same solution, but some methods are more appropriate for some systems of equations than others.

Graphing Functions

Different types of functions behave in different ways. A function is defined to be increasing over a subset of its domain if for all $x_1 \geq x_2$ in that interval, $f(x_1) \geq f(x_2)$. Also, a function is decreasing over an interval if for all $x_1 \geq x_2$ in that interval, $f(x_1) \leq f(x_2)$. A point in which a function changes from increasing to decreasing can also be labeled as the *maximum value* of a function if it is the largest point the graph reaches on the y-axis. A point in which a function changes from decreasing to increasing can be labeled as the minimum value of a function if it is the smallest point the graph reaches on the y-axis. Maximum values are also known as *extreme values*. The graph of a continuous function does not have any breaks or jumps in the graph. This description is not true of all functions. A radical function, for example, $f(x) = \sqrt{x}$, has a restriction for the domain and range because there are no real negative inputs or outputs for this function. The domain can be stated as $x \geq 0$, and the range is $y \geq 0$.

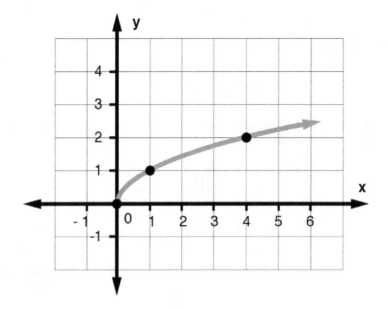

A piecewise-defined function also has a different appearance on the graph. In the following function, there are three equations defined over different intervals. It is a function because there is only one y-value for each x-value, passing the Vertical Line Test. The domain is all real numbers less than or equal to 6. The range is all real numbers greater than zero. From left to right, the graph decreases to zero, then increases to almost 4, and then jumps to 6.

From input values greater than 2, the input decreases just below 8 to 4, and then stops.

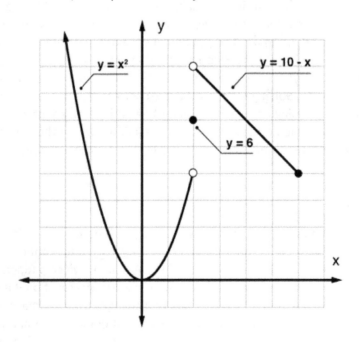

Logarithmic and exponential functions also have different behavior than other functions. These two types of functions are inverses of each other. The *inverse* of a function can be found by switching the place of x and y, and solving for y. When this is done for the exponential equation, $y = 2^x$, the function $y = \log_2 x$ is found. The general form of a *logarithmic function* is $y = \log_b x$, which says b raised to the y power equals x.

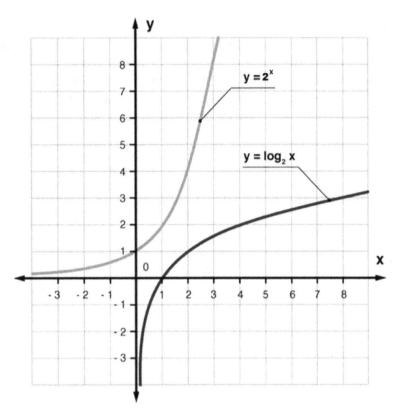

The thick black line on the graph above represents the logarithmic function $y = \log_2 x$. This curve passes through the point $(1, 0)$, just as all log functions do, because any value $b^0 = 1$. The graph of this logarithmic function starts very close to zero, but does not touch the y-axis. The output value will never be zero by the definition of logarithms. The thinner gray line seen above represents the exponential function $y = 2^x$. The behavior of this function is opposite the logarithmic function because the graph of an inverse function is the graph of the original function flipped over the line $y = x$. The curve passes through the point $(0, 1)$ because any number raised to the zero power is one. This curve also gets very close to the x-axis but never touches it because an exponential expression never has an output of zero. The x-axis on this graph is called a horizontal asymptote. An *asymptote* is a line that represents a boundary for a function. It shows a value that the function will get close to, but never reach.

Three common functions used to model different relationships between quantities are linear, quadratic, and exponential functions. Linear functions are the simplest of the three, and the independent variable x has an exponent of 1. Written in the most common form, $y = mx + b$, the coefficient of x indicates how fast the function grows at a constant rate, and the b-value denotes the starting point. A quadratic function has an exponent of 2 on the independent variable x. Standard form for this type of function is $y = ax^2 + bx + c$, and the graph is a parabola. These type functions grow at a changing rate. An exponential function has an independent variable in the exponent $y = ab^x$. The graph of these types of functions is described as *growth* or *decay*, based on whether the base, b, is greater than or less than 1.

These functions are different from quadratic functions because the base stays constant. A common base is base *e*.

The following three functions model a linear, quadratic, and exponential function respectively: $y = 2x$, $y = x^2$, and $y = 2^x$. Their graphs are shown below. The first graph, modeling the linear function, shows that the growth is constant over each interval. With a horizontal change of 1, the vertical change is 2. It models a constant positive growth. The second graph shows the quadratic function, which is a curve that is symmetric across the y-axis. The growth is not constant, but the change is mirrored over the axis. The last graph models the exponential function, where the horizontal change of 1 yields a vertical change that increases more and more. The exponential graph gets very close to the *x*-axis, but never touches it, meaning there is an asymptote there. The y-value can never be zero because the base of 2 can never be raised to an input value that yields an output of zero.

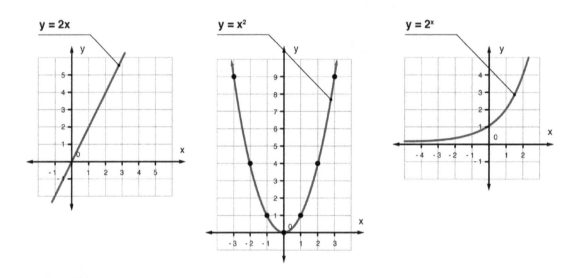

The three tables below show specific values for three types of functions. The third column in each table shows the change in the y-values for each interval. The first table shows a constant change of 2 for each equal interval, which matches the slope in the equation $y = 2x$. The second table shows an increasing change, but it also has a pattern. The increase is changing by 2 more each time, so the change is quadratic. The third table shows the change as factors of the base, 2. It shows a continuing pattern of factors of the base.

$y = 2x$		
x	y	Δy
1	2	
2	4	2
3	6	2
4	8	2
5	10	2

$y = x^2$		
x	y	Δy
1	1	
2	4	3
3	9	5
4	16	7
5	25	9

$y = 2^x$		
x	y	Δy
1	2	
2	4	2
3	8	4
4	16	8
5	32	16

Given a table of values, the type of function can be determined by observing the change in y over equal intervals. For example, the tables below model two functions. The changes in interval for the x-values is 1 for both tables. For the first table, the y-values increase by 5 for each interval. Since the change is constant, the situation can be described as a linear function. The equation would be $y = 5x + 3$. For the second table, the change for y is 5, 20, 100, and 500, respectively. The increases are multiples of 5, meaning the situation can be modeled by an exponential function. The equation $y = 5^x + 3$ models this situation.

x	y		x	y
0	3		0	3
1	8		1	8
2	13		2	28
3	18		3	128
4	23		4	628

Quadratic equations can be used to model real-world area problems. For example, a farmer may have a rectangular field that he needs to sow with seed. The field has length $x + 8$ and width $2x$. The formula for area should be used: $A = lw$. Therefore:

$$A = (x + 8) \times 2x = 2x^2 + 16x$$

The possible values for the length and width can be shown in a table, with input x and output A. If the equation was graphed, the possible area values can be seen on the y-axis for given x-values.

Exponential growth and decay can be found in real-world situations. For example, if a piece of notebook paper is folded 25 times, the thickness of the paper can be found. To model this situation, a table can be used. The initial point is one-fold, which yields a thickness of 2 papers. For the second fold, the thickness is 4. Since the thickness doubles each time, the table below shows the thickness for the next few folds. Notice the thickness changes by the same factor each time. Since this change for a constant interval of folds is a factor of 2, the function is exponential. The equation for this is $y = 2^x$. For twenty-five folds, the thickness would be 33,554,432 papers.

x (folds)	y (paper thickness)
0	1
1	2
2	4
3	8
4	16
5	32

One exponential formula that is commonly used is the *interest formula*: $A = Pe^{rt}$. In this formula, interest is compounded continuously. A is the value of the investment after the time, t, in years. P is the initial amount of the investment, r is the interest rate, and e is the constant equal to approximately 2.718. Given an initial amount of $200 and a time of 3 years, if interest is compounded continuously at a rate of 6%, the total investment value can be found by plugging each value into the formula. The invested value at the end is $239.44. In more complex problems, the final investment may be given, and the rate may be the unknown. In this case, the formula becomes $239.44 = 200e^{r3}$. Solving for r requires isolating the exponential expression on one side by dividing by 200, yielding the equation $1.20 = e^{r3}$. Taking the natural log of both sides results in $\ln(1.2) = r3$. Using a calculator to evaluate the logarithmic expression, $r = 0.06 = 6\%$.

When working with logarithms and exponential expressions, it is important to remember the relationship between the two. In general, the logarithmic form is $y = log_b x$ for an exponential form $b^y = x$. Logarithms and exponential functions are inverses of each other.

Finding the zeros of polynomial functions is the same process as finding the solutions of polynomial equations. These are the points at which the graph of the function crosses the x-axis. As stated previously, factors can be used to find the zeros of a polynomial function. The degree of the function shows the number of possible zeros. If the highest exponent on the independent variable is 4, then the degree is 4, and the number of possible zeros is 4. If there are complex solutions, the number of roots is less than the degree.

Given the function $y = x^2 + 7x + 6$, y can be set equal to zero, and the polynomial can be factored. The equation turns into $0 = (x + 1)(x + 6)$, where $x = -1$ and $x = -6$ are the zeros. Since this is a quadratic equation, the shape of the graph will be a parabola. Knowing that zeros represent the points where the parabola crosses the x-axis, the maximum or minimum point is the only other piece needed to sketch a rough graph of the function. By looking at the function in standard form, the coefficient of x is positive; therefore, the parabola opens *up*. Using the zeros and the minimum, the following rough sketch of the graph can be constructed:

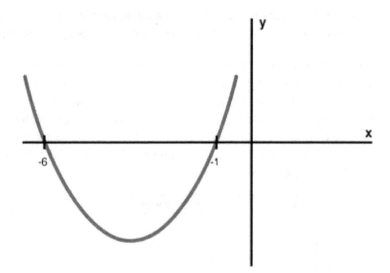

A quadratic function can be written in the standard form:

$$y = ax^2 + bx + c$$

It can be represented by a U-shaped graph called a parabola. The graph can either open up or open down (upside down U). The graph is symmetric about a vertical line, called the axis of symmetry. Corresponding points on the parabola are directly across from each other (same y-value) and are the same distance from the axis of symmetry (on either side). The axis of symmetry intersects the parabola at its vertex. For a quadratic function where the value of a is positive, as the inputs increase, the outputs increase until a certain value (maximum of the function) is reached. As inputs increase past the value that corresponds with the maximum output, the relationship reverses, and the outputs decrease. For a quadratic function where a is negative, as the inputs increase, the outputs (1) decrease, (2) reach a maximum, and (3) then increase.

Creating Equations and Inequalities

An algebraic expression is a statement about unknown quantities expressed in mathematical symbols. The statement *five times a number added to forty* is expressed as $5x + 40$. An equation is a statement in which two expressions (with at least one containing a variable) are equal to one another. The statement *five times a number added to forty is equal to ten* is expressed as $5x + 40 = 10$.

Real world scenarios can also be expressed mathematically. Suppose a job pays its employees $300 per week and $40 for each sale made. The weekly pay is represented by the expression $40x + 300$ where x is the number of sales made during the week.

Consider the following scenario: Bob had $20 and Tom had $4. After selling 4 ice cream cones to Bob, Tom has as much money as Bob. The cost of an ice cream cone is an unknown quantity and can be represented by a variable (x). The amount of money Bob has after his purchase is four times the cost of an ice cream cone subtracted from his original $20 → $20 - 4x$. The amount of money Tom has after his sale is four times the cost of an ice cream cone added to his original $4 → $4x + 4$. After the sale, the amount of money that Bob and Tom have is equal → $20 - 4x = 4x + 4$.

When expressing a verbal or written statement mathematically, it is vital to understand words or phrases that can be represented with symbols. The following are examples:

Symbol	Phrase
+	Added to; increased by; sum of; more than
−	Decreased by; difference between; less than; take away
×	Multiplied by; 3(4,5…) times as large; product of
÷	Divided by; quotient of; half (third, etc.) of
=	Is; the same as; results in; as much as; equal to
x,t,n, etc.	A number; unknown quantity; value of; variable

Use of Formulas

Formulas are mathematical expressions that define the value of one quantity, given the value of one or more different quantities. Formulas look like equations because they contain variables, numbers, operators, and an equal sign. All formulas are equations, but not all equations are formulas. A formula must have more than one variable. For example, $2x + 7 = y$ is an equation and a formula (it relates the unknown quantities x and y). However, $2x + 7 = 3$ is an equation but not a formula (it only expresses the value of the unknown quantity x).

Formulas are typically written with one variable alone (or isolated) on one side of the equal sign. This variable can be thought of as the *subject* in that the formula is stating the value of the *subject* in terms of the relationship between the other variables. Consider the distance formula: $distance = rate \times time$ or $d = rt$. The value of the subject variable d (distance) is the product of the variable r and t (rate and time). Given the rate and time, the distance traveled can easily be determined by substituting the values into the formula and evaluating.

The formula $P = 2l + 2w$ expresses how to calculate the perimeter of a rectangle (P) given its length (l) and width (w). To find the perimeter of a rectangle with a length of 3ft and a width of 2ft, these values are substituted into the formula for l and w: $P = 2(3ft) + 2(2ft)$. Following the order of operations, the perimeter is determined to be 10ft. When working with formulas such as these, including units is an important step.

Given a formula expressed in terms of one variable, the formula can be manipulated to express the relationship in terms of any other variable. In other words, the formula can be rearranged to change which variable is the *subject*. To solve for a variable of interest by manipulating a formula, the equation may be solved as if all other variables were numbers. The same steps for solving are followed, leaving operations in terms of the variables instead of calculating numerical values. For the formula $P = 2l + 2w$, the perimeter is the subject expressed in terms of the length and width. To write a formula to calculate the width of a rectangle, given its length and perimeter, the previous formula relating the three variables is solved for the variable w. If P and l were numerical values, this is a two-step linear equation solved by subtraction and division. To solve the equation $P = 2l + 2w$ for w, $2l$ is first subtracted from both sides: $P - 2l = 2w$. Then both sides are divided by 2: $\frac{P-2l}{2} = w$.

Functions and Function Notation, Interpreting Key Features of Graphs and Tables in Terms of Quantities

A relation is a set of input and output values that can be written as ordered pairs. A function is a relation in which each input is paired with exactly one output. A *function* is defined as a relationship between inputs and outputs where there is only one output value for a given input. As an example, the following function is in function notation: $f(x) = 3x - 4$. The $f(x)$ represents the output value for an input of x. If $x = 2$, the equation becomes:

$$f(2) = 3(2) - 4 = 6 - 4 = 2$$

The input of 2 yields an output of 2, forming the ordered pair $(2, 2)$. The following set of ordered pairs corresponds to the given function: $(2, 2), (0, -4), (-2, -10)$. The set of all possible inputs of a function is its *domain*, and all possible outputs is called the *range*. By definition, each member of the domain is paired with only one member of the range.

Functions can also be defined recursively. In this form, they are not defined explicitly in terms of variables. Instead, they are defined using previously-evaluated function outputs, starting with either $f(0)$ or $f(1)$. An example of a recursively-defined function is:

$$f(1) = 2, f(n) = 2f(n - 1) + 2n, n > 1$$

The domain of this function is the set of all integers.

Functions can also be described as being even, odd, or neither. If $f(-x) = f(x)$, the function is even. For example, the function $f(x) = x^2 - 2$ is even. Plugging in $x = 2$ yields an output of $y = 2$. After

changing the input to $x = -2$, the output is still $y = 2$. The output is the same for opposite inputs. Another way to observe an even function is by the symmetry of the graph. If the graph is symmetrical about the axis, then the function is even. If the graph is symmetric about the origin, then the function is odd. Algebraically, if $f(-x) = -f(x)$, the function is odd.

Also, a function can be described as periodic if it repeats itself in regular intervals. Common periodic functions are trigonometric functions. For example, $y = \sin x$ is a periodic function with period 2π because it repeats itself every 2π units along the x-axis.

Functions can be built out of the context of a situation. For example, the relationship between the money paid for a gym membership and the months that someone has been a member can be described through a function. If the one-time membership fee is $40 and the monthly fee is $30, then the function can be written $f(x) = 30x + 40$. The x-value represents the number of months the person has been part of the gym, while the output is the total money paid for the membership. The table below shows this relationship. It is a representation of the function because the initial cost is $40 and the cost increases each month by $30.

x (months)	y (money paid to gym)
0	40
1	70
2	100
3	130

Functions can also be built from existing functions. For example, a given function $f(x)$ can be transformed by adding a constant, multiplying by a constant, or changing the input value by a constant. The new function $g(x) = f(x) + k$ represents a vertical shift of the original function. In $f(x) = 3x - 2$, a vertical shift 4 units up would be:

$$g(x) = 3x - 2 + 4 = 3x + 2$$

Multiplying the function times a constant k represents a vertical stretch, based on whether the constant is greater than or less than 1. The function

$$g(x) = kf(x) = 4(3x - 2) = 12x - 8$$

represents a stretch. Changing the input x by a constant forms the function:

$$g(x) = f(x + k) = 3(x + 4) - 2 = 3x + 12 - 2 = 3x + 10$$

and this represents a horizontal shift to the left 4 units. If $(x - 4)$ was plugged into the function, it would represent a vertical shift.

A composition function can also be formed by plugging one function into another. In function notation, this is written:

$$(f \circ g)(x) = f(g(x))$$

For two functions $f(x) = x^2$ and $g(x) = x - 3$, the composition function becomes:

$$f(g(x)) = (x - 3)^2 = x^2 - 6x + 9$$

The composition of functions can also be used to verify if two functions are inverses of each other. Given the two functions $f(x) = 2x + 5$ and $g(x) = \frac{x-5}{2}$, the composition function can be found $(f \circ g)(x)$. Solving this equation yields:

$$f(g(x)) = 2\left(\frac{x-5}{2}\right) + 5 = x - 5 + 5 = x$$

It also is true that $g(f(x)) = x$. Since the composition of these two functions gives a simplified answer of x, this verifies that $f(x)$ and $g(x)$ are inverse functions. The domain of $f(g(x))$ is the set of all x-values in the domain of $g(x)$ such that $g(x)$ is in the domain of $f(x)$. Basically, both $f(g(x))$ and $g(x)$ have to be defined.

To build an inverse of a function, $f(x)$ needs to be replaced with y, and the x and y values need to be switched. Then, the equation can be solved for y. For example, given the equation $y = e^{2x}$, the inverse can be found by rewriting the equation $x = e^{2y}$. The natural logarithm of both sides is taken down, and the exponent is brought down to form the equation:

$$\ln(x) = \ln(e)\, 2y$$

ln (e)=1, which yields the equation $\ln(x) = 2y$. Dividing both sides by 2 yields the inverse equation

$$\frac{\ln(x)}{2} = y = f^{-1}(x)$$

The domain of an inverse function is the range of the original function, and the range of an inverse function is the domain of the original function. Therefore, an ordered pair (x, y) on either a graph or a table corresponding to $f(x)$ means that the ordered pair (y, x) exists on the graph of $f^{-1}(x)$. Basically, if $f(x) = y$, then $f^{-1}(y) = x$. For a function to have an inverse, it must be one-to-one. That means it must pass the *Horizontal Line Test*, and if any horizontal line passes through the graph of the function twice, a function is not one-to-one. The domain of a function that is not one-to-one can be restricted to an interval in which the function is one-to-one, to be able to define an inverse function.

Functions can also be formed from combinations of existing functions.

Given $f(x)$ and $g(x)$, the following can be built:

$$f + g$$

$$f - g$$

$$fg$$

$$\frac{f}{g}$$

The domains of $f + g, f - g,$ and fg are the intersection of the domains of f and g. The domain of $\frac{f}{g}$ is the same set, excluding those values that make $g(x) = 0$.

For example, if:

$$f(x) = 2x + 3$$

$$g(x) = x + 1$$

then

$$\frac{f}{g} = \frac{2x + 3}{x + 1}$$

Its domain is all real numbers except -1.

Domain and Range of a Function

The domain and range of a function can be found visually by its plot on the coordinate plane. In the function $f(x) = x^2 - 3$, for example, the domain is all real numbers because the parabola stretches as far left and as far right as it can go, with no restrictions. This means that any input value from the real number system will yield an answer in the real number system. For the range, the inequality $y \geq -3$ would be used to describe the possible output values because the parabola has a minimum at $y = -3$.

This means there will not be any real output values less than -3 because -3 is the lowest value it reaches on the y-axis.

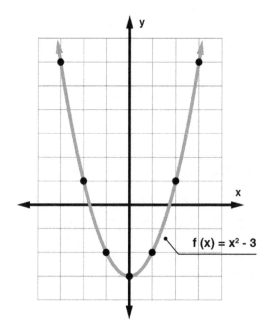

These same answers for domain and range can be found by observing a table. The table below shows that from input values $x = -1$ to $x = 1$, the output results in a minimum of -3. On each side of $x = 0$, the numbers increase, showing that the range is all real numbers greater than or equal to -3.

x (domain/input)	y (range/output)
-2	1
-1	-2
0	-3
-1	-2
2	1

Steps in Solving a Simple Equation

All equations, from the simple to more complex, involve a series of steps that build upon the solution found in the prior step. Sometimes, subsequent steps further manipulate the equation or apply an additional formula, while other times, steps simplify the solution obtained in the prior step or convert its units or presentation in one way or another. In the latter cases, solutions to the two steps are actually equivalent, but presented in different forms. In all situations, it is important to understand and be able to logically explain the reason behind each step involved in finding a solution to a given problem and why the given procedure was followed. To that end, one should verify that his or her obtained answer is reasonable for the provided problem and can defend its accuracy.

For example, when solving a linear equation, the desired result requires determining a numerical value for the unknown variable. If given a linear equation involving addition, subtraction, multiplication, or division, working backwards isolates the variable. Addition and subtraction are inverse operations, as are multiplication and division. Therefore, they can be used to cancel each other out.

The first steps to solving linear equations are distributing, if necessary, and combining any like terms on the same side of the equation. Sides of an equation are separated by an *equal* sign. Next, the equation is manipulated to show the variable on one side. Whatever is done to one side of the equation must be done to the other side of the equation to remain equal. Inverse operations are then used to isolate the variable and undo the order of operations backwards. Addition and subtraction are undone, then multiplication and division are undone.

For example, solve $4(t - 2) + 2t - 4 = 2(9 - 2t)$

Distributing: $4t - 8 + 2t - 4 = 18 - 4t$

Combining like terms: $6t - 12 = 18 - 4t$

Adding $4t$ to each side to move the variable: $10t - 12 = 18$

Adding 12 to each side to isolate the variable: $10t = 30$

Dividing each side by 10 to isolate the variable: $t = 3$

The answer can be checked by substituting the value for the variable into the original equation, ensuring that both sides calculate to be equal.

Calculating and Interpreting the Average Rate of Change of a Function

Rate of change for any line calculates the steepness of the line over a given interval. Rate of change is also known as the slope or rise/run. The rates of change for nonlinear functions vary depending on the interval being used for the function. The rate of change over one interval may be zero, while the next interval may have a positive rate of change. The equation plotted on the graph below, $y = x^2$, is a quadratic function and non-linear. The average rate of change from points $(0, 0)$ to $(1, 1)$ is 1 because the vertical change is 1 over the horizontal change of 1. For the next interval, $(1, 1)$ to $(2, 4)$, the average rate of change is 3 because the slope is $\frac{3}{1}$.

You can see that here:

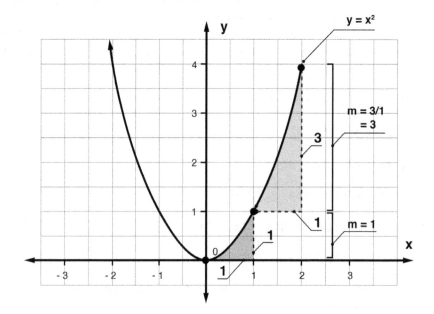

The rate of change for a linear function is constant and can be determined based on a few representations. One method is to place the equation in slope-intercept form: $y = mx + b$. Thus, m is the slope, and b is the y-intercept. In the graph below, the equation is $y = x + 1$, where the slope is 1 and the y-intercept is 1. For every vertical change of 1 unit, there is a horizontal change of 1 unit. The x-intercept is -1, which is the point where the line crosses the x-axis.

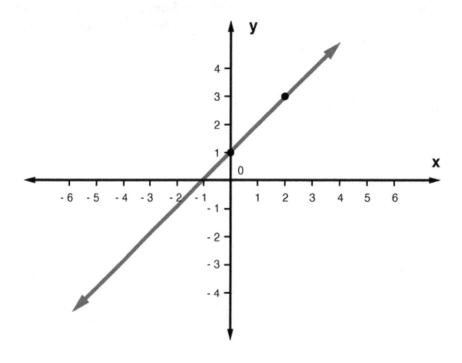

Practice Questions

1. Which of the following is the result of simplifying the expression: $\frac{4a^{-1}b^3}{a^4b^{-2}} \times \frac{3a}{b}$?

    a. $12a^3b^5$

    b. $12\frac{b^4}{a^4}$

    c. $\frac{12}{a^4}$

    d. $7\frac{b^4}{a}$

    e. $4\frac{7b}{a}$

2. What is the product of two irrational numbers?

    a. Irrational
    b. Rational
    c. Contradictory
    d. Complex and imaginary
    e. Irrational or rational

3. The graph shows the position of a car over a 10-second time interval. Which of the following is the correct interpretation of the graph for the interval 1 to 3 seconds?

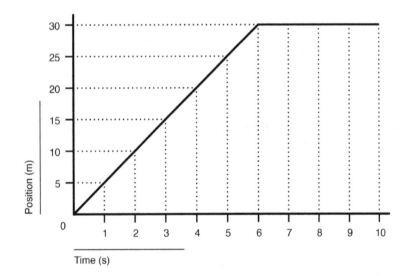

    a. The car remains in the same position.
    b. The car is traveling at a speed of 5m/s.
    c. The car is traveling up a hill.
    d. The car is traveling at 5 mph.
    e. The car is traveling at a speed of 10m/s.

4. Being as specific as possible, how is the number -4 classified?
    a. Real, rational, integer, whole, natural
    b. Real, rational, integer, natural
    c. Real, rational, integer
    d. Real, irrational, complex
    e. Real, irrational, whole

5. What are the zeros of the function: $f(x) = x^3 + 4x^2 + 4x$?
    a. -2
    b. 0, -2
    c. 2
    d. 0, 2
    e. 0, 0

6. If $g(x) = x^3 - 3x^2 - 2x + 6$ and $f(x) = 2$, then what is $g(f(x))$?
    a. -26
    b. 6
    c. $2x^3 - 6x^2 - 4x + 12$
    d. -2
    e. $2^2 + 3x + 6$

7. What is the solution to the following system of equations?

$$x^2 - 2x + y = 8$$
$$x - y = -2$$

    a. $(-2, 3)$
    b. There is no solution.
    c. $(-2, 0) (1, 3)$
    d. $(-2, 0) (3, 5)$
    e. $(-2, 3) (3, 5)$

8. Which of the following shows the correct result of simplifying the following expression:
$(7n + 3n^3 + 3) + (8n + 5n^3 + 2n^4)$?
    a. $9n^4 + 15n - 2$
    b. $2n^4 + 5n^3 + 15n - 2$
    c. $9n^4 + 8n^3 + 15n$
    d. $2n^4 + 8n^3 + 15n + 3$
    e. $3n^4 + 4n^3 + 15n - 4$

9. What is the product of the following expression?

$$(4x - 8)(5x^2 + x + 6)$$

    a. $20x^3 - 36x^2 + 16x - 48$
    b. $6x^3 - 41x^2 + 12x + 15$
    c. $20x^4 + 11x^2 - 37x - 12$
    d. $2x^3 - 11x^2 - 32x + 20$
    e. $10x^3 + 2x^2 - 8x + 48$

10. How could the following equation be factored to find the zeros?

$$y = x^3 - 3x^2 - 4x$$

a. $0 = x^2(x - 4), x = 0, 4$
b. $0 = 3x(x + 1)(x + 4), x = 0, -1, -4$
c. $0 = x(x + 1)(x + 6), x = 0, -1, -6$
d. $0 = x^2(x - 1)(x - 4), x = 0, -1, 4$
e. $0 = x(x + 1)(x - 4), x = 0, -1, 4$

11. What is the simplified quotient of $\frac{5x^3}{3x^2y} \div \frac{25}{3y^9}$?

a. $\frac{125x}{9y^{10}}$

b. $\frac{x}{5y^8}$

c. $\frac{5}{xy^8}$

d. $\frac{xy^8}{5}$

e. $\frac{xy^2}{5}$

12. What is the solution for the following equation?

$$\frac{x^2 + x - 30}{x - 5} = 11$$

a. $x = -6$
b. There is no solution.
c. $x = 16$
d. $x = 5$
e. $x = 6$

13. Mom's car drove 72 miles in 90 minutes. How fast did she drive in feet per second?
a. 0.8 feet per second
b. 48.9 feet per second
c. 0.009 feet per second
d. 70.4 feet per second
e. 55 feet per second

14. How do you solve $V = lwh$ for h?

    a. $lwV = h$

    b. $h = \dfrac{V}{lw}$

    c. $h = \dfrac{Vl}{w}$

    d. $h = \dfrac{Vw}{l}$

    e. $h = \dfrac{wl}{V}$

15. What is the domain for the function $y = \sqrt{x}$?

    a. All real numbers
    b. $x \geq 0$
    c. $x > 0$
    d. $y \geq 0$
    e. $y < 0$

16. If Sarah reads at an average rate of 21 pages in four nights, how long will it take her to read 140 pages?

    a. 6 nights
    b. 26 nights
    c. 8 nights
    d. 12 nights
    e. 27 nights

17. The phone bill is calculated each month using the equation $c = 50g + 75$. The cost of the phone bill per month is represented by c, and g represents the gigabytes of data used that month. What is the value and interpretation of the slope of this equation?

    a. 75 dollars per day
    b. 75 gigabytes per day
    c. 50 dollars per day
    d. 50 dollars per gigabyte
    e. 25 dollars per day

18. What are the zeros of $f(x) = x^2 + 4$?

    a. $x = -4$
    b. $x = \pm 2i$
    c. $x = \pm 2$
    d. $x = \pm 4i$
    e. $x = \pm 7i$

19. Twenty is 40 percent of what number?

    a. 500
    b. 8
    c. 200
    d. 5000
    e. 50

20. What is the simplified form of the expression $1.2 \times 10^{12} \div 3.0 \times 10^{8}$?

    a. 0.4×10^{4}

    b. 4.0×10^{4}

    c. 4.0×10^{3}

    d. 3.6×10^{20}

    e. 3.8×10^{30}

21. You measure the width of your door to be 36 inches. The true width of the door is 35.75 inches. What is the relative error in your measurement?

    a. 0.7%

    b. 0.007%

    c. 0.99%

    d. 0.1%

    e. 0.77%

22. What is the y-intercept for $y = x^2 + 3x - 4$?

    a. $y = 1$

    b. $y = -4$

    c. $y = 3$

    d. $y = 4$

    e. $y = 8$

23. Is the following function even, odd, neither, or both?

$$y = \frac{1}{2}x^4 + 2x^2 - 6$$

    a. Even

    b. Odd

    c. Neither

    d. Both

    e. There is no way to tell.

24. Which equation is not a function?

    a. $y = |x|$

    b. $y = \sqrt{x}$

    c. $x = 3$

    d. $y = 4$

    e. $y = 5$

25. How could the following function be rewritten to identify the zeros?

$$y = 3x^3 + 3x^2 - 18x$$

    a. $y = 3x(x + 3)(x - 2)$

    b. $y = x(x - 2)(x + 3)$

    c. $y = 3x(x - 3)(x + 2)$

    d. $y = (x + 3)(x - 2)$

    e. $y = (x + 1)(x - 1)$

26. If the volume of a sphere is 288π cubic meters, what are the radius and surface area of the same sphere?

a. Radius 6 meters and surface area 144π square meters
b. Radius 36 meters and surface area 144π square meters
c. Radius 6 meters and surface area 12π square meters
d. Radius 36 meters and surface area 12π square meters
e. Radius 30 meters and surface area 24π square meters

27. What is the type of function that is modeled by the values in the following table?

X	f(x)
1	2
2	4
3	8
4	16
5	32

a. Linear
b. Exponential
c. Quadratic
d. Cubic
e. Elliptic

28. A sample data set contains the following values: 1, 3, 5, 7. What's the standard deviation of the set?

a. 2.58
b. 4
c. 6.23
d. 1.1
e. 3.67

29. A ball is drawn at random from a ball pit containing 8 red balls, 7 yellow balls, 6 green balls, and 5 purple balls. What's the probability that the ball drawn is yellow?

a. $\frac{1}{26}$

b. $\frac{19}{26}$

c. $\frac{7}{26}$

d. 1

e. $\frac{5}{26}$

30. Two cards are drawn from a shuffled deck of 52 cards. What's the probability that both cards are Kings if the first card isn't replaced after it's drawn?

    a. $\frac{1}{169}$

    b. $\frac{1}{221}$

    c. $\frac{1}{13}$

    d. $\frac{4}{13}$

    e. $\frac{1}{15}$

31. What's the probability of rolling a 6 at least once in two rolls of a die?

    a. $\frac{1}{3}$

    b. $\frac{1}{36}$

    c. $\frac{1}{6}$

    d. $\frac{11}{36}$

    e. $\frac{11}{43}$

32. For a group of 20 men, the median weight is 180 pounds and the range is 30 pounds. If each man gains 10 pounds, which of the following would be true?

    a. The median weight will increase, and the range will remain the same.

    b. The median weight and range will both remain the same.

    c. The median weight will stay the same, and the range will increase.

    d. The median weight and range will both increase.

    e. The median weight will decrease, and the range will increase.

33. For the following similar triangles, what are the values of x and y (rounded to one decimal place)?

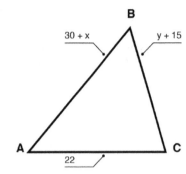 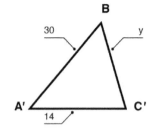

a. $x = 16.5, y = 25.1$
b. $x = 19.5, y = 24.1$
c. $x = 17.1, y = 26.3$
d. $x = 26.3, y = 17.1$
e. $x = 28.9, y = 19.1$

34. What are the center and radius of a circle with equation $4x^2 + 4y^2 - 16x - 24y + 51 = 0$?
a. Center (3, 2) and radius $\frac{1}{2}$
b. Center (2, 3) and radius $\frac{1}{2}$
c. Center (3, 2) and radius $\frac{1}{4}$
d. Center (2, 3) and radius $\frac{1}{4}$
e. Center (1, 3) and radius $\frac{1}{3}$

35. If the ordered pair $(-3, -4)$ is reflected over the x-axis, what's the new ordered pair?
a. $(-3, -4)$
b. $(3, -4)$
c. $(3, 4)$
d. $(-3, 4)$
e. $(4, 3)$

36. Dwayne has received the following scores on his math tests: 78, 92, 83, and 97. What score must Dwayne get on his next math test to have an overall average of 90?
a. 89
b. 98
c. 95
d. 94
e. 100

37. A line passes through the origin and through the point (-3, 4). What is the slope of the line?

    a. $-\frac{4}{3}$

    b. $-\frac{3}{4}$

    c. $\frac{4}{3}$

    d. $\frac{3}{4}$

    e. $\frac{1}{3}$

38. An equilateral triangle has a perimeter of 18 feet. If a square whose sides have the same length as one side of the triangle is built, what will be the area of the square?

    a. 6 square feet
    b. 36 square feet
    c. 256 square feet
    d. 1000 square feet
    e. 324 square feet

39. On Monday, Robert mopped the floor in 4 hours. On Tuesday, he did it in 3 hours. If on Monday, his average rate of mopping was p sq. ft. per hour, what was his average rate on Tuesday?

    a. $\frac{4}{3}p$ sq. ft. per hour

    b. $\frac{3}{4}p$ sq. ft. per hour

    c. $\frac{5}{4}p$ sq. ft. per hour

    d. $p + 1$ sq. ft. per hour

    e. $\frac{1}{3}p$ sq. ft. per hour

40. Which of the following inequalities is equivalent to $3 - \frac{1}{2}x \geq 2$?

    a. $x \geq 2$
    b. $x \leq 2$
    c. $x \geq 1$
    d. $x \leq 1$
    e. $x \leq -2$

41. If $\sqrt{1 + x} = 4$, what is x?

    a. 10
    b. 15
    c. 20
    d. 25
    e. 36

42. A line passes through the point (1, 2) and crosses the y-axis at $y = 1$. Which of the following is an equation for this line?
    a. $y = 2x$
    b. $y = x + 1$
    c. $x + y = 1$
    d. $y = \frac{x}{2} - 2$
    e. $y = x - 1$

43. $x^4 - 16$ can be simplified to which of the following?
    a. $(x^2 - 4)(x^2 + 4)$
    b. $(x^2 + 4)(x^2 + 4)$
    c. $(x^2 - 4)(x^2 - 4)$
    d. $(x^2 - 2)(x^2 + 4)$
    e. $(x^2 - 2)(x^2 + 2)$

44. What is the solution to $4 \times 7 + (25 - 21)^2 \div 2$?
    a. 512
    b. 36
    c. 60.5
    d. 22
    e. 16

45. Johnny earns $2334.50 from his job each month. He pays $1437 for monthly expenses. Johnny is planning a vacation in 3 months that he estimates will cost $1750 total. How much will Johnny have left over from three months of saving once he pays for his vacation?
    a. $948.50
    b. $584.50
    c. $852.50
    d. $942.50
    e. $848.50

46. What is the volume of a cube with the side equal to 5 centimeters?
    a. 10 cm³
    b. 15 cm³
    c. 50 cm³
    d. 25 cm³
    e. 125 cm³

47. A pizzeria owner regularly creates jumbo pizzas, each with a radius of 9 inches. She is mathematically inclined and wants to know the area of the pizza to purchase the correct boxes and know how much she is feeding her customers. What is the area of the circle, in terms of π, with a radius of 9 inches?
    a. 3π in²
    b. 18π in²
    c. 90π in²
    d. 9π in²
    e. 81π in²

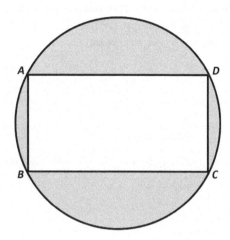

48. Rectangle *ABCD* is inscribed in the circle above. The length of side *AB* is 9 inches and the length of side *BC* is 12 inches. What is the area of the shaded region?

    a. 64.4 sq. in.

    b. 68.6 sq. in.

    c. 62.8 sq. in.

    d. 61.3 sq. in.

    e. 64.6 sq. in.

49. The following graph compares the various test scores of the top three students in each of these teacher's classes. Based on the graph, which teacher's students had the smallest range of test scores?

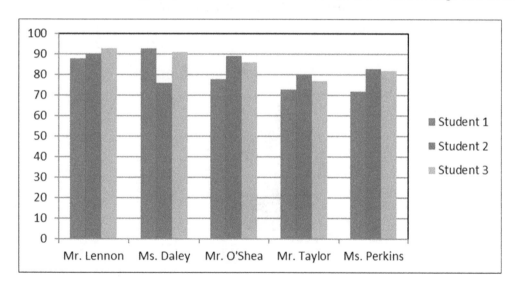

    a. Mr. Lennon

    b. Mr. O'Shea

    c. Mr. Taylor

    d. Ms. Daley

    e. Ms. Perkins

50. Kimberley earns $10 an hour babysitting, and after 10 p.m., she earns $12 an hour, with the amount paid being rounded to the nearest hour accordingly. On her last job, she worked from 5:30 p.m. to 11 p.m. In total, how much did Kimberley earn on her last job?

    a. $45

    b. $57

    c. $62

    d. $42

    e. $55

51. What is the length of the hypotenuse of a right triangle with one leg equal to 3 centimeters and the other leg equal to 4 centimeters?

    a. 7 cm

    b. 5 cm

    c. 25 cm

    d. 12 cm

    e. 6 cm

52. What is the overall median of Dwayne's current scores: 78, 92, 83, 97?

    a. 19

    b. 85

    c. 83

    d. 94.5

    e. 87.5

53. This chart indicates how many sales of CDs, vinyl records, and MP3 downloads occurred over the last year. Approximately what percentage of the total sales was from CDs?

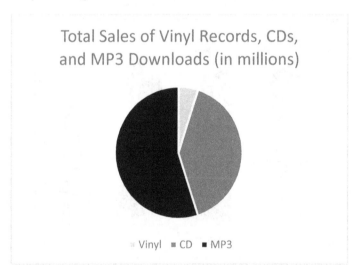

Total Sales of Vinyl Records, CDs, and MP3 Downloads (in millions)

    Vinyl ■ CD ■ MP3

    a. 55%

    b. 25%

    c. 40%

    d. 5%

    e. 75%

54. What is the slope of this line?

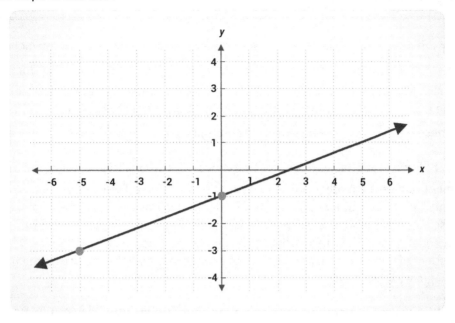

a. 2

b. $\frac{5}{2}$

c. $\frac{1}{2}$

d. $\frac{2}{5}$

e. $-\frac{1}{2}$

55. Which of the following equations best represents the problem below?

The width of a rectangle is 2 centimeters less than the length. If the perimeter of the rectangle is 44 centimeters, then what are the dimensions of the rectangle?

a. $2l + 2(l - 2) = 44$
b. $(l + 2) + (l + 2) + l = 48$
c. $l \times (l - 2) = 44$
d. $(l + 2) + (l + 2) + l = 44$
e. $2l + 2(w - 2) = 44$

Answer Explanations

1. B: To simplify the given equation, the first step is to make all exponents positive by moving them to the opposite place in the fraction. This expression becomes $\frac{4b^3b^2}{a^1a^4} \times \frac{3a}{b}$. Then the rules for exponents can be used to simplify. Multiplying the same bases means the exponents can be added. Dividing the same bases means the exponents are subtracted. Thus, after multiplying the exponents in the first fraction the equation becomes $\frac{4b^5}{a^5} \times \frac{3a}{b}$. Therefore, we can first multiply to get $\frac{12ab^5}{a^5b}$. Then, dividing yields $12\frac{b^4}{a^4}$.

2. E: The product of two irrational numbers can be rational or irrational. Sometimes, the irrational parts of the two numbers cancel each other out, leaving a rational number. For example, $\sqrt{2} \times \sqrt{2} = 2$ because the roots cancel each other out. Technically, the product of two irrational numbers can be complex because complex numbers can have either the real or imaginary part (in this case, the imaginary part) equal zero and still be considered a complex number. However, Choice *D* is incorrect because the product of two irrational numbers is not an imaginary number so saying the product is complex *and* imaginary is incorrect.

3. B: The car is traveling at a speed of five meters per second. On the interval from one to three seconds, the position changes by ten meters. By making this change in position over time into a rate, the speed becomes ten meters in two seconds or five meters in one second.

4. C: The number negative four is classified as a real number because it exists and is not imaginary. It is rational because it does not have a decimal that never ends. It is an integer because it does not have a fractional component. The next classification would be whole numbers, for which negative four does not qualify because it is negative. Choices *D* and *E* are wrong because -4 is not considered an irrational number because it does not have a never-ending decimal component.

5. B: There are two zeros for the function $x = 0, -2$. The zeros can be found several ways, but this particular equation can be factored into:

$$f(x) = x(x^2 + 4x + 4) = x(x + 2)(x + 2)$$

By setting each factor equal to zero and solving for x, there are two solutions. On a graph, these zeros can be seen where the line crosses the x-axis.

6. D: This problem involves a composition function, where one function is plugged into the other function. In this case, the $f(x)$ function is plugged into the $g(x)$ function for each x-value. The composition equation becomes:

$$g\big(f(x)\big) = 2^3 - 3(2^2) - 2(2) + 6$$

Simplifying the equation gives the answer:

$$g\big(f(x)\big) = 8 - 3(4) - 2(2) + 6$$

$$8 - 12 - 4 + 6 = -2$$

7. D: This system of equations involves one quadratic function and one linear function, as seen from the degree of each equation. One way to solve this is through substitution. Solving for y in the second equation yields $y = x + 2$. Plugging this equation in for the y of the quadratic equation yields:

$$x^2 - 2x + x + 2 = 8$$

Simplifying the equation, it becomes $x^2 - x + 2 = 8$. Setting this equal to zero and factoring, it becomes:

$$x^2 - x - 6 = 0$$

$$(x - 3)(x + 2) = 0$$

Solving these two factors for x gives the zeros $x = 3, -2$. To find the y-value for the point, each number can be plugged in to either original equation. Solving each one for y yields the points $(3, 5)$ and $(-2, 0)$.

8. D: The expression is simplified by collecting like terms. Terms with the same variable and exponent are like terms, and their coefficients can be added.

Since the two sets of parentheses are being added, the parentheses are actually not needed. Like terms can be added together even if they are in different sets of parentheses.

9. A: Finding the product means distributing one polynomial onto the other. Each term in the first must be multiplied by each term in the second. Then, like terms can be collected. Multiplying the factors yields the expression:

$$20x^3 + 4x^2 + 24x - 40x^2 - 8x - 48$$

Collecting like terms means adding the x^2 terms and adding the x terms. The final answer after simplifying the expression is:

$$20x^3 - 36x^2 + 16x - 48$$

10. E: Finding the zeros for a function by factoring is done by setting the equation equal to zero, then completely factoring. Since there was a common x for each term in the provided equation, that would be factored out first. Then the quadratic that was left could be factored into two binomials, which are $(x + 1)(x - 4)$. Setting each factor equal to zero and solving for x yields three zeros.

11. D: Dividing rational expressions follows the same rule as dividing fractions. The division is changed to multiplication by the reciprocal of the second fraction. This turns the expression into:

$$\frac{5x^3}{3x^2y} \times \frac{3y^9}{25}$$

Multiplying across and simplifying, the final expression is $\frac{xy^8}{5}$.

12. B: The equation can be solved by factoring the numerator into $(x + 6)(x - 5)$. Since that same factor exists on top and bottom, that factor $(x - 5)$ cancels. This leaves the equation $x + 6 = 11$. Solving the equation gives the answer $x = 5$. When this value is plugged into the equation, it yields a zero in the denominator of the fraction. Since this is undefined, there is no solution.

13. D: This problem can be solved by using unit conversion. The initial units are miles per minute. The final units need to be feet per second. Converting miles to feet uses the equivalence statement 1 mile equals 5,280 feet. Converting minutes to seconds uses the equivalence statement 1 minute equals 60 seconds. Setting up the ratios to convert the units is shown in the following equation:

$$\frac{72 \text{ mi}}{90 \text{ min}} \times \frac{1 \text{ min}}{60 \text{ s}} \times \frac{5280 \text{ ft}}{1 \text{ mi}} = 70.4 \frac{\text{ft}}{\text{s}}$$

The initial units cancel out, and the new units are left.

14. B: The formula can be manipulated by dividing both the length, *l*, and the width, *w*, on both sides. The length and width will cancel on the right, leaving height by itself.

15. B: The domain is all possible input values, or *x*-values. For this equation, the domain is every number greater than or equal to zero. There are no negative numbers in the domain because taking the square root of a negative number results in an imaginary number.

16. E: This problem can be solved by setting up a proportion involving the given information and the unknown value. The proportion is:

$$\frac{21 \text{ } pages}{4 \text{ } nights} = \frac{140 \text{ } pages}{x \text{ } nights}$$

Solving the proportion by cross-multiplying, the equation becomes $21x = 4 \times 140$, where $x = 26.67$. Since it is not an exact number of nights, the answer is rounded up to 27 nights. Twenty-six nights would not give Sarah enough time.

17. D: The slope from this equation is 50, and it is interpreted as the cost per gigabyte used. Since the *g*-value represents number of gigabytes and the equation is set equal to the cost in dollars, the slope relates these two values. For every gigabyte used on the phone, the bill goes up 50 dollars.

18. B: The zeros of this function can be found by using the quadratic formula:

$$x = \frac{-b \pm \sqrt{b^2 - 4ac}}{2a}$$

Identifying a, b, and c can also be done from the equation because it is in standard form. The formula becomes:

$$x = \frac{0 \pm \sqrt{0^2 - 4(1)(4)}}{2(1)} = \frac{\sqrt{-16}}{2}$$

Since there is a negative underneath the radical, the answer is a complex number:

$$x = \pm 2i$$

19. E: Setting up a proportion is the easiest way to represent this situation. The proportion becomes $\frac{20}{x} = \frac{40}{100}$, where cross-multiplication can be used to solve for x. The answer can also be found by observing the two fractions as equivalent, knowing that twenty is half of forty, and fifty is half of one-hundred.

20. C: Scientific notation division can be solved by grouping the first terms together and grouping the tens together. The first terms can be divided, and the tens terms can be simplified using the rules for exponents. The initial expression becomes 0.4×10^4. This is not in scientific notation because the first number is not between 1 and 10. Shifting the decimal and subtracting one from the exponent, the answer becomes 4.0×10^3./121. A: The relative error can be found by finding the absolute error and

making it a percent of the true value. The absolute error is $36 - 35.75 = 0.25$. This error is then divided by 35.75—the true value—to find 0.7%.

22. B: The y-intercept of an equation is found where the x-value is zero. Plugging zero into the equation for x, the first two terms cancel out, leaving -4.

23. A: The equation is *even* because $f(-x) = f(x)$. Plugging in a negative value will result in the same answer as when plugging in the positive of that same value. The function:

$$f(-2) = \frac{1}{2}(-2)^4 + 2(-2)^2 - 6$$

$$8 + 8 - 6 = 10$$

yields the same value as:

$$f(2) = \frac{1}{2}(2)^4 + 2(2)^2 - 6$$

$$8 + 8 - 6 = 10$$

24. C: The equation $x = 3$ is not a function because it does not pass the vertical line test. This test is made from the definition of a function, where each x-value must be mapped to one and only one y-value. This equation is a vertical line, so the x-value of 3 is mapped with an infinite number of y-values.

25. A: The function can be factored to identify the zeros. First, the term $3x$ is factored out to the front because each term contains $3x$. Then, the quadratic is factored into $(x + 3)(x - 2)$.

26. A: Because the volume of the given sphere is 288π cubic meters, this means $\frac{4}{3}\pi r^3 = 288\pi$. This equation is solved for r to obtain a radius of 6 meters. The formula for the surface area of a sphere is $4\pi r^2$, so if $r = 6$ in this formula, the surface area is 144π square meters.

27. B: The table shows values that are increasing exponentially. The differences between the inputs are the same, while the differences in the outputs are changing by a factor of 2. The values in the table can be modeled by the equation $f(x) = 2^x$.

28. A: First, the sample mean must be calculated.

$$\bar{x} = \frac{1}{4}(1 + 3 + 5 + 7) = 4$$

The standard deviation of the data set is:

$$\sigma = \sqrt{\frac{\sum(x - \bar{x})^2}{n - 1}}$$

and $n = 4$ represents the number of data points.

Therefore, $\sigma =$

$$\sqrt{\frac{1}{3}[(1-4)^2 + (3-4)^2 + (5-4)^2 + (7-4)^2]}$$

$$\sqrt{\frac{1}{3}(9+1+1+9)} = 2.58.$$

29. C: The sample space is made up of $8 + 7 + 6 + 5 = 26$ balls. The probability of pulling each individual ball is $\frac{1}{26}$. Since there are 7 yellow balls, the probability of pulling a yellow ball is $\frac{7}{26}$.

30. B: For the first card drawn, the probability of a King being pulled is $\frac{4}{52}$. Since this card isn't replaced, if a King is drawn first the probability of a King being drawn second is $\frac{3}{51}$. The probability of a King being drawn in both the first and second draw is the product of the two probabilities: $\frac{4}{52} \times \frac{3}{51} = \frac{12}{2652}$. This fraction, when divided by 12, equals $\frac{1}{221}$.

31. D: The addition rule is necessary to determine the probability because a 6 can be rolled on either roll of the die. The rule used is:

$$P(A \text{ or } B) = P(A) + P(B) - P(A \text{ and } B)$$

The probability of a 6 being individually rolled is $\frac{1}{6}$ and the probability of a 6 being rolled twice is:

$$\frac{1}{6} \times \frac{1}{6} = \frac{1}{36}$$

Therefore, the probability that a 6 is rolled at least once is:

$$\frac{1}{6} + \frac{1}{6} - \frac{1}{36} = \frac{11}{36}$$

32. A: If each man gains 10 pounds, every original data point will increase by 10 pounds. Therefore, the man with the original median will still have the median value, but that value will increase by 10. The smallest value and largest value will also increase by 10 and, therefore, the difference between the two won't change. The range does not change in value and, thus, remains the same.

33. C: Because the triangles are similar, the lengths of the corresponding sides are proportional. Therefore:

$$\frac{30 + x}{30} = \frac{22}{14} = \frac{y + 15}{y}$$

This results in the equation:

$$14(30 + x) = 22 \times 30$$

When solved, gives:

$$x = 17.1$$

The proportion also results in the equation:

$$14(y + 15) = 22y$$

When solved, gives:

$$y = 26.3$$

34. B: The technique of completing the square must be used to change $4x^2 + 4y^2 - 16x - 24y + 51 = 0$ into the standard equation of a circle. First, the constant must be moved to the right-hand side of the equal sign, and each term must be divided by the coefficient of the x^2 term (which is 4). The x and y terms must be grouped together to obtain:

$$x^2 - 4x + y^2 - 6y = -\frac{51}{4}$$

Now the process of completing the square must be completed for each variable. This gives:

$$(x^2 - 4x + 4) + (y^2 - 6y + 9) = -\frac{51}{4} + 4 + 9$$

The equation can be written as:

$$(x - 2)^2 + (y - 3)^2 = \frac{1}{4}$$

Therefore, the center of the circle is (2, 3) and the radius is:

$$\sqrt{\frac{1}{4}} = \frac{1}{2}$$

35. D: When an ordered pair is reflected over an axis, the sign of one of the coordinates must change. When it's reflected over the x-axis, the sign of the y-coordinate must change. The x-value remains the same. Therefore, the new ordered pair is $(-3, 4)$.

36. E: To find the average of a set of values, add the values together and then divide by the total number of values. In this case, include the unknown value of what Dwayne needs to score on his next test, in order to solve it.

$$\frac{78 + 92 + 83 + 97 + x}{5} = 90$$

Add the unknown value to the new average total, which is 5. Then multiply each side by 5 to simplify the equation, resulting in:

$$78 + 92 + 83 + 87 + x = 450$$

$$350 + x = 450$$

$$x = 100$$

Dwayne would need to get a perfect score of 100 in order to get an average of at least 90.

Test this answer by substituting back into the original formula.

$$\frac{78 + 92 + 83 + 97 + 100}{5} = 90$$

37. A: The slope is given by:

$$m = \frac{y_2 - y_1}{x_2 - x_1} = \frac{0 - 4}{0 - (-3)} = -\frac{4}{3}$$

38. B: An equilateral triangle has three sides of equal length, so if the total perimeter is 18 feet, each side must be 6 feet long. A square with sides of 6 feet will have an area of $6^2 = 36$ square feet.

39. A: Robert accomplished his task on Tuesday in $\frac{3}{4}$ the time compared to Monday. He must have worked $\frac{4}{3}$ as fast.

40. B: To simplify this inequality, subtract 3 from both sides to get $-\frac{1}{2}x \geq -1$. Then, multiply both sides by -2 (remembering this flips the direction of the inequality) to get $x \leq 2$.

41. B: Start by squaring both sides to get $1 + x = 16$. Then subtract 1 from both sides to get $x = 15$.

42. B: From the slope-intercept form, $y = mx + b$, it is known that b is the y-intercept, which is 1. Compute the slope as $\frac{2-1}{1-0} = 1$, so the equation should be $y = x + 1$.

43. A: This has the form $t^2 - y^2$, with $t = x^2$ and $y = 4$. It's also known that $t^2 - y^2 = (t + y)(t - y)$, and substituting the values for t and y into the right-hand side gives $(x^2 - 4)(x^2 + 4)$.

44. B: To solve this correctly, keep in mind the order of operations with the mnemonic PEMDAS (Please Excuse My Dear Aunt Sally). This stands for Parentheses, Exponents, Multiplication, Division, Addition, Subtraction. Taking it step by step, solve inside the parentheses first:

$$4 \times 7 + 4^2 \div 2$$

Then, apply the exponent:

$$4 \times 7 + 16 \div 2$$

Multiplication and division are both performed next:

$$28 + 8 = 36$$

Addition and subtraction are done last. The solution is 36.

45. D: First, subtract $1437 from $2334.50 to find Johnny's monthly savings; this equals $897.50. Then, multiply this amount by 3 to find out how much he will have (in three months) before he pays for his vacation: this equals $2692.50. Finally, subtract the cost of the vacation ($1750) from this amount to find how much Johnny will have left: $942.50.

46. E: The volume of a cube is the length of the side cubed, and 5 centimeters cubed is 125 cm³. Choice A is not the correct answer because that is 2 × 5 centimeters. Choice B is not the correct answer because that is 3 × 5 centimeters. Choice C is not the correct answer because that is 5 × 10 centimeters and Choice D is not correct because that is 5 × 5 inches.

47. E: The formula for the area of the circle is πr^2 and 9 squared is 81. Choice A is not the correct answer because that takes the square root of the radius instead of squaring the radius. Choice B is not the correct answer because that is 2 × 9. Choice C is not the correct answer because that is 9 × 10. Choice D is not the correct answer because that is simply the value of the radius.

48. B: The inscribed rectangle is 9 × 12 inches. First find the length of AC using the Pythagorean Theorem. So, $9^2 + 12^2 = c^2$, where c is the length of AC in this case. This means that $AC = 15$ inches. This means the diameter of the circle is 15 inches. This can be used to find the area of the entire circle. The formula is πr^2. So, $3.14(7.5)^2 = 176.6$ sq. inches. Then, subtract the area of the rectangle to find just the just the area of the shaded region. This is $176.6 - 108 = 68.6$.

49. A: To calculate the range in a set of data, subtract the lowest value from the highest value. In this graph, the range of Mr. Lennon's students is 5, which can be seen physically in the graph as having the smallest difference between the highest value and the lowest value compared with the other teachers.

50. C: Kimberley worked 4.5 hours at the rate of $10/h and 1 hour at the rate of $12/h. The problem states that her pay is rounded to the nearest hour, so the 4.5 hours would round up to 5 hours at the rate of $10/h.

$$(5h)\left(\frac{\$10}{h}\right) + (1h)\left(\frac{\$12}{h}\right) = \$50 + \$12 = \$62$$

51. B: This answer is correct because $3^2 + 4^2$ is $9 + 16$, which is 25. Taking the square root of 25 is 5. Choice A is not the correct answer because that is $3 + 4$. Choice C is not the correct answer because that is stopping at $3^2 + 4^2$ is $9 + 16$, which is 25. Choice D is not the correct answer because that is 3×4. Choice E is incorrect because the square root of 25 is 5 not 6.

52. E: For an even number of total values, the *median* is calculated by finding the *mean* or average of the two middle values once all values have been arranged in ascending order from least to greatest. In this case, $(83 + 92) \div 2$ would equal the median 87.5, Choice E.

53. C: The sum total percentage of a pie chart must equal 100%. Since the CD sales take up less than half of the chart and more than a quarter (25%), it can be determined to be 40% overall. This can also be measured with a protractor. The angle of a circle is 360°. Since 25% of 360° would be 90° and 50% would be 180°, the angle percentage of CD sales falls in between; therefore, it would be Choice C.

54. D: The slope is given by the change in y divided by the change in x. Specifically, it's:

$$slope = \frac{y_2 - y_1}{x_2 - x_1}$$

The first point is (-5, -3) and the second point is (0, -1). Work from left to right when identifying coordinates. Thus, the point on the left is point 1 (-5, -3) and the point on the right is point 2 (0, -1).

Now we need to just plug those numbers into the equation:

$$slope = \frac{-1 - (-3)}{0 - (-5)}$$

It can be simplified to:

$$slope = \frac{-1 + 3}{0 + 5}$$

$$slope = \frac{2}{5}$$

55. A: The first step is to determine the unknown, which is in terms of the length, l.

The second step is to translate the problem into the equation using the perimeter of a rectangle, $P = 2l + 2w$. The width is the length minus 2 centimeters. The resulting equation is $2l + 2(l - 2) = 44$. The equation can be solved as follows:

$2l + 2l - 4 = 44$	Apply the distributive property on the left side of the equation
$4l - 4 = 44$	Combine like terms on the left side of the equation
$4l = 48$	Add 4 to both sides of the equation
$l = 12$	Divide both sides of the equation by 4

The length of the rectangle is 12 centimeters. The width is the length minus 2 centimeters, which is 10 centimeters. Checking the answers for length and width forms the following equation:

$$44 = 2(12) + 2(10)$$

The equation can be solved using the order of operations to form a true statement: $44 = 44$.

Science

Life Science

Understand Organisms, Their Environments, and Their Life Cycles

Cell Differentiation

Cell differentiation refers to the process of a cell transforming into another type of cell. It most commonly involves a less specialized cell transforming into a more specialized cell.

The human body contains a vast array of cells which undergo division and differentiation to compose each unique human being. The trillions of cells composing the human body are derived from one cell, a fertilized egg called a zygote. The zygote not only divides, but also differentiates into cells that perform specific tasks.

Genes control the process of cell differentiation during human development. The zygote divides through mitosis into a blastula and then into a gastrula. At this stage, the three embryonic germ layers (endoderm, mesoderm, and ectoderm) are formed. Most of the human body systems develop from one or more of the embryonic germ layers. For example, the digestive system develops from the endoderm, or innermost germ layer; the cardiovascular system develops from the mesoderm, or middle germ layer; and the nervous system develops from the ectoderm, or outer germ layer.

Mitosis and Meiosis

Mitosis

Mitosis, or asexual reproduction, produces two new cells that are genetically identical to the parent cell. It can happen in virtually every healthy adult cell, although some cells like red blood cells and neurons do not divide in general. When a cell is not undergoing cell division, it is in a stage called *interphase* which is characterized by growth, typical maintenance, and DNA synthesis in the nucleus. Each healthy human cell nucleus typically has 46 chromosomes and is said to be *diploid* (2n), as this count comes from 23 pairs of *homologous chromosomes*. Homologous chromosomes are pairs of chromatids with similar sections that correspond to similar genes, as in pairs of chromosome 1 or pairs of chromosome 21.

Mitosis is divided into the following events:

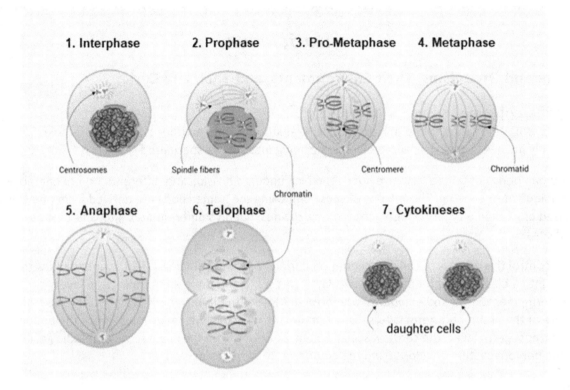

Prophase: The already duplicated chromatin condenses to form chromosomes. Each new chromosome is made up of two identical sister chromatids joined by a structure called a *centromere*. The nuclear envelope is degraded and spindle fibers form and attach to structures called *centrioles*. The centrioles separate and proceed to opposite poles of the cell.

Pro-metaphase: The centrioles build spindle fibers and attach them to the chromosomes.

Metaphase: Using tension from spindle fibers, the chromosomes align in the middle of the cell.

Anaphase: The spindle fibers contract and separate the chromosomes at their centromere. The single chromatids, pulled by the spindle fibers, begin migrating to opposite poles of the cell.

Telophase: The chromatids arrive at opposite poles of the cell. The spindle fibers disappear, the nuclear envelope reforms, and the chromosomes uncoil back into chromatin.

Cytokinesis: The process refers to the cleaving of the cytoplasm to form two daughter cells genetically identical to the parent cell. In animal cells, this happens via a *cleavage furrow*; a cleavage furrow is a pinching of the cell membrane near the center that deepens until it reaches the point that the cell membrane can recombine and split the entity into two separate cells.

Meiosis

Meiosis, or sexual division, happens only in specialized sex cells and produces four cells called gametes. In humans, sex cells are found in the ovaries and in the testes, and are in contrast to somatic cells which constitute the rest of the body of the organism. Each gamete contains half the number of chromosomes of a normal cell, and each is said to be *haploid* (n) rather than *diploid* (2n). In humans, a gamete has 23

chromosomes instead of the 46 which are typically found in somatic cells. The female gamete is called an *egg* and the male gamete is called a *sperm*.

Preceding meiosis, the DNA is synthesized, and the chromatin coalesces into chromosomes, as in mitosis. However, the pairs of sister chromatids that are homologous combine together, joining their centromeres into a single *chiasma* and forming a *tetrad*. At this point, sections of the different chromatids may break off and rejoin, possibly in another place. Half of a leg of one chromatid may swap with that of another chromatid; the chromatids essentially exchange some of their genes with one another. This process, called *crossing over* or *genetic recombination*, happens in prophase I and leads to greater genetic diversity.

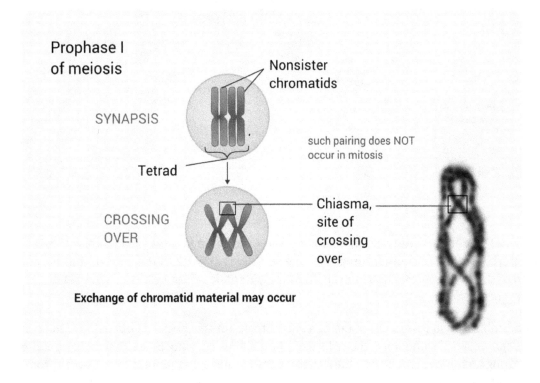

Like mitosis, meiosis is divided into stages of prophase, metaphase, anaphase, telophase, and cytokinesis. However, as the end products have half the genetic material as the end products of mitosis, another round of division, so meiosis is first partitioned into meiosis I and meiosis II, each round similar in scope to mitosis. During meiosis I, homologous chromosome pairs are separated into two daughter cells. Each daughter cell is haploid (n) because, although each cell at the end of meiosis I has 46 chromatids, half of them are duplicates of the other and not considered unique genetic material.

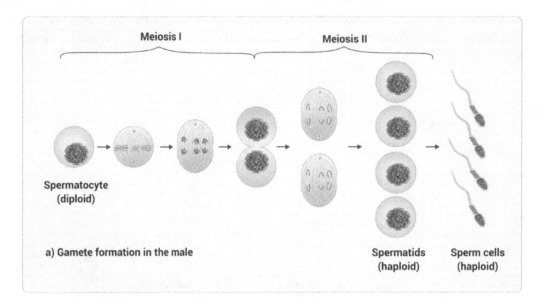

a) Gamete formation in the male

After cytokinesis I, the daughter cells immediately enter prophase II, rather than duplicating DNA or entering interphase. The nucleus disintegrates, the centrioles migrate to the ends of the cell, and the next round of divisions begins. This results in four haploid (n) daughter cells, the gametes of egg and sperm referenced earlier.

A common problem that arises in both meiosis and mitosis but is especially noticeable in meiosis is that of *nondisjunction*. Nondisjunction is the failure of homologous chromosomes or sister chromatids to separate during anaphase. This causes the daughter cells to have one more or one fewer chromosomes than usual and can ultimately result in genetic conditions like Down's syndrome when a meiotic egg with nondisjunction is fertilized.

Photosynthesis and Cellular Respiration

Photosynthesis

Photosynthesis refers to the process used by plants, some algae, and some bacteria to convert sunlight into chemical energy. It combines water and carbon dioxide using light energy to produce the sugar glucose and oxygen. Glucose is converted to adenosine triphosphate (ATP) through cellular respiration. ATP is the molecule that provides energy for all cellular activities and can be thought of as a sort of energy currency.

Photosynthesis is the most prolific process to create useable energy for life on Earth. In plants, photosynthesis takes place in chloroplasts, organelles containing photosynthetic structures and chemicals. *Thylakoids*, the structural units of photosynthesis, are found within chloroplasts and use *chlorophyll*, a green pigment, to harness sunlight in photosynthesis.

Photosynthesis has two types of reactions: light-dependent and light-independent. Light-dependent reactions produce ATP and release oxygen into the atmosphere. Light-independent reactions utilize ATP to produce glucose.

$$6CO_2 + 6H_2O + Energy \rightarrow C_6H_{12}O_6 + 6O_2$$

In photosynthesis, solar energy is used to fix carbon dioxide into glucose, oxidizing water to oxygen.

$$6CO_2 + 6H_2O + Energy \rightarrow C_6H_{12}O_6 + 6O_2$$

In aerobic respiration, the opposite happens, and glucose and oxygen are broken down into carbon dioxide and water to liberate energy in the form of ATP.

$$C_6H_{12}O_6 + 6O_2 \rightarrow 6CO_2 + 6H_2O + Energy$$

Heterotrophs are organisms that eat other living things to obtain organic compounds. This is usually simplified as the thought of eating other animals for food to gain energy, as most animals are unable to produce their own food. As a result, heterotrophs are completely dependent on autotrophs and other heterotrophs as food sources. All animals are considered heterotrophs.

Autotrophs are able to fix carbon dioxide into useable organic compounds. Most of them produce their own food by harnessing the power of sunlight and employing photosynthesis. As a result, autotrophs are not dependent on other organisms for sources of carbon. All plants are considered autotrophs.

Cellular Respiration
Cellular respiration is the pathway used by cells to release energy stored in food molecules. It culminates in the production of ATP (adenosine triphosphate), which is the energy currency for all cellular activities.

The two types of cellular respiration are *aerobic* respiration and *anaerobic* respiration. Aerobic respiration requires oxygen while anaerobic respiration does not, but aerobic respiration produces significantly more ATP and is the principal mode of respiration carried out by human cells.

In aerobic respiration, a molecule of glucose is broken down in the cytoplasm into pyruvate during *glycolysis* before further being broken down to a pair of carbon atoms called an *acetyl group*. This acetyl group is ferried by acetyl coenzyme A to the mitochondria where it enters the *Krebs cycle* (also known as the *tricaboxylic acid cycle* and the *TCA cycle*). In the Krebs cycle, the acetyl group is further broken down into carbon dioxide (C2→CO2) while the energy of its bonds are translated into molecules of NADH (reduced nicotinamide adenine dinucleotide).

These molecules of NADH are the source of electrons and hydrogen ions that are used in the last phase of aerobic respiration to generate ATP so that the energy from the sugar can actually be used. The electrons and hydrogens from the NADH enter the *electron transport chain* (also called the ETC or *oxidative phosphorylation*) where many complicated reactions occur before finally reducing oxygen to water (O2→H2O).

If no oxygen is available, aerobic respiration can't continue to completion. The cell stops aerobic respiration after the glucose is broken down to pyruvate, and instead converts it to lactate (or lactic acid). This is *fermentation*, and it generates ATP, but it does so only at a much lower rate than aerobic respiration. Because it doesn't use oxygen, fermentation is a type of anaerobic respiration. Many types of bacteria use it as their primary source of energy, but it is also a last defense for human cells when deprived of oxygen for long periods of time.

Nucleic Acids

Nucleic acids have two important duties in the body. As monomers, they are crucial for energy transfer. As polymers, they are a fundamental component of genetic material. Monomers form the building blocks of macromolecules, while polymers are formed when monomers link together in chains, forming larger macromolecules.

Nucleotides are the monomers that link together to form nucleic acids. Nucleotides have three components: a nitrogenous base and a phosphate functional group both attached to a five-carbon (pentose) sugar. There are two classes of nitrogenous bases: purines and pyrimidines. The two types of purines are guanine (G) and adenine (A), while the three types of pyrimidines are thymine (T), cytosine (C), and uracil (U). The two types of pentose sugars are deoxyribose and ribose. Nucleotides containing deoxyribose are termed deoxyribonucleic acids (DNA) and utilize guanine, adenine, cytosine, and thymine as their nitrogen bases. Nucleotides containing ribose are termed ribonucleic acids (RNA) and utilize guanine, adenine, cytosine, and uracil as their nitrogenous bases.

Mutations

Mutations are permanent alterations to an organism's genetic DNA sequence. Mutations can result from DNA failing to replicate accurately. They can also result from environmental influences such as radiation or chemicals. Mutations occur randomly and spontaneously at low rates.

Mutations can occur in reproductive and non-reproductive cells. Those occurring in non-reproductive cells are termed somatic mutations. Those occurring in reproductive cells (eggs or sperm) are termed germ line mutations. Although somatic mutations cannot be transmitted to offspring, germ line mutations are transmitted to offspring and can be advantageous, neutral, or disadvantageous.

In general, single-nucleotide alterations to DNA, or point mutations, can be either silent or same-sense (so that an identical amino acid sequence is encoded), missense (so that a different amino acid sequence is encoded), or nonsense (so that a new stop codon is encoded and the resultant protein is truncated). They can also either be transitions (a purine base to another purine base or vice versa) or transversions (a purine base to a pyrimidine base or vice versa).

The four common multi-nucleotide mutations are:

- Insertions: One or more nitrogen bases are added or inserted into the typical DNA sequence.

- Deletions: One or more nitrogen bases are removed or deleted from the typical DNA sequence.

- Inversions: A length of DNA is removed and reattached in reverse order from the typical DNA sequence.

- Translocations: A length of DNA is removed and reattached in an alternate place or chromosome than where found in the typical DNA sequence.

Cell Replication

Cell replication in eukaryotes involves duplicating the genetic material (DNA) and then dividing to yield two daughter cells, which are clones of the parent cell. The cell cycle is a series of stages leading to the growth and division of a cell. The cell cycle helps to replenish damaged or depleted cells. On average, eukaryotic cells go through a complete cell cycle every 24 hours. Some cells such as epithelial, or skin, cells are constantly dividing, while other cells such as mature nerve cells do not divide. Prior to mitosis, cells exist in a nondivisional stage of the cell cycle called interphase. During interphase, the cell begins to

prepare for division by duplicating DNA and its cytoplasmic contents. Interphase is divided into three phases: gap 1 (G_1), synthesis (S), and gap 2 (G_2).

DNA Replication
Replication refers to the process during which DNA makes copies of itself. Enzymes govern the major steps of DNA replication.

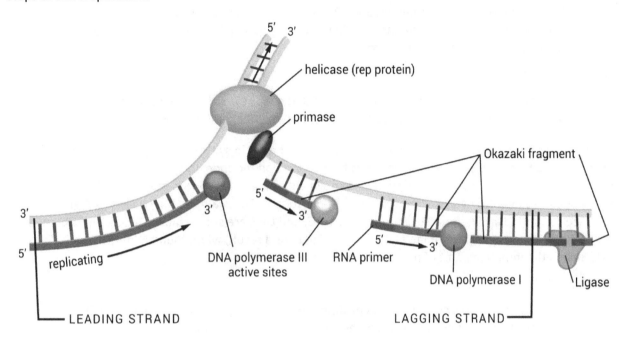

The process begins with the uncoiling of the double helix of DNA. *Helicase*, an enzyme, accomplishes this task by breaking the weak hydrogen bonds uniting base pairs. The uncoiling of DNA gives rise to the replication fork, which has a Y-shape. Each separated strand of DNA will act as a template for the production of a new molecule of DNA. The strand of DNA oriented toward the replication fork is called the *leading strand* and the strand oriented away from the replication fork is named the *lagging strand*.

Replication of the leading strand is continuous. *DNA polymerase*, an enzyme, binds to the leading strand and adds complementary bases. Replication of the lagging strand of DNA on the other hand is discontinuous. DNA polymerase produces discontinuous segments, called *Okazaki fragments*, which are later joined together by another enzyme, *DNA ligase*. To start the DNA synthesis on the lagging strand, the protein *primase* lays down a strip of RNA, called an *RNA primer*, to which the DNA polymerase can bind. As a result, two clones of the original DNA emerge from this process. DNA replication is considered *semiconservative* due to the fact that half of the new molecule is old and the other half is new.

Genes and Heredity
Genes are the basis of heredity. The German scientist Gregor Mendel first suggested the existence of genes in 1866. A gene can be pinpointed to a locus, or a particular position, on DNA. It is estimated that humans have approximately 20,000 to 25,000 genes. For any particular gene, a human inherits one copy from each parent for a total of two.

Chromosomes, Genes, Proteins, RNA, and DNA

Chromosomes are composed of hundreds to thousands of genes. Human cells contain 23 pairs of chromosomes for a total of 46 chromosomes. As explained above, genes are inherited in pairs, one from each parent.

Proteins are made of long chains of amino acids. In total, there are 20 amino acids, 11 of which humans can synthesize on their own and the remaining 9 of which are procured through diet. DNA contains the information for the synthesis of proteins, but that information on DNA has to undergo transcription and translation by RNA in order to produce proteins.

Codons

A codon represents a sequence of three nucleotides that codes for either one specific amino acid or a stop signal during protein synthesis. Codons are found on messenger RNA (mRNA).

Twenty essential amino acids are utilized in the process of protein synthesis. The full set of codons encompasses 64 possible combinations and is termed the genetic code. In the genetic code, 61 codons represent amino acids and three codons are stop signals. The genetic code is redundant due to the fact that a single amino acid may be produced by multiple codons. For example, the codons AAA and AAG produce the amino acid lysine. The codons UAA, UAG, and UGA are stop signals. The codon AUG codes for both the amino acid methionine and the start signal. As a result, when found in mRNA, the codon AUG marks the initiation point of protein translation.

RNA

Ribonucleic acid (RNA) plays crucial roles in protein synthesis and gene regulation. RNA is made of nucleotides consisting of ribose (a sugar), a phosphate group, and one of four possible nitrogen bases— adenine (A), cytosine (C), guanine (G), and uracil (U). RNA utilizes the nitrogen base uracil in place of the base thymine found in DNA. Another difference between RNA and DNA is that RNA is typically found as a single-stranded structure, while DNA typically exists in a double-stranded structure.

RNA can be categorized into three major groups—messenger RNA (mRNA), ribosomal RNA (rRNA), and transfer RNA (tRNA). Messenger RNA (mRNA) transports instructions from DNA in the nucleus of a cell to the areas responsible for protein synthesis in the cytoplasm of a cell. This process is known as transcription. Transfer RNA (tRNA) deciphers the amino acid sequence for the construction of proteins found in mRNA. Both tRNA and ribosomal RNA (rRNA) are found in the ribosomes of cells. Ribosomes are responsible for protein synthesis. The process is also known as translation and both tRNA and rRNA play crucial roles. Both translation and transcription are further described below.

DNA

Deoxyribonucleic acid, or DNA, contains the genetic material that is passed from parents to offspring. It contains specific instructions for the development and function of a unique eukaryotic organism. The vast majority of cells in any eukaryotic organism contains the same DNA.

The majority of DNA can be found in the cell's nucleus and is referred to as nuclear DNA. A small amount of DNA can be located in the mitochondria and is referred to as mitochondrial DNA. Mitochondria are the site of aerobic energy production for the cell. All offspring inherit mitochondrial DNA from their mother. James Watson, an American geneticist, and Frances Crick, a British molecular biologist, first outlined the structure of DNA in 1953.

The structure of DNA visually approximates a twisting ladder, and is described as a double helix. DNA is made of nucleotides consisting of deoxyribose (a sugar), a phosphate group, and one of four possible nitrogenous bases—thymine (T), adenine (A), cytosine (C), and guanine (G). It is estimated that human DNA contains three billion bases. The sequence of these bases dictates the instructions contained in the DNA, making each species singular. The bases in DNA pair in a particular manner—thymine (T) with adenine (A) and guanine (G) with cytosine (C). Weak hydrogen bonds between the nitrogenous bases ensures easy uncoiling of DNA's double helical structure in preparation for replication.

Transcription

Transcription refers to a portion of DNA being copied into RNA, specifically mRNA. It represents the first crucial step in gene expression. The process begins with the enzyme RNA polymerase binding to the promoter region of DNA, which initiates transcription of a specific gene. RNA polymerase then untwists the double helix of DNA by breaking weak hydrogen bonds between its nucleotides. Once DNA is untwisted, RNA polymerase travels down the strand reading the DNA sequence and adding complementary nitrogen bases. With the assistance of RNA polymerase, the pentose sugar and phosphate functional group are added to the nitrogen base to form a nucleotide. Lastly, the weak hydrogen bonds uniting the DNA-RNA complex are broken to free the newly formed mRNA. The mRNA travels from the nucleus of the cell out to the cytoplasm of the cell where translation occurs.

Translation

Translation refers to the process of ribosomes synthesizing proteins. It represents the second crucial step in gene expression. The instructions encoding specific proteins to be made are contained in codons on mRNA, which have previously been transcribed from DNA. Each codon represents a specific amino acid or stop signal in the genetic code.

Amino acids are the building blocks of proteins. Ribosomes contain transfer RNA (tRNA) and ribosomal RNA (rRNA). Translation occurs in ribosomes located in the cytoplasm of cells and consists of the following three phases:

Initiation: The ribosome gathers at a target point on the mRNA, and tRNA attaches at the start codon (AUG), which is also the codon for the amino acid methionine.

Elongation: A new tRNA reads the next codon on the mRNA and links the two amino acids together with a peptide bond. The process is repeated until a polypeptide, or long chain of amino acids, is formed.

Termination: The ribosome disengages from the mRNA when it encounters a stop codon (UAA, UAG, or UGA). The event releases the polypeptide molecule. Proteins are made of one or more polypeptide molecules.

Genotypes and Phenotypes

Genotype refers to the genetic makeup of an individual within a species. Phenotype refers to the visible characteristics and observable behavior of an individual within a species.

Genotypes are written with pairs of letters that represent alleles. Alleles are different versions of the same gene, and, in simple systems, each gene has one dominant allele and one recessive allele. The letter of the dominant trait is capitalized, while the letter of the recessive trait is not capitalized. An individual can be homozygous dominant, homozygous recessive, or heterozygous for a particular gene. Homozygous means that the individual inherits two alleles of the same type, while heterozygous means inheriting one dominant allele and one recessive allele.

If an individual has homozygous dominant alleles or heterozygous alleles, the dominant allele is expressed. If an individual has homozygous recessive alleles, the recessive allele is expressed. For example, imagine a species of bird can develop either white or black feathers. The white feathers are the dominant allele, or trait (A), while the black feathers are the recessive allele (a). Homozygous dominant (AA) and heterozygous (Aa) birds will develop white feathers. Homozygous recessive (aa) birds will develop black feathers.

Genotype (genetic makeup)	Phenotype (observable traits)
AA	white feathers
Aa	white feathers
aa	black feathers

The genetic material (DNA) inherited from an individual's parents determines genotype. Natural selection leads to adaptations within a species, which affects the phenotype. Over time, individuals within a species with the most advantageous phenotypes will survive and reproduce. As result of reproduction, the subsequent generation of phenotypes receives the fittest genotype. Eventually, the individuals within a species with genetic fitness flourish and those without it are erased from the environment. As explained above, this is also referred to as the concept of "survival of the fittest." When this process is duplicated over numerous generations, the outcome is offspring with a level of genetic fitness that meets or exceeds that of their parents.

Mendel's Laws of Genetics and Punnett Squares

Mendel's first law of genetics is the principle of *segregation* and states that alleles will segregate into different cells during the formation of gametes in meiosis. Mendel's second law of genetics is the principle of *independent assortment* and states that genes for different traits will be assigned to different gametes independent of the others. Together, these two laws state the assumptions on which genetic probabilities are based.

Punnett squares are simple graphic representations of all the possible genotypes of offspring, given the genotypes of the parent organisms. For example, in the above example with the species of bird with black or white feathers, A represents a dominant allele and determines white colored feathers on a bird. The recessive allele a determines black colored feathers on a bird. If both parents are heterozygous (Aa, the x and y-axis of the square), the offspring will have the possible genotypes AA, Aa, Aa, and aa. Phenotypically, three offspring would have white feathers and one would have black feathers, as shown in the following Punnett square:

	A	a
A	White	White
a	White	Black

Monohybrid and Dihybrid Genetic Crosses

Genetic crosses represent all possible permutations of gene combinations, or alleles. A monohybrid cross investigates the inheritance pattern of a single gene such as in the above example of the birds with black or white feathers. Both parents must have heterozygous gene pairs in a monohybrid cross.

The phenotypic ratio for a monohybrid cross is 3:1 (AA, Aa, Aa, aa), in favor of the dominant gene. A dihybrid cross investigates the inheritance patterns of two genes that are related, for example A and B. A dihybrid cross has a phenotypic ratio of 9:3:3:1, with nine offspring inheriting both dominant genes,

six offspring inheriting a single dominant and a single recessive gene, and one offspring inheriting both recessive genes.

Natural Selection and Adaptation

The theory of natural selection is one of the fundamental tenets of evolution. It affects the phenotype, or visible characteristics, of individuals in a species, which ultimately affects the genotype, or genetic makeup, of those same individuals. Charles Darwin was the first to explain the theory of natural selection, and it is described by Herbert Spencer as favoring *survival of the fittest*. Natural selection encompasses three assumptions:

- A species has heritable traits: All traits have some likelihood of being propagated to offspring.

- The traits of a species vary: Some traits are more advantageous than others.

- Individuals of a species are subject to differing rates of reproduction: Some individuals of a species may not get the opportunity to reproduce while others reproduce frequently.

Over time, certain variations in traits may increase both the survival and reproduction of certain individuals within a species. The desirable heritable traits are passed on from generation to generation. Eventually, the desirable traits will become more common and permeate the entire species.

Adaptation

The theory of *adaptation* is defined as an alteration in a species that causes it to become more well-suited to its environment. It increases the probability of survival, thus increasing the rate of successful reproduction. As a result, an adaptation becomes more common within the population of that species.

For examples, bats use reflected sound waves (echolocation) to prey on insects, and chameleons change colors to blend in with their surroundings to evade detection by its prey and predators. Adaptations are brought about by natural selection.

Adaptive radiation refers to rapid diversification within a species into an array of unique forms. It may occur as a result of changes in a habitat creating new challenges, ecological niches, or natural resources.

Darwin's finches are often thought of as an example of the theory of adaptive radiation. Charles Darwin documented 13 varieties of finches on the Galapagos Islands. Each island in the chain presented a unique and changing environment, which was believed to cause rapid adaptive radiation among the finches. There was also diversity among finches inhabiting the same island. Darwin believed that as a result of natural selection, each variety of finch developed adaptations to fit into its native environment.

A major difference in Darwin's finches had to do with the size and shapes of beaks. The variation in beaks allowed the finches to access different foods and natural resources, which decreased competition and preserved resources. As a result, various finches of the same species were allowed to coexist, thrive, and diversify. Finches had:

- Short beaks, which were suited for foraging for seeds
- Thin, sharp beaks, which were suited for preying on insects
- Long beaks, which were suited for probing for food inside plants

Darwin believed that the finches on the Galapagos Islands resulted from chance mutations in genes transmitted from generation to generation.

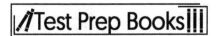

Understanding the Interdependence of Organisms

The study of biological relationships is called *ecology*, which explores how organisms affect one another. These relationships can be defined as either *intraspecific* (within the same species) or *interspecific* (between different species) interactions. For example, a man being friendly to his neighbor can be considered an intraspecific interaction, while a man interacting with a dog is an interspecific interaction. These can also be categorized in terms of the effects they have on participating organisms:

- Competition: This is the basis for Darwin's theory of natural selection and survival of the fittest. *Competition* occurs when the organisms' interactions are mutually harmful, as they compete with one another in order to benefit themselves.

- Amensalism: Occurs when an organism harms another, but the organism inflicting harm does not benefit from the interaction.

- Antagonism: One individual or species hunts another (inflicting harm) for the benefit of food. The harmed species is not working toward the same goal, differentiating amensalism from competition. Classic examples of antagonism are predation and parasitism.

- Neutralism: Occurs when different individuals or species interact with one another and neither species is harmed or benefits from the relationship.

- Facilitation: Describes interactions in which one participant benefits without harming the other.

- Commensalism: Occurs when only one participant receives benefit from the interaction.

- Mutualism: Occurs when both parties benefit from the interaction.

- Symbiosis: A close mutualistic relationship between organisms of different species. A popular example is the relationship between a clown fish and a sea anemone. The clown fish cleans the sea anemone, and in return, it is protected from predators by the anemone's stinging arms.

Another large area of ecological studies is the *food chain*—the network of links between organisms. Producer organisms like grass or algae make up the bottom of the chain, while the top is comprised of apex predators such as lions and tigers. The food chain is useful for depicting the flow of energy, or nutrition, from bottom to top. Terms commonly used while discussing the food chain include:

- Autotroph: A producer in the food chain with the ability to produce organic compounds from the environment. Photosynthetic plants are examples of autotrophs.

- Heterotroph: A consumer in the food chain that must consume organic materials to thrive. Humans are examples of heterotrophs.

- Lithotroph: An organism with the ability to use inorganic material to survive. Lithotrophs are exclusively microbes, such as those that survive deep underwater, where they utilize inorganic sulfur that escapes from volcanic vents.

Biological Classification System

Biological classification, or taxonomy, is the science of grouping organisms based on common characteristics. The father of modern taxonomy is the Swedish botanist Carl Linnaeus. There are eight levels of organization in the biological classification system including:

- Domain: Domains classify organisms by common characteristics such as cellular chemistry and structure. The concept of domains was introduced by Carl Woese in the late 1970s, but not widely adopted by biologists until 1990. The three domains are Bacteria, Eukarya, and Archaea.

- Kingdom: Kingdoms group organisms according to nutritional strategies and developmental features. Most scientists recognize five kingdoms: *Animalia*, *Plantae*, *Monera*, *Protista*, and *Fungi*. However, since the two divisions of *Monera* (*Bacteria* and *Archaea*) are treated as their own domains, this organizational scheme is under fire. Additionally, scientists have recently begun debates over the inclusion of viruses as a separate domain.

- Phylum: Within the animal kingdom, phyla of note include *Chordata*, *Mollusca*, *Cnidaria* (jellyfish), *Porifera* (sponges), and *Arthropoda* (insects).

- Class: Within the phylum *Chordata*, classes of note include *Chondrichthyes* (bony fish), *Amphibia*, *Reptilia*, *Aves*, and *Mammalia*. These classes also fall in the subphylum *Vertebrata*, which is distinguished by its unique notochord, dorsal hollow nerve cord, pharyngeal slits, and post-anal tail.

- Order: Within the class *Mammalia*, orders of note include *Lagomorpha* (rabbits), *Rodentia*, *Carnivora*, *Chiroptera* (bats), and *Primates*. It also includes the monotremes (like platypi and echidnas) and the marsupials (like opossums and kangaroos). This class is specially known for its teeth's enamel coating and the presence of a middle ear to sense sound.

- Family: Within the order *Primates*, the most commonly known family is *Hominidae*, to which the great apes and humans belong. This order is typically divided into the prosimians and anthropoids, and it includes monkeys, lemurs, and gibbons.

- Genus: Within the family *Hominidae*, there are four genera which encompass orangutans, gorillas, chimpanzees, and humans. Humans belong to the genus *Homo*.

- Species: Finally, within the genus *Homo*, humans are recognized by the species *sapiens*, but modern humans are further specified by the subspecies *sapiens* again. By conventional reference, modern humans are classified as *Homo sapiens sapiens*.

Moving down the list, each subsequent level becomes more specific. Each organism receives a scientific two-part name, referred to as binomial nomenclature. The first part of the name is the genus and the first letter is capitalized, while the second part of the name is the species and the first letter is lowercased. By convention, the name must be Latin and italicized. The two-part name may also be abbreviated by shortening the genus to its first letter with a period coming after. For example, the scientific two-part name for humans is *Homo sapiens* (abbreviated *H. sapiens*). Biological classification has been practiced for more than 300 years and is constantly evolving.

Biological Concepts

Biological concepts include all the terms previously discussed as well as the following:

- Territoriality: The behavior of an organism that claims a specific sociographical area/location and defends it, usually because its food or sexual mates are in that area.
- Community: A collection of individual organisms living and interacting within a specific area.
- Niche: The role of a species in a community.
- Dominance: Refers to the most prevalent species in a community.
- Altruism: A biological interaction in which one individual or species provides benefit to another at a cost to itself. A wealthy man is considered altruistic when he donates money to the poor.
- Threat display: Behavior of an organism with the intent to scare other individuals.
- Competitive exclusion: Also called Gause's Law, which dictates that when two species within the same community are competing for the same resource, they cannot stably coexist. In other words, one will dominate the other.
- Species diversity: Refers to the number of different species in a community.
- Biotic: Living organisms in a community, such as other species.
- Abiotic: Non-living factors such as terrain and weather.
- Ecosystem: Encompasses a community and all its environmental factors.
- Biomass: Mass of organic matter within an ecosystem.
- Mimicry: An organism that adapts the appearance or actions of a different species in order to avoid predators or harm.

Population Growth and Decline

A population's size can grow or decline based on fluctuating birth rates, death rates, immigration, and emigration. The *birth rate* is defined as the total number of live births per 1,000 individuals in a defined population in a year. The *death rate*, or *mortality rate*, is defined as the total number of deaths per 1,000 individuals in a defined population in a year. *Immigration* refers to a person or organism coming from another population to the one currently being examined, and *emigration* refers a person or organism leaving that population to settle elsewhere. Because these are the chief factors on a population's size, the rate of its growth can be determined from them.

Other variables such as food availability, adequate shelter, and water supply can also affect population. These resources are finite and help determine the carrying capacity of a particular geographic area. *Carrying capacity* is defined as the maximum number of individuals that can be sustained indefinitely in a particular habitat. As a population of humans grows, other factors such as government, education, economics, healthcare, and cultural values will also begin to influence population.

As time goes on, certain countries may experience population growth while others experience its decline. Population growth, especially if unchecked, can be troubling when a community grows to a point that exceeds its carrying capacity. Ultimately, levels of emigration may rise as individuals leave in search of more favorable conditions. Conversely, population decline can also be threatening due to the diminished pools of individuals available for labor and reproduction. Ultimately, rising levels of immigration will be needed to remedy this stress on the population.

Demographic Transition Model

The Demographic Transition Model (DTM) explains changes in two key areas, birth rate and death rate, and their effect on the total population of a country as it undergoes economic and industrial development. This model was introduced in 1929. Birth rate would be defined as the number of live

births per 1,000 individuals in a population over the course of an entire year, while death rate would be defined as the total number of deaths per 1,000 individuals in a population over the course of an entire year.

As a general rule, a country will progress through stages as it undergoes economic and industrial development. Very few countries are in stage 1 of the model. Most of the developing countries are grouped in stage 2 or 3 of the model, while most of the industrialized countries are categorized in stage 3 or 4 of the model.

The Demographic Transition Model has the following stages:

Stage 1—High Stationary: Both a high birth and a high death rate characterize stage 1 of the DTM. These factors combine to produce a constant, relatively low total population. High birth rates may be accounted for by factors such as poor family planning, high infant and child death rates, and child labor requirements for farming and manufacturing. High death rates may be accounted for by factors such as famine, poor sanitation, poor health care, and epidemics of disease. Most of the world's countries were categorized as stage 1 pre-Industrial Revolution. Today, only the least economically developed countries would be classified as this stage.

Stage 2—Early Expanding: A high birth rate and a rapidly decreasing death rate characterize stage 2 of the DTM. These factors converge to produce a rapid increase in the total population. Rapidly decreasing death rates may be explained by falling infant and child death rates, improved sanitation, improvements in healthcare, and better nutrition. Today, the African countries of Ethiopia, Kenya, and Egypt are examples of countries classified in this stage.

Stage 3—Late Expanding: A decreasing birth rate along with a continued falling (but less rapid) death rate characterizes stage 3 of the DTM. These factors continue to produce an increase in total population, but at a less rapid rate than seen in stage 2. Child welfare laws, a desire for smaller families, and the changing role of women in the workplace may explain the effect of falling birth rates on the total population. Current examples of countries grouped in this stage are Brazil, South Korea, and India.

Stage 4—Low Stationary: Both a low birth rate and death rate characterize stage 4 of the DTM. As a result, the factors combine to have a stabilizing effect on total population. Current examples of countries classified in this stage are the United States, Canada, and Great Britain.

Factors Influencing Birth and Fertility Rates

Fertility rate is defined as the average lifetime number of children a woman will bear for a given population. By convention, the reproductive lifetime of a woman ranges from about 15 to 45 years old. The fertility rate is an artificial measure, and does not represent the real fertility rate of any particular group of women, nor does it take into account child mortality. *Fertility rate* is not the same as *birth rate*, and it is instead defined as the total number of live births per 1,000 individuals in a defined population in a year.

Key factors influencing birth and fertility rates of a particular country include:

- Urbanization: Rates are lower amongst individuals living in urban areas because they tend to have smaller families than those living in rural areas.

- Child labor: Rates are lower in developed countries because fewer children are required to work at a young age and help cultivate the land.

- Education costs: Rates are lower in developed countries because these costs are higher due to the fact children enter the work force in their late teens to early twenties.

- Women labor: Rates are lower in developed countries because women have access to education and paid employment outside of the household.

- Infant mortality: Rates are lower in locales with low infant mortality rates because individuals tend to have fewer children if those children don't die at an early age.

- Age of marriage: Rates are lower in countries where the average age at which women marry or have their first child is greater than 25. Women who marry or have children later in life tend to have fewer children.

- Pension systems: Rates are lower in developed countries as pensions and social security eliminate the burden of children providing for aging parents.

- Abortions: Rates are lower in countries where women have access to legal elective abortions.

- Contraception: Rates are lower in developed countries, because individuals have access to reliable birth control methods.

- Cultural norms and religious beliefs: Rates are lower in countries that don't hold particular norms and beliefs that strongly oppose abortion or birth control.

Cell Structure and Function

The cell is the main functional and structural component of all living organisms. Robert Hooke, an English scientist, coined the term "cell" in 1665. Hooke's discovery laid the groundwork for the cell theory, which is composed of three principals:

- All organisms are composed of cells.
- All existing cells are created from other living cells.
- The cell is the most fundamental unit of life.

Organisms can be unicellular (composed of one cell) or multicellular (composed of many cells). All cells must be bounded by a cell membrane, be filled with cytoplasm of some sort, and be coded by a genetic sequence.

The cell membrane separates a cell's internal and external environments. It is a selectively permeable membrane, which usually only allows the passage of certain molecules by diffusion. Phospholipids and proteins are crucial components of all cell membranes. The cytoplasm is the cell's internal environment and is aqueous, or water-based. The genome represents the genetic material inside the cell that is passed on from generation to generation.

Prokaryotes and Eukaryotes

Prokaryotic cells are much smaller than eukaryotic cells. The majority of prokaryotes are unicellular, while the majority of eukaryotes are multicellular. Prokaryotic cells have no nucleus, and their genome is found in an area known as the nucleoid. They also do not have membrane-bound organelles, which are "little organs" that perform specific functions within a cell.

Eukaryotic cells have a proper nucleus containing the genome. They also have numerous membrane-bound organelles such as lysosomes, endoplasmic reticula (rough and smooth), Golgi complexes, and mitochondria. The majority of prokaryotic cells have cell walls, while most eukaryotic cells do not have cell walls. The DNA of prokaryotic cells is contained in a single circular chromosome, while the DNA of eukaryotic cells is contained in multiple linear chromosomes. Prokaryotic cells divide using binary fission, while eukaryotic cells divide using mitosis. Examples of prokaryotes are bacteria and archaea while examples of eukaryotes are animals and plants.

Nuclear Parts of a Cell

Nucleus (plural nuclei): Houses a cell's genetic material, deoxyribonucleic acid (DNA), which is used to form chromosomes. A single nucleus is the defining characteristic of eukaryotic cells. The nucleus of a cell controls gene expression. It ensures genetic material is transmitted from one generation to the next.

Chromosomes: Complex thread-like arrangements composed of DNA that is found in a cell's nucleus. Humans have 23 pairs of chromosomes for a total of 46.

Chromatin: An aggregate of genetic material consisting of DNA and proteins that forms chromosomes during cell division.

Nucleolus (plural nucleoli): The largest component of the nucleus of a eukaryotic cell. With no membrane, the primary function of the nucleolus is the production of ribosomes, which are crucial to the synthesis of proteins.

Cell Membranes

Cell membranes encircle the cell's cytoplasm, separating the intracellular environment from the extracellular environment. They are selectively permeable, which enables them to control molecular traffic entering and exiting cells. Cell membranes are made of a double layer of phospholipids studded with proteins. Cholesterol is also dispersed in the phospholipid bilayer of cell membranes to provide stability. The proteins in the phospholipid bilayer aid the transport of molecules across cell membranes.

Scientists use the term "fluid mosaic model" to refer to the arrangement of phospholipids and proteins in cell membranes. In that model, phospholipids have a head region and a tail region. The head region of the phospholipids is attracted to water (hydrophilic), while the tail region is repelled by it (hydrophobic). Because they are hydrophilic, the heads of the phospholipids are facing the water, pointing inside and outside of the cell. Because they are hydrophobic, the tails of the phospholipids are oriented inward between both head regions. This orientation constructs the phospholipid bilayer.

Cell membranes have the distinct trait of selective permeability. The fact that cell membranes are amphiphilic (having hydrophilic and hydrophobic zones) contributes to this trait. As a result, cell membranes are able to regulate the flow of molecules in and out of the cell.

Factors relating to molecules such as size, polarity, and solubility determine their likelihood of passage across cell membranes. Small molecules are able to diffuse easily across cell membranes compared to large molecules. Polarity refers to the charge present in a molecule. Polar molecules have regions, or poles, of positive and negative charge and are water soluble, while nonpolar molecules have no charge and are fat-soluble. Solubility refers to the ability of a substance, called a solute, to dissolve in a solvent. A soluble substance can be dissolved in a solvent, while an insoluble substance cannot be dissolved in a solvent. Nonpolar, fat-soluble substances have a much easier time passing through cell membranes compared to polar, water-soluble substances.

Passive Transport Mechanisms

Passive transport refers to the migration of molecules across a cell membrane that does not require energy. The three types of passive transport include simple diffusion, facilitated diffusion, and osmosis.

Simple diffusion relies on a concentration gradient, or differing quantities of molecules inside or outside of a cell. During simple diffusion, molecules move from an area of high concentration to an area of low concentration. Facilitated diffusion utilizes carrier proteins to transport molecules across a cell membrane. Osmosis refers to the transport of water across a selectively permeable membrane. During osmosis, water moves from a region of low solute concentration to a region of high solute concentration.

Active Transport Mechanisms

Active transport refers to the energy-requiring migration of molecules across a cell membrane. It's a useful way to move molecules from an area of low concentration to an area of high concentration. Adenosine triphosphate (ATP), the currency of cellular energy, is needed to work against the concentration gradient.

Active transport can involve carrier proteins that cross the cell membrane to pump molecules and ions across the membrane, like in facilitated diffusion. The difference is that active transport uses the energy from ATP to drive this transport, as typically the ions or molecules are going against their concentration gradient. For example, glucose pumps in the kidney pump all of the glucose into the cells from the lumen of the nephron even though there is a higher concentration of glucose in the cell than in the lumen. This is because glucose is a precious fuel source, and the body wants to conserve as much as possible. Pumps can either send a molecule in one direction, multiple molecules in the same direction (symports), or multiple molecules in different directions (antiports).

Active transport can also involve the movement of membrane-bound particles, either into a cell (endocytosis) or out of a cell (exocytosis). The three major forms of endocytosis are: pinocytosis, where the cell is *drinking* and intakes only small molecules; phagocytosis, where the cell is *eating* and intakes large particles or small organisms; and receptor-mediated endocytosis, where the cell's membrane splits off to form an internal vesicle as a response to molecules activating receptors on its surface. Exocytosis is the inverse of endocytosis, and the membranes of the vesicle join to that of the cell's surface while the molecules inside the vesicle are released outside. This is common in nervous and muscle tissue for the release of neurotransmitters and in endocrine cells for the release of hormones. The two major categories of exocytosis are excretion and secretion. Excretion is defined as the removal of waste from a cell. Secretion is defined as the transport of molecules, such as hormones or enzymes, from a cell.

Structure and Function of Cellular Organelles

Organelles are specialized structures that perform specific tasks in a cell. The term literally means "little organ." Most organelles are membrane bound and serve as sites for the production or degradation of chemicals. The following are organelles found in eukaryotic cells:

Nucleus: A cell's nucleus contains genetic information in the form of DNA. The nucleus is surrounded by the nuclear envelope. A single nucleus is the defining characteristic of eukaryotic cells. The nucleus is also the most important organelle of the cell. It contains the nucleolus, which manufactures ribosomes (another organelle) that are crucial in protein synthesis (also called gene expression).

Mitochondria: Mitochondria are oval-shaped and have a double membrane. The inner membrane has multiple folds called cristae. Mitochondria are responsible for the production of a cell's energy in the

form of adenosine triphosphate (ATP). ATP is the principal energy transfer molecule in eukaryotic cells. Mitochondria also participate in cellular respiration.

Rough Endoplasmic Reticulum: The rough endoplasmic reticulum (RER) is composed of linked membranous sacs called cisternae with ribosomes attached to their external surfaces. The RER is responsible for the production of proteins that will eventually get shipped out of the cell.

Smooth Endoplasmic Reticulum: The smooth endoplasmic reticulum (SER) is composed of linked membranous sacs called cisternae without ribosomes, which distinguishes it from the RER. The SER's main function is the production of carbohydrates and lipids, which can be created expressly for the cell, or to modify the proteins from the RER that will eventually get shipped out of the cell.

Golgi Apparatus: The Golgi apparatus is located next to the SER. Its main function is the final modification, storage, and shipping of products (proteins, carbohydrates, and lipids) from the endoplasmic reticulum.

Lysosomes: Lysosomes are specialized vesicles that contain enzymes capable of digesting food, surplus organelles, and foreign invaders such as bacteria and viruses. They often destroy dead cells in order to recycle cellular components. Lysosomes are found only in animal cells.

Secretory Vesicles: Secretory vesicles transport and deliver molecules into or out of the cell via the cell membrane. Endocytosis refers to the movement of molecules into a cell via secretory vesicles. Exocytosis refers to the movement of molecules out of a cell via secretory vesicles.

Ribosomes: Ribosomes are not membrane bound. They are responsible for the production of proteins as specified from DNA instructions. Ribosomes can be free or bound.

Cilia and Flagella: Cilia are specialized hair-like projections on some eukaryotic cells that aid in movement, while flagella are long, whip-like projections that are used in the same capacity.

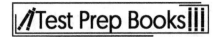

Here is an illustration of the cell:

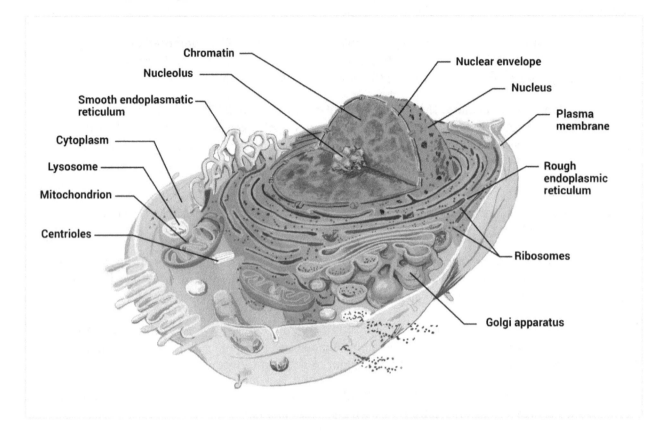

The following organelles are not found in animal cells:

Cell Walls: Cell walls can be found in plants, bacteria, and fungi, and are made of cellulose, peptidoglycan, and lignin, depending on the organism it surrounds. Each of these materials is a type of sugar recognized as a structural carbohydrate. The carbohydrates form rigid structures located outside of the cell membrane. Cell walls protect the cell, help maintain a cell's shape, and provide structural support.

Vacuoles: Plant cells have central vacuoles, which are essentially a membrane surrounding a body of water. They may store nutrients or waste products. Since vacuoles are large, they also help to support the structure of plant cells.

Chloroplasts: Chloroplasts are membrane-bound organelles that perform photosynthesis. They contain structural units called thylakoids. Chlorophyll, a green pigment that circulates within the thylakoids, harnesses light energy (sunlight) and helps convert it into chemical energy (glucose).

Human Body Systems

Anatomy is the structural makeup of an organism. The study of anatomy may be divided into microscopic/fine anatomy and macroscopic/gross anatomy. Fine anatomy concerns itself with viewing the features of the body with the aid of a microscope, while gross anatomy concerns itself with viewing the features of the body with the naked eye. Physiology refers to the functions of an organism, and it examines the chemical or physical functions that help the body function appropriately.

Levels of Organization of the Human Body

All the parts of the human body are built of individual units called *cells*. Groups of similar cells are arranged into *tissues,* different tissues are arranged into *organs,* and organs working together form entire *organ systems.* The human body has twelve organ systems that govern circulation, digestion, immunity, hormones, movement, support, coordination, urination & excretion, reproduction (male and female), respiration, and general protection.

Body Cavities

The body is partitioned into different hollow spaces that house organs. The human body contains the following cavities:

- Cranial cavity: The cranial cavity is surrounded by the skull and contains organs such as the brain and pituitary gland.

- Thoracic cavity: The thoracic cavity is encircled by the sternum (breastbone) and ribs. It contains organs such as the lungs, heart, trachea (windpipe), esophagus, and bronchial tubes.

- Abdominal cavity: The abdominal cavity is separated from the thoracic cavity by the diaphragm. It contains organs such as the stomach, gallbladder, liver, small intestines, and large intestines. The abdominal organs are held in place by a membrane called the peritoneum.

- Pelvic cavity: The pelvic cavity is enclosed by the pelvis, or bones of the hip. It contains organs such as the urinary bladder, urethra, ureters, anus, and rectum. It contains the reproductive organs as well. In females, the pelvic cavity also contains the uterus.

- Spinal cavity: The spinal cavity is surrounded by the vertebral column. The vertebral column has five regions: cervical, thoracic, lumbar, sacral, and coccygeal. The spinal cord runs through the middle of the spinal cavity.

Human Tissues

Human tissues can be grouped into four categories:

Muscle

Muscle tissue supports the body and allows it to move, and muscle cells have the ability to contract. There are three distinct types of muscle tissue: skeletal, smooth, and cardiac. Skeletal muscle is voluntary, or under conscious control, and is usually attached to bones. Most body movement is directly caused by the contraction of skeletal muscle. Smooth muscle is typically involuntary, or not under conscious control, and it is found in blood vessels, the walls of hollow organs, and the urinary bladder. Cardiac muscle is involuntary and found in the heart, which helps pump blood throughout the body.

Nervous

Nervous tissue is unique in that it is able to coordinate information from sensory organs as well as communicate the proper behavioral responses. Neurons, or nerve cells, are the workhorses of the nervous system. They communicate via action potentials (electrical signals) and neurotransmitters (chemical signals).

Epithelial

Epithelial tissue covers the external surfaces of organs and lines many of the body's cavities. Epithelial tissue helps to protect the body from invasion by microbes (bacteria, viruses, parasites), fluid loss, and injury. Epithelial cell shapes can be:

- Squamous: cells with a flat shape
- Cuboidal: cells with a cubed shape
- Columnar: cells shaped like a column

Epithelial cells can be arranged in four patterns:

- Simple: a type of epithelium composed solely from a single layer of cells
- Stratified: a type of epithelium composed of multiple layers of cells
- Pseudostratified: a type of epithelium that appears to be stratified but actually consists of only one layer of cells
- Transitional: a type of epithelium noted for its ability to expand and contract

Connective

Connective tissue supports and connects the tissues and organs of the body. Connective tissue is composed of cells dispersed throughout a matrix which can be gel, liquid, protein fibers, or salts. The primary protein fibers in the matrix are collagen (for strength), elastin (for flexibility), and reticulum (for support). Connective tissue can be categorized as either *loose* or *dense*. Examples of connective tissue include bones, cartilage, ligaments, tendons, blood, and adipose (fat) tissue.

Three Primary Body Planes

A plane is an imaginary flat surface. The three primary planes of the human body are frontal, sagittal, and transverse. The coronal plane is a vertical plane that divides the body or organ into front (anterior) and back (posterior) portions. The sagittal, or lateral, plane is a vertical plane that divides the body or organ into right and left sides. The transverse plane is a horizontal plane that divides the body or organ into upper and lower portions.

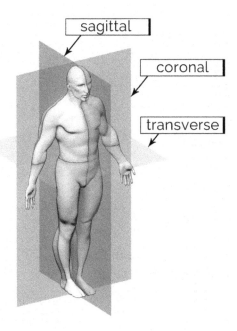

Terms of Direction

- *Medial* refers to a structure being closer to the midline of the body. For example, the nose is medial to the eyes.

- *Lateral* refers to a structure being farther from the midline of the body, and it is the opposite of *medial*. For example, the eyes are lateral to the nose.

- *Proximal* refers to a structure or body part located near an attachment point. For example, the elbow is proximal to the wrist.

- *Distal* refers to a structure or body part located far from an attachment point, and it is the opposite of *proximal*. For example, the wrist is distal to the elbow.

- *Anterior* means toward the front in humans. For example, the lips are anterior to the teeth. The term *ventral* can be used in place of *anterior*.

- *Posterior* means toward the back in humans, and it is the opposite of *anterior*. For example, the teeth are posterior to the lips. The term *dorsal* can be used in place of *posterior*.

- *Superior* means above and refers to a structure closer to the head. For example, the head is superior to the neck. The terms *cephalic* or *cranial* may be used in place of *superior*.

- *Inferior* means below and refers to a structure farther from the head, and it is the opposite of *superior*. For example, the neck is inferior to the head. The term *caudal* may be used in place of *inferior*.

- *Superficial* refers to a structure closer to the surface. For example, the muscles are superficial because they are just beneath the surface of the skin.

- *Deep* refers to a structure farther from the surface, and it is the opposite of *superficial*. For example, the femur is a deep structure lying beneath the muscles.

Body Regions

Terms for general locations on the body include:

- Cervical: relating to the neck
- Clavicular: relating to the clavicle, or collarbone
- Ocular: relating to the eyes
- Acromial: relating to the shoulder
- Cubital: relating to the elbow
- Brachial: relating to the arm
- Carpal: relating to the wrist
- Thoracic: relating to the chest
- Abdominal: relating to the abdomen
- Pubic: relating to the groin
- Pelvic: relating to the pelvis, or bones of the hip
- Femoral: relating to the femur, or thigh bone
- Geniculate: relating to the knee
- Pedal: relating to the foot
- Palmar: relating to the palm of the hand
- Plantar: relating to the sole of the foot

Abdominopelvic Regions and Quadrants

The abdominopelvic region may be defined as the combination of the abdominal and the pelvic cavities. The region's upper border is the breasts and its lower border is the groin region.

The region is divided into the following nine sections:

- Right hypochondriac: region below the cartilage of the ribs
- Epigastric: region above the stomach between the hypochondriac regions
- Left hypochondriac: region below the cartilage of the ribs
- Right lumbar: region of the waist
- Umbilical: region between the lumbar regions where the umbilicus, or belly button (navel), is located
- Left lumbar: region of the waist
- Right inguinal: region of the groin
- Hypogastric: region below the stomach between the inguinal regions
- Left inguinal: region of the groin

A simpler way to describe the abdominopelvic area is to divide it into the following quadrants:

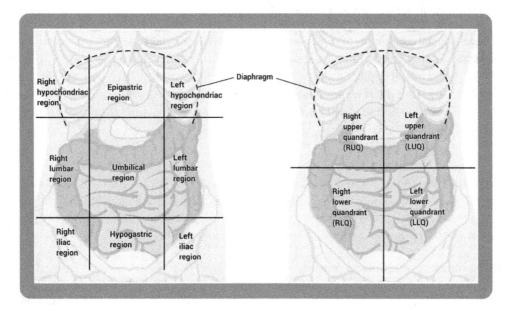

- Right upper quadrant (RUQ): Encompasses the right hypochondriac, right lumbar, epigastric, and umbilical regions.

- Right lower quadrant (RLQ): Encompasses the right lumbar, right inguinal, hypogastric, and umbilical regions.

- Left upper quadrant (LUQ): Encompasses the left hypochondriac, left lumbar, epigastric, and umbilical regions.

- Left lower quadrant (LLQ): Encompasses the left lumbar, left inguinal, hypogastric, and umbilical regions.

Anatomy and Physiology of Various Systems of the Body
Circulatory System
The circulatory system is a network of organs and tubes that transport blood, hormones, nutrients, oxygen, and other gases to cells and tissues throughout the body. It is also known as the cardiovascular system. The major components of the circulatory system are the blood vessels, blood, and heart.

Blood Vessels
In the circulatory system, blood vessels are responsible for transporting blood throughout the body. The three major types of blood vessels in the circulatory system are arteries, veins, and capillaries. Arteries carry blood from the heart to the rest of the body. Veins carry blood from the body back to the heart. Capillaries connect arteries to veins and form networks that exchange materials between the blood and the cells.

In general, arteries are stronger and thicker than veins, as they withstand high pressures exerted by the blood as the heart pumps it through the body. Arteries control blood flow through either vasoconstriction (narrowing of the blood vessel's diameter) or vasodilation (widening of the blood vessel's diameter). The blood in veins is under much lower pressures, so veins have valves to prevent the backflow of blood.

Most of the exchange between the blood and tissues takes place through the capillaries. There are three types of capillaries: continuous, fenestrated, and sinusoidal.

Continuous capillaries are made up of epithelial cells tightly connected together. As a result, they limit the types of materials that pass into and out of the blood. Continuous capillaries are the most common type of capillary. Fenestrated capillaries have openings that allow materials to be freely exchanged between the blood and tissues. They are commonly found in the digestive, endocrine, and urinary systems. Sinusoidal capillaries have larger openings and allow proteins and blood cells through. They are found primarily in the liver, bone marrow, and spleen.

Blood

Blood is vital to the human body. It is a liquid connective tissue that serves as a transport system for supplying cells with nutrients and carrying away their wastes. The average adult human has five to six quarts of blood circulating through their body. Approximately 55% of blood is plasma (the fluid portion), and the remaining 45% is composed of solid cells and cell parts.

There are three major types of blood cells:

- Red blood cells, or erythrocytes, transport oxygen throughout the body. They contain a protein called hemoglobin that allows them to carry oxygen. The iron in the hemoglobin gives the cells and the blood their red colors.

- White blood cells, or leukocytes, are responsible for fighting infectious diseases and maintaining the immune system. There are five types of white blood cells: neutrophils, lymphocytes, eosinophils, monocytes, and basophils.

- Platelets are cell fragments that play a central role in the blood clotting process.

All blood cells in adults are produced in the bone marrow—red blood cells and most white blood cells are produced in the red marrow, and some white blood cells are produced in the yellow bone marrow.

Heart

The heart is a two-part, muscular pump that forcefully pushes blood throughout the human body. The human heart has four chambers—two upper atria and two lower ventricles separated by a partition

called the septum. There is a pair on the left and a pair on the right. Anatomically, *left* and *right* correspond to the sides of the body that the patient themselves would refer to as left and right.

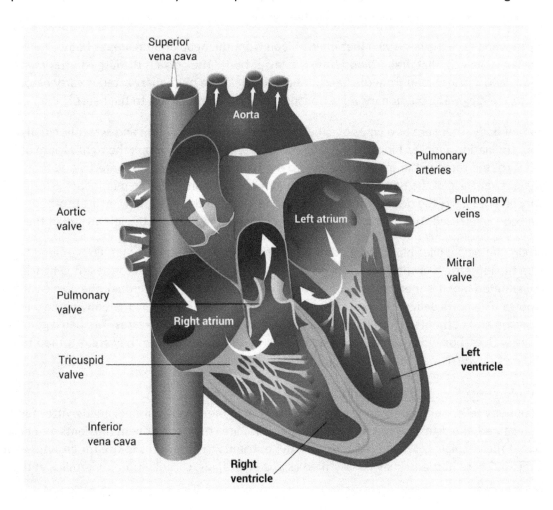

Four valves help to section off the chambers from one another. Between the right atrium and ventricle, the three flaps of the tricuspid valve keep blood from flowing backwards from the ventricle to the atrium, similar to how the two flaps of the mitral valve work between the left atrium and ventricle. As these two valves lie between an atrium and a ventricle, they are referred to as atrioventricular (AV) valves. The other two valves are semilunar (SL) and control blood flow into the two great arteries leaving the ventricles. The pulmonary valve connects the right ventricle to the pulmonary artery, while the aortic valve connects the left ventricle to the aorta.

Cardiac Cycle

A cardiac cycle is one complete sequence of cardiac activity. The cardiac cycle represents the relaxation and contraction of the heart and can be divided into two phases: diastole and systole.

Diastole is the phase during which the heart relaxes and fills with blood. It gives rise to the diastolic blood pressure (DBP), which is the bottom number of a blood pressure reading. Systole is the phase during which the heart contracts and discharges blood. It gives rise to the systolic blood pressure (SBP), which is the top number of a blood pressure reading. The heart's electrical conduction system coordinates the cardiac cycle.

Types of Circulation

Five major blood vessels manage blood flow to and from the heart: the superior and inferior venae cava, the aorta, the pulmonary artery, and the pulmonary vein.

The superior vena cava is a large vein that drains blood from the head and the upper body. The inferior vena cava is a large vein that drains blood from the lower body. The aorta is the largest artery in the human body and carries blood from the heart to body tissues. The pulmonary arteries carry blood from the heart to the lungs. The pulmonary veins transport blood from the lungs to the heart.

In the human body, there are two types of circulation: pulmonary circulation and systemic circulation. Pulmonary circulation supplies blood to the lungs. Deoxygenated blood enters the right atrium of the heart and is routed through the tricuspid valve into the right ventricle. Deoxygenated blood then travels from the right ventricle of the heart through the pulmonary valve and into the pulmonary arteries. The pulmonary arteries carry the deoxygenated blood to the lungs. In the lungs, oxygen is absorbed, and carbon dioxide is released. The pulmonary veins carry oxygenated blood to the left atrium of the heart.

Systemic circulation supplies blood to all other parts of the body, except the lungs. Oxygenated blood flows from the left atrium of the heart through the mitral, or bicuspid, valve into the left ventricle of the heart. Oxygenated blood is then routed from the left ventricle of the heart through the aortic valve and into the aorta. The aorta delivers blood to the systemic arteries, which supply the body tissues. In the tissues, oxygen and nutrients are exchanged for carbon dioxide and other wastes. The deoxygenated blood along with carbon dioxide and wastes enter the systemic veins, where they are returned to the right atrium of the heart via the superior and inferior vena cava.

Digestive System Structure and Function

The human body relies completely on the digestive system to meet its nutritional needs. After food and drink are ingested, the digestive system breaks them down into their component nutrients and absorbs them so that the circulatory system can transport the nutrients to other cells to use for growth, energy, and cell repair. These nutrients may be classified as proteins, lipids, carbohydrates, vitamins, and minerals.

The digestive system is thought of chiefly in two parts: the digestive tract (also called the alimentary tract or gastrointestinal tract) and the accessory digestive organs. The digestive tract is the pathway in which food is ingested, digested, absorbed, and excreted. It is composed of the mouth, pharynx, esophagus, stomach, small and large intestines, rectum, and anus. *Peristalsis*, or wave-like contractions of smooth muscle, moves food and wastes through the digestive tract. The accessory digestive organs are the salivary glands, liver, gallbladder, and pancreas.

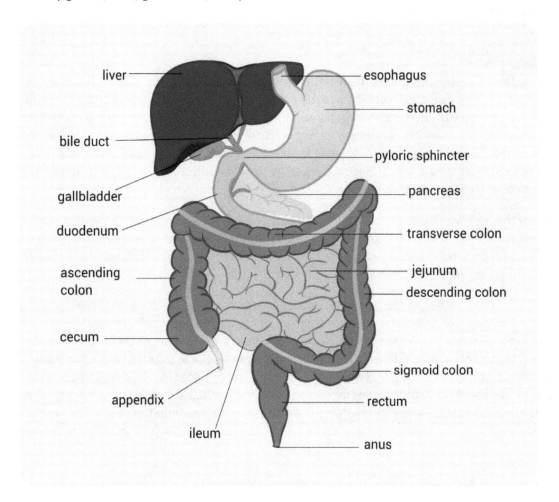

Mouth and Stomach

The mouth is the entrance to the digestive system. Here, the mechanical and chemical digestion of the food begins. The food is chewed mechanically by the teeth and shaped into a *bolus* by the tongue so that it can be more easily swallowed by the esophagus. The food also becomes more watery and pliable with the addition of saliva secreted from the salivary glands, the largest of which are the parotid glands. The glands also secrete amylase in the saliva, an enzyme which begins chemical digestion and breakdown of the carbohydrates and sugars in the food.

The food then moves through the pharynx and down the muscular esophagus to the stomach.

The stomach is a large, muscular sac-like organ at the distal end of the esophagus. Here, the bolus is subjected to more mechanical and chemical digestion. As it passes through the stomach, it is physically squeezed and crushed while additional secretions turn it into a watery nutrient-filled liquid that exits into the small intestine as *chyme*.

The stomach secretes many substances into the *lumen* of the digestive tract. Some cells produce gastrin, a hormone that prompts other cells in the stomach to secrete a gastric acid composed mostly of hydrochloric acid (HCl). The HCl is at such a high concentration and low pH that it denatures most proteins and degrades a lot of organic matter. The stomach also secretes mucous to form a protective film that keeps the corrosive acid from dissolving its own cells; gaps in this mucous layer can lead to peptic ulcers. Finally, the stomach also uses digestive enzymes like proteases and lipases to break down proteins and fats; although there are some gastric lipases here, the stomach mostly breaks down proteins.

Small Intestine

The chyme from the stomach enters the first part of the small intestine, the *duodenum*, through the *pyloric sphincter*, and its extreme acidity is partly neutralized by sodium bicarbonate secreted along with mucous. The presence of chyme in the duodenum triggers the secretion of the hormones secretin and cholecystokinin (CCK). Secretin acts on the pancreas to dump more sodium bicarbonate into the small intestine so that the pH is kept at a reasonable level, while CCK acts on the gallbladder to release the *bile* that it has been storing. Bile, a substance produced by the liver and stored in the gallbladder, helps to emulsify or dissolve fats and lipids.

Because of the bile, which aids in lipid absorption, and the secreted lipases, which break down fats, the duodenum is the chief site of fat digestion in the body. The duodenum also represents the last major site of chemical digestion in the digestive tract, as the other two sections of the small intestine (the *jejunum* and *ileum*) are instead heavily involved in absorption of nutrients.

The small intestine reaches 40 feet in length, and its cells are arranged in small finger-like projections called villi. This is due to its key role in the absorption of nearly all nutrients from the ingested and digested food, effectively transferring them from the lumen of the GI tract to the bloodstream, where they travel to the cells that need them. These nutrients include simple sugars like glucose from carbohydrates, amino acids from proteins, emulsified fats, electrolytes like sodium and potassium, minerals like iron and zinc, and vitamins like D and B12. Vitamin B12's absorption, though it takes place in the intestines, is actually aided by *intrinsic factor* that was released into the chyme back in the stomach.

Large Intestine

The leftover parts of food that remain unabsorbed or undigested in the lumen of the small intestine next travel through the large intestine, which is also referred to as the large bowel or colon. The large intestine is mainly responsible for water absorption. As the chyme at this stage no longer has any useful nutrients that can be absorbed by the body, it is now referred to as *waste*, and it is stored in the large intestine until it can be excreted from the body. Removing the liquid from the waste transforms it from liquid to solid stool, or feces.

This waste first passes from the small intestine to the cecum, a pouch that forms the first part of the large intestine. In herbivores, it provides a place for bacteria to digest cellulose, but in humans most of it is vestigial and is known as the appendix. The appendix has no known function other than arbitrarily becoming inflamed. From the cecum, waste next travels up the ascending colon, across the transverse colon, down the descending colon, and through the sigmoid colon to the rectum. The rectum is responsible for the final storage of waste before it is expelled through the anus. The anal canal is a small portion of the rectum leading through to the anus and the outside of the body.

Pancreas

The pancreas has endocrine and exocrine functions. The endocrine function works to regulate blood sugar levels. It involves releasing the hormone insulin, which decreases blood sugar (glucose) levels, or glucagon, which increases blood sugar (glucose) levels, directly into the bloodstream. Both hormones are produced in the islets of Langerhans, insulin in the beta cells and glucagon in the alpha cells.

The major part of the gland has an exocrine function, which consists of acinar cells secreting inactive digestive enzymes (zymogens) into the main pancreatic duct. The main pancreatic duct joins the common bile duct, which empties into the small intestine (specifically the duodenum). The digestive enzymes are then activated and take part in the digestion of carbohydrates, proteins, and fats within chyme (the mixture of partially digested food and digestive juices).

Endocrine System Structure and Function

The endocrine system is made of the ductless tissues and glands that secrete hormones into the interstitial fluids of the body. Interstitial fluid is the solution that surrounds tissue cells within the body. This system works closely with the nervous system to regulate the physiological activities of the other systems of the body to maintain homeostasis. While the nervous system provides quick, short-term responses to stimuli, the endocrine system acts by releasing hormones into the bloodstream that get distributed to the whole body. The response is slow but long-lasting, ranging from a few hours to a few weeks.

Hormones are chemical substances that change the metabolic activity of tissues and organs. While regular metabolic reactions are controlled by enzymes, hormones can change the type, activity, or quantity of the enzymes involved in the reaction. They bind to specific cells and start a biochemical chain of events that changes the enzymatic activity. Hormones can regulate development and growth, digestive metabolism, mood, and body temperature, among other things. Often small amounts of hormone will lead to large changes in the body.

The endocrine system has the following major glands:

- Hypothalamus: A part of the brain, the hypothalamus connects the nervous system to the endocrine system via the pituitary gland. Although it is considered part of the nervous system, it plays a dual role in regulating endocrine organs.

- Pituitary Gland: A pea-sized gland found at the bottom of the hypothalamus. It has two lobes, called the anterior and posterior lobes. It plays an important role in regulating the function of other endocrine glands. The hormones released control growth, blood pressure, certain functions of the sex organs, salt concentration of the kidneys, internal temperature regulation, and pain relief.

- Thyroid Gland: This gland releases hormones, such as thyroxine, that are important for metabolism, growth and development, temperature regulation, and brain development during infancy and childhood. Thyroid hormones also monitor the amount of circulating calcium in the body.

- Parathyroid Glands: These are four pea-sized glands located on the posterior surface of the thyroid. The main hormone secreted is called parathyroid hormone (PTH) and helps with the thyroid's regulation of calcium in the body.

- Thymus Gland: The thymus is located in the chest cavity, embedded in connective tissue. It produces several hormones important for development and maintenance of normal immunological defenses. One hormone promotes the development and maturation of lymphocytes, which strengthens the immune system.

- Adrenal Gland: One adrenal gland is attached to the top of each kidney. It produces adrenaline and is responsible for the "fight or flight" reactions in the face of danger or stress. The hormones epinephrine and norepinephrine cooperate to regulate states of arousal.

- Pancreas: The pancreas is an organ that has both endocrine and exocrine functions. The endocrine functions are controlled by the pancreatic islets of Langerhans, which are groups of beta cells scattered throughout the gland that secrete insulin to lower blood sugar levels in the body. Neighboring alpha cells secrete glucagon to raise blood sugar.

- Pineal Gland: The pineal gland secretes melatonin, a hormone derived from the neurotransmitter serotonin. Melatonin can slow the maturation of sperm, oocytes, and reproductive organs. It also regulates the body's circadian rhythm, which is the natural awake/asleep cycle. It also serves an important role in protecting the CNS tissues from neural toxins.

- Testes and Ovaries: These glands secrete testosterone and estrogen, respectively, and are responsible for secondary sex characteristics, as well as reproduction.

Immune System Structure and Function

The immune system is the body's defense against invading microorganisms (bacteria, viruses, fungi, and parasites) and other harmful, foreign substances. It is capable of limiting or preventing infection.

There are two general types of immunity: innate immunity and acquired immunity. Innate immunity uses physical and chemical barriers to block the entry of microorganisms into the body. The skin forms a physical barrier that blocks microorganisms from entering underlying tissues. Mucous membranes in the digestive, respiratory, and urinary systems secrete mucus to block and remove invading microorganisms. Saliva, tears, and stomach acids are examples of chemical barriers intended to block infection with microorganisms. In addition, macrophages and other white blood cells can recognize and eliminate foreign objects through phagocytosis or direct lysis.

Acquired immunity refers to a specific set of events used by the body to fight a particular infection. Essentially, the body accumulates and stores information about the nature of an invading microorganism. As a result, the body can mount a specific attack that is much more effective than innate immunity. It also provides a way for the body to prevent future infections by the same microorganism.

Acquired immunity is divided into a primary response and a secondary response. The primary immune response occurs the first time a particular microorganism enters the body, where macrophages engulf the microorganism and travel to the lymph nodes. In the lymph nodes, macrophages present the invader to helper T lymphocytes, which then activate humoral and cellular immunity. Humoral immunity refers to immunity resulting from antibody production by B lymphocytes. After being activated by helper T lymphocytes, B lymphocytes multiply and divide into plasma cells and memory cells. Plasma cells are B lymphocytes that produce immune proteins called antibodies, or immunoglobulins. Antibodies then bind the microorganism to flag it for destruction by other white blood cells. Cellular immunity refers to

the immune response coordinated by T lymphocytes. After being activated by helper T lymphocytes, other T lymphocytes attack and kill cells that cause infection or disease.

The secondary immune response takes place during subsequent encounters with a known microorganism. Memory cells respond to the previously encountered microorganism by immediately producing antibodies. Memory cells are B lymphocytes that store information to produce antibodies. The secondary immune response is swift and powerful because it eliminates the need for the time-consuming macrophage activation of the primary immune response. Suppressor T lymphocytes also take part to inhibit the immune response as an overactive immune response could cause damage to healthy cells.

Active and Passive Immunity

Immunization is the process of inducing immunity. *Active immunization* refers to immunity gained by exposure to infectious microorganisms or viruses and can be *natural* or *artificial*. Natural immunization refers to an individual being exposed to an infectious organism as a part of daily life. For example, it was once common for parents to expose their children to childhood diseases such as measles or chicken pox. Artificial immunization refers to therapeutic exposure to an infectious organism as a way of protecting an individual from disease. Today, the medical community relies on artificial immunization as a way to induce immunity.

Vaccines are used for the development of active immunity. A vaccine contains a killed, weakened, or inactivated microorganism or virus that is administered through injection, by mouth, or by aerosol. Vaccinations are administered to prevent an infectious disease but do not always guarantee immunity.

Passive immunity refers to immunity gained by the introduction of antibodies. This introduction can be natural or artificial. The process occurs when antibodies from the mother's bloodstream are passed on to the bloodstream of the developing fetus. Breast milk can also transmit antibodies to a baby. Babies are born with passive immunity, which provides protection against general infection for approximately the first six months of its life.

Integumentary System (Skin) Structure and Function

Skin consists of three layers: epidermis, dermis, and the hypodermis. There are four types of cells that make up the keratinized stratified squamous epithelium in the epidermis. They are keratinocytes, melanocytes, Merkel cells, and Langerhans cells. Skin is composed of many layers, starting with a basement membrane. On top of that sits the stratum germinativum, the stratum spinosum, the stratum granulosum, the stratum lucidum, and then the stratum corneum at the outer surface. Skin can be classified as thick or thin. These descriptions refer to the epidermis layer. Most of the body is covered with thin skin, but areas such as the palm of the hands are covered with thick skin. The dermis consists of a superficial papillary layer and a deeper reticular layer. The papillary layer is made of loose connective tissue, containing capillaries and the axons of sensory neurons. The reticular layer is a meshwork of tightly packed irregular connective tissue, containing blood vessels, hair follicles, nerves, sweat glands, and sebaceous glands. The hypodermis is a loose layer of fat and connective tissue. Since it is the third layer, if a burn reaches this third degree, it has caused serious damage.

Sweat glands and sebaceous glands are important exocrine glands found in the skin. Sweat glands regulate temperature, and remove bodily waste by secreting water, nitrogenous waste, and sodium salts to the surface of the body. Some sweat glands are classified as apocrine glands. Sebaceous glands are holocrine glands that secrete sebum, which is an oily mixture of lipids and proteins. Sebum protects the skin from water loss, as well as bacterial and fungal infections.

The three major functions of skin are protection, regulation, and sensation. Skin acts as a barrier and protects the body from mechanical impacts, variations in temperature, microorganisms, and chemicals. It regulates body temperature, peripheral circulation, and fluid balance by secreting sweat. It also contains a large network of nerve cells that relay changes in the external environment to the body.

Lymphatic System Structure and Function

The lymphatic system is one of the major systems that is benefited by proper massage. This system, like the circulatory system, is a network of vessels and organs that move fluid—in this case lymph—throughout the body. The lymphatic system works in concert with the immune system to help the body process toxins and waste. Lymph has a high concentration of white blood cells, which help attack viruses and bacteria throughout body cells and tissues. Lymph is filtered in nodes along the vessels; the body has 600 to 700 lymph nodes, which may be superficial (like those in the armpit and groin) or deep (such as those around the heart and lungs). The spleen is the largest organ of the lymphatic system and it helps produce the lymphocytes (white blood cells) to control infections. It also controls the number of red blood cells in the body. Other lymphatic organs include the tonsils, adenoids, and thymus.

Muscular System Structure and Function

The muscular system of the human body is responsible for all movement that occurs. There are approximately 700 muscles in the body that are attached to the bones of the skeletal system and that make up half of the body's weight. Muscles are attached to the bones through tendons. Tendons are made up of dense bands of connective tissue and have collagen fibers that firmly attach to the bone on one side and the muscle on the other. Their fibers are actually woven into the coverings of the bone and muscle so they can withstand the large forces that are put on them when muscles are moving. There are three types of muscle tissue in the body: Skeletal muscle tissue pulls on the bones of the skeleton and causes body movement; cardiac muscle tissue helps pump blood through veins and arteries; and smooth muscle tissue helps move fluids and solids along the digestive tract and contributes to movement in other body systems. All of these muscle tissues have four important properties in common: They are excitable, meaning they respond to stimuli; contractile, meaning they can shorten and pull on connective tissue; extensible, meaning they can be stretched repeatedly, but maintain the ability to contract; and elastic, meaning they rebound to their original length after a contraction.

Muscles begin at an origin and end at an insertion. Generally, the origin is proximal to the insertion and the origin remains stationary while the insertion moves. For example, when bending the elbow and moving the hand up toward the head, the part of the forearm that is closest to the wrist moves and the part closer to the elbow is stationary. Therefore, the muscle in the forearm has an origin at the elbow and an insertion at the wrist.

Body movements occur by muscle contraction. Each contraction causes a specific action. Muscles can be classified into one of three muscle groups based on the action they perform. Primary movers, or agonists, produce a specific movement, such as flexion of the elbow. Synergists are in charge of helping the primary movers complete their specific movements. They can help stabilize the point of origin or provide extra pull near the insertion. Some synergists can aid an agonist in preventing movement at a joint. Antagonists are muscles whose actions are the opposite of that of the agonist. If an agonist is contracting during a specific movement, the antagonist is stretched. During flexion of the elbow, the biceps' brachii muscle contracts and acts as an agonist, while the triceps' brachii muscle on the opposite side of the upper arm acts as an antagonist and stretches.

Skeletal muscle tissue has several important functions. It causes movement of the skeleton by pulling on tendons and moving the bones. It maintains body posture through the contraction of specific muscles

responsible for the stability of the skeleton. Skeletal muscles help support the weight of internal organs and protect these organs from external injury. They also help to regulate body temperature within a normal range. Muscle contractions require energy and produce heat, which heats the body when cold.

Nervous System Structure and Function

The human nervous system coordinates the body's response to stimuli from inside and outside the body. There are two major types of nervous system cells: neurons and neuroglia. Neurons are the workhorses of the nervous system and form a complex communication network that transmits electrical impulses termed action potentials, while neuroglia connect and support the neurons.

Although some neurons monitor the senses, some control muscles, and some connect the brain to other neurons, all neurons have four common characteristics:

- Dendrites: These receive electrical signals from other neurons across small gaps called *synapses*.
- Nerve cell body: This is the hub of processing and protein manufacture for the neuron.
- Axon: This transmits the signal from the cell body to other neurons.
- Terminals: These bridge the neuron to dendrites of other neurons and deliver the signal via chemical messengers called neurotransmitters.

Here is an illustration of a neuron:

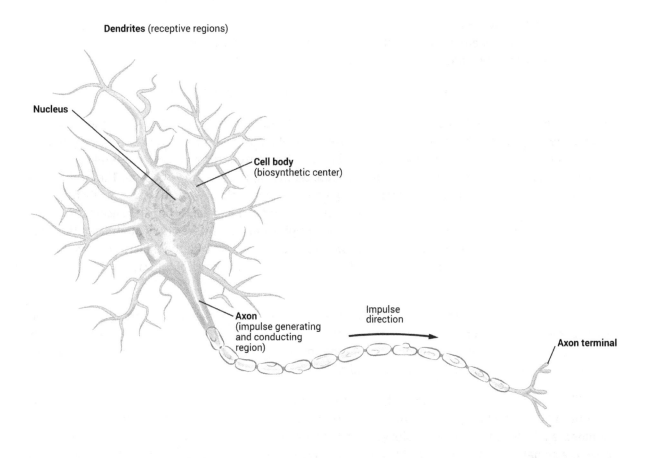

There are two major divisions of the nervous system: central and peripheral.

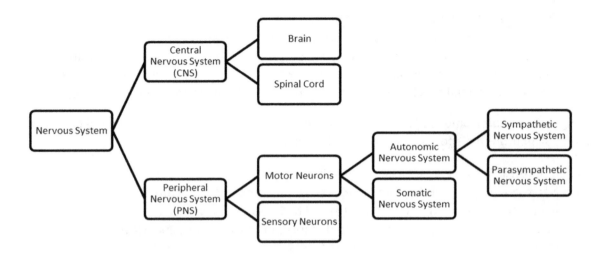

Central Nervous System

The central nervous system (CNS) consists of the brain and spinal cord. Three layers of membranes called the meninges cover and separate the CNS from the rest of the body.

The major divisions of the brain are the forebrain, the midbrain, and the hindbrain.

The *forebrain* consists of the cerebrum, the thalamus and hypothalamus, and the rest of the limbic system. The *cerebrum* is the largest part of the brain, and its most well-researched part is the outer cerebral cortex. The cerebrum is divided into right and left hemispheres, and each cerebral cortex hemisphere has four discrete areas, or lobes: frontal, temporal, parietal, and occipital. The frontal lobe governs duties such as voluntary movement, judgment, problem solving, and planning, while the other lobes are more sensory. The temporal lobe integrates hearing and language comprehension, the parietal lobe processes sensory input from the skin, and the occipital lobe processes visual input from the eyes. For completeness, the other two senses, smell and taste, are processed via the olfactory bulbs. The thalamus helps organize and coordinate all of this sensory input in a meaningful way for the brain to interpret.

The hypothalamus controls the endocrine system and all of the hormones that govern long-term effects on the body. Each hemisphere of the limbic system includes a hippocampus (which plays a vital role in memory), an amygdala (which is involved with emotional responses like fear and anger), and other small bodies and nuclei associated with memory and pleasure.

The midbrain is in charge of alertness, sleep/wake cycles, and temperature regulation, and it includes the substantia nigra which produces melatonin to regulate sleep patterns. The notable components of the hindbrain include the medulla oblongata and cerebellum. The medulla oblongata is located just above the spinal cord and is responsible for crucial involuntary functions such as breathing, swallowing, and the regulation of heart rate and blood pressure. Together with other parts of the hindbrain, the midbrain and medulla oblongata form the brain stem. The connects the spinal cord to the rest of the brain. To the rear of the brain stem sits the cerebellum, which plays key roles in posture, balance, and muscular coordination. The spinal cord itself, which is encapsulated by the protective bony spinal column, carries sensory information to the brain and motor information to the body.

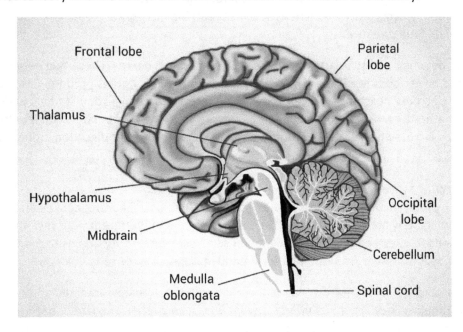

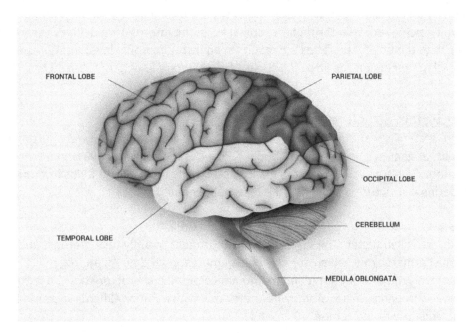

Peripheral Nervous System

The peripheral nervous system (PNS) includes all nervous tissue besides the brain and spinal cord. The PNS consists of the sets of cranial and spinal nerves and relays information between the CNS and the rest of the body. The PNS has two divisions: the autonomic nervous system and the somatic nervous system.

Autonomic Nervous System

The autonomic nervous system (ANS) governs involuntary, or reflexive, body functions. Ultimately, the autonomic nervous system controls functions such as breathing, heart rate, digestion, body temperature, and blood pressure.

The ANS is split between parasympathetic nerves and sympathetic nerves. These two nerve types are antagonistic and have opposite effects on the body. Parasympathetic nerves predominate resting conditions, and decrease heart rate, decrease breathing rate, prepare digestion, and allow urination and excretion. Sympathetic nerves, on the other hand, become active when a person is under stress or excited, and they increase heart rate, increase breathing rates, and inhibit digestion, urination, and excretion.

Somatic Nervous System and the Reflex Arc

The somatic nervous system (SNS) governs the conscious, or voluntary, control of skeletal muscles and their corresponding body movements. The SNS contains afferent and efferent neurons. Afferent neurons carry sensory messages from the skeletal muscles, skin, or sensory organs to the CNS. Efferent neurons relay motor messages from the CNS to skeletal muscles, skin, or sensory organs.

The SNS also has a role in involuntary movements called reflexes. A reflex is defined as an involuntary response to a stimulus. They are transmitted via what is termed a reflex arc, where a stimulus is sensed by a receptor and its afferent neuron, interpreted and rerouted by an interneuron, and delivered to effector muscles by an efferent neuron where they respond to the initial stimulus. A reflex is able to bypass the brain by being rerouted through the spinal cord; the interneuron decides the proper course of action rather than the brain. The reflex arc results in an instantaneous, involuntary response. For example, a physician tapping on the knee produces an involuntary knee jerk referred to as the patellar tendon reflex.

Reproductive System Structure and Function

The reproductive system is responsible for producing, storing, nourishing, and transporting functional reproductive cells, or gametes, in the human body. It includes the reproductive organs, also known as gonads, the reproductive tract, the accessory glands and organs that secrete fluids into the reproductive tract, and the perineal structures, which are the external genitalia.

The Male System

The male gonads are called testes. The testes secrete androgens, mainly testosterone, and produce and store 500 million spermatocytes, which are the male gametes, each day. An androgen is a steroid hormone that controls the development and maintenance of male characteristics. Once the sperm are mature, they move through a duct system, where they mix with additional fluids secreted by accessory glands, forming a mixture called semen.

The Female System

The female gonads are the ovaries. Ovaries generally produce one immature gamete, an egg or oocyte, per month. They are also responsible for secreting the hormones estrogen and progesterone. When the oocyte is released from the ovary, it travels along the uterine tubes, or Fallopian tubes, and then into the uterus. The uterus opens into the vagina. When sperm cells enter the vagina, they swim through the uterus and may fertilize the oocyte in the Fallopian tubes. The resulting zygote travels down the tube and implants into the uterine wall. The uterus protects and nourishes the developing embryo for nine months until it is ready for the outside environment. If the oocyte is not fertilized, it is released in the

uterine, or menstrual, cycle. The menstrual cycle occurs monthly and involves the shedding of the functional part of the uterine lining.

Human Reproduction

Humans procreate through sexual reproduction. Sexual reproduction involves the fusion of gametes, one from each parent. A gamete is a reproductive cell that contains half the chromosomes of a normal cell. Chromosomes are found in the nucleus of cells and contain DNA. The female gamete is called an ovum, or egg, and the male gamete is called a sperm. In sexual reproduction, the gametes fuse through a process called fertilization. As a result, sexual reproduction often produces offspring with varying characteristics.

Gametes are created by the human reproductive systems. In women, the ovaries produce eggs, the female gamete. The ovaries produce on average one mature egg per month, which is referred to as the menstrual cycle. The release of an egg from the ovaries is termed ovulation. The female menstrual cycle is under the control of hormones such as luteinizing hormone (LH), follicle stimulating hormone (FSH), estrogen, and progesterone. In men, the testes produce sperm, the male gamete, and they produce millions of sperm at a time. The hormones LH and testosterone regulate the production of sperm in the testes. Leydig cells in the testes produce testosterone, while sperm is manufactured in the seminiferous tubules of the testes.

The fusion of the gametes (egg and sperm) is termed *fertilization*, and the resulting fusion creates a zygote. The zygote takes approximately seven days to travel through the fallopian tube and implant itself into the uterus. Upon implantation, it has developed into a blastocyst and will next grow into a gastrula. It is during this stage that the embryological germ layers are formed. The three germ layers are the ectoderm (outer layer), mesoderm (middle layer), and endoderm (inner layer). All of the human body systems develop from one or more of the germ layers. The gastrula further develops into an embryo which then matures into a fetus. The entire process takes approximately nine months and culminates in labor and birth.

Respiratory System Structure and Function

The respiratory system mediates the exchange of gas between the air and the blood, mainly through the act of breathing. This system is divided into the upper respiratory system and the lower respiratory system. The upper system comprises the nose, the nasal cavity and sinuses, and the pharynx. The lower respiratory system comprises the larynx (voice box), the trachea (windpipe), the small passageways leading to the lungs, and the lungs. The upper respiratory system is responsible for filtering, warming, and humidifying the air that gets passed to the lower respiratory system, protecting the lower respiratory system's more delicate tissue surfaces. The process of breathing in is referred to as *inspiration* while the process of breathing out is referred to as *expiration*.

The Lungs

Bronchi are tubes that lead from the trachea to each lung, and are lined with cilia and mucus that collect dust and germs along the way. The bronchi, which carry air into the lungs, branch into bronchioles and continue to divide into smaller and smaller passageways, until they become alveoli, which are the smallest passages. Most of the gas exchange in the lungs occurs between the blood-filled pulmonary capillaries and the air-filled alveoli. Within the lungs, oxygen and carbon dioxide are exchanged between the air in the alveoli and the blood in the pulmonary capillaries. Oxygen-rich blood returns to the heart and is pumped through the systemic circuit. Carbon dioxide-rich air is exhaled from the body. Together, the lungs contain approximately 1,500 miles of airway passages, and this extremely high amount is due to the enormous amount of branching.

Breathing is possible due to the muscular diaphragm pulling on the lungs, increasing their volume and decreasing their pressure. Air flows from the external high-pressure system to the low-pressure system inside the lungs. When breathing out, the diaphragm releases its pressure difference, decreases the lung volume, and forces the stale air back out.

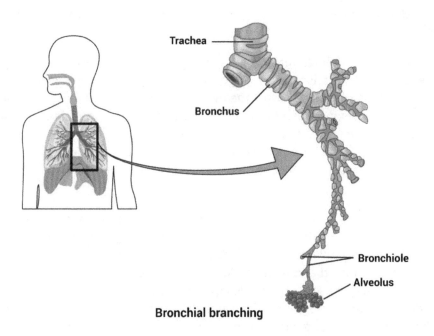

Bronchial branching

Functions of the Respiratory System

The respiratory system has many functions. Most importantly, it provides a large area for gas exchange between the air and the circulating blood. It protects the delicate respiratory surfaces from environmental variations and defends them against pathogens. It is responsible for producing the sounds that the body makes for speaking and singing, as well as for non-verbal communication. It also helps regulate blood volume and blood pressure by releasing vasopressin, and it is a regulator of blood pH due to its control over carbon dioxide release, as the aqueous form of carbon dioxide is the chief buffering agent in blood.

Skeletal System Structure and Function

The skeletal system consists of the 206 bones that make up the skeleton, as well as the cartilage, ligaments, and other connective tissues that stabilize them. Bone is made of collagen fibers and calcium inorganic minerals, mostly in the form of hydroxyapatite, calcium carbonate, and phosphate salts. The inorganic minerals are strong but brittle, and the collagen fibers are weak but flexible, so the combination makes bone resistant to shattering. There are two types of bone: compact and spongy. Compact bone has a basic functional unit, called the Haversian system. Osteocytes, or bone cells, are arranged in concentric circles around a central canal, called the Haversian canal, which contains blood vessels. While Haversian canals run parallel to the surface of the bone, perforating canals, also known as the canals of Volkmann, run perpendicularly between the central canal and the surface of the bone. The concentric circles of bone tissue that surround the central canal within the Haversian system are called

lamellae. The spaces that are found between the lamellae are called lacunae. The Haversian system is a reservoir for calcium and phosphorus for blood. Spongy bone, in contrast to compact bone, is lightweight and porous. It has a branching network of parallel lamellae, called trabeculae. Although spongy bone forms an open framework inside the compact bone, it is still quite strong. Different bones have different ratios of compact-to-spongy bone, depending on their functions. The outside of the bone is covered by a periosteum, which has four major functions. It isolates and protects bones from the surrounding tissue; provides a place for attachment of the circulatory and nervous system structures; participates in growth and repair of the bone; and attaches the bone to the deep fascia. An endosteum is found inside the bone, covers the trabeculae of the spongy bone, and lines the inner surfaces of the central canals.

One major function of the skeletal system is to provide structural support for the entire body. It provides a framework for the soft tissues and organs to attach to. The skeletal system also provides a reserve of important nutrients, such as calcium and lipids. Normal concentrations of calcium and phosphate in body fluids are partly maintained by the calcium salts stored in bone. Lipids that are stored in yellow bone marrow can be used as a source of energy. Yellow bone marrow also produces some white blood cells. Red bone marrow produces red blood cells, most white blood cells, and platelets that circulate in the blood. Certain groups of bones form protective barriers around delicate organs. The ribs, for example, protect the heart and lungs, the skull encloses the brain, and the vertebrae cover the spinal cord.

Special Senses Structure and Function

The special senses include vision, hearing and balance, smell, and taste. They are distinguished from general senses in that special senses have *special somatic afferents* and *special visceral afferents*, both a type of nerve fiber relaying information to the CNS, as well as specialized organs devoted to their function. Touch is the other sense that is typically discussed, but unlike the special senses, it relays information to the CNS from all over the body and not just one particular organ; skin, the largest organ of the body, is the largest contributor to tactile information, but touch receptors also include mechanoreceptors, nociceptors for pain, and thermoreceptors for heat. Tactile messages are carried via *general somatic afferents* and *general visceral afferents.* Massage therapists should be familiar with the various touch receptors such as the following:

- Pacinian corpuscles: detect rapid vibration in the skin and fascia
- Meissner's corpuscles: respond to light touch and slower vibrations
- Merkel's discs: respond to sustained pressure
- Ruffini endings: detect deep touch and tension in the skin and fascia

Urinary System Structure and Function

The urinary system includes the kidneys, ureters, urinary bladder, and the urethra. It is the main system responsible for getting rid of the organic waste products, excess water, and electrolytes are generated by the body's other systems. The kidneys are responsible for producing urine, which is a fluid waste product containing water, ions, and small soluble compounds. The urinary system has many important functions related to waste excretion. It regulates the concentrations of sodium, potassium, chloride, calcium, and other ions in the plasma by controlling the amount of each that is excreted in urine. This also contributes to the maintenance of blood pH. It regulates blood volume and pressure by controlling the amount of water lost in the urine. It eliminates toxic substances, drugs, and organic waste products, such as urea and uric acid.

The Kidneys

Under normal circumstances, humans have two functioning kidneys. They are the main organs responsible for filtering waste products out of the blood and transferring them to urine. Kidneys are made of millions of tiny filtering units called *nephrons*. Nephrons have two parts: a glomerulus, which is the filter, and a tubule. As blood enters the kidneys, the glomerulus allows fluid and waste products to pass through it and enter the tubule. Blood cells and large molecules, such as proteins, do not pass through and remain in the blood. The filtered fluid and waste then pass through the tubule, where any final essential minerals are sent back to the bloodstream. The final product at the end of the tubule is urine.

Waste Excretion

Once urine accumulates, it leaves the kidneys. The urine travels through the ureters into the urinary bladder, a muscular organ that is hollow and elastic. As more urine enters the urinary bladder, its walls stretch and become thinner, so there is no significant difference in internal pressure. The urinary bladder stores the urine until the body is ready for urination, at which time the muscles contract and force the urine through the urethra and out of the body.

Tissue Injury and Repair

Healing tissue injuries is a complicated process, mediated by specialized cells. The immune system initiates the response and sends out macrophages, which are white blood cells that scavenge for damaged cells and foreign particles. Fibroblasts help repair injured cells, laying down scar tissue, which is fibrous connective tissue that forms over the injured area. It is denser than healthy tissue. Vitamin C is necessary for tissue repair and Vitamin K plays an instrumental role in blood clotting during initial injury.

Immediately after the injury, there is often bleeding and swelling. The first phase of healing is called the inflammatory stage. It takes place from 2 to 3 days post-injury to 2 to 3 weeks, depending on the injury and the health status of the individual. Bleeding stops during this time and the tissue swells, as the immune system mobilizes the macrophages to the area to clean up debris and prevent infection if the area is open. During the repair and regeneration phase, the fibroblasts begin to organize around the site of injury and create new tissue. This occurs anywhere from 2 to 3 days to six weeks after the initial injury. After this point and up to a year or so later, the remodeling phase occurs, where the scar tissue forms fully over the site of injury.

Physical Science

Recognizing Physical Properties

Physical properties of matter can be observed and used for identification of substances, classification, observation, experimental design, among other things. Depending on the substance and property of interest, these properties may or may not be modifiable. For example, the size of a rock specimen in a lab is not going to grow, although a large sedimentary rock formation in a canyon may increase in size over thousands of years as layers of sediment deposit on top of the existing surface. Other properties, such as hardness or flexibility, are not modifiable, but rather are inherent to the material. In many cases, observable properties, such as *color*, may be variables in scientific experiments, as researchers look at how manipulations to the environment affect the properties of the constituents. For example, students studying chemical reactions may investigate how the *color* of copper pennies changes when soaked in vinegar or sodium hydroxide.

Size can be measured with rulers or tape measures, while *weight* can be measured on a scale. It is important to differentiate between mass and weight, with weight being affected by gravity. Thermometers can be used to measure *temperature*, typically either in degrees Celsius, Fahrenheit, or Kelvin. Depending on the state of an object—solid, liquid, or gas—various physical properties may be different as well.

Physical Properties vs. Chemical Properties
Both physical and chemical properties are used to sort and classify objects:

- Physical properties: refers to the appearance, mass, temperature, state, size, or color of an object or fluid; a physical change indicates a change in the appearance, mass, temperature, state, size or color of an object or fluid.

- Chemical properties: refers to the chemical makeup of an object or fluid; a chemical change refers to an alteration in the makeup of an object or fluid and forms a new solution or compound.

Reversible Change vs. Non-Reversible Change
Reversible change (physical change) is the changing of the size or shape of an object without altering its chemical makeup. Examples include the heating or cooling of water, change of state (solid, liquid, gas), the freezing of water into ice, or cutting a piece of wood in half.

When two or more materials are combined, it is called a mixture. Generally, a mixture can be separated out into the original components. When one type of matter is dissolved into another type of matter (a solid into a liquid or a liquid into another liquid), and cannot easily be separated back into its original components, it is called a solution.

States of matter refer to the form substances take such as solid, liquid, gas, or plasma. Solid refers to a rigid form of matter with a flexed shape and a fixed volume. Liquid refers to the fluid form of matter with no fixed shape and a fixed volume. Gas refers to an easily compressible fluid form of matter with no fixed shape that expands to fill any space available. Finally, plasma refers to an ionized gas where electrons flow freely from atom to atom.

> Examples: A rock is a solid because it has a fixed shape and volume. Water is considered to be a liquid because it has a set volume, but not a set shape; therefore, you could pour it into different containers of different shapes, as long as they were large enough to contain the existing volume of the water. Oxygen is considered to be a gas. Oxygen does not have a set volume or a set shape; therefore, it could expand or contract to fill a container or even a room. Gases in fluorescent lamps become plasma when electric current is applied to them.

Matter can change from one state to another in many ways, including through heating, cooling, or a change in pressure.

Changes of state are identified as:

- Melting: solid to liquid
- Sublimation: solid to gas
- Evaporation: liquid to gas
- Freezing: liquid to solid
- Condensation: gas to liquid

- Non-reversible change (chemical change): When one or more types of matter change and it results in the production of new materials. Examples include burning, rusting, and combining solutions. If a piece of paper is burned it cannot be turned back into its original state. It has forever been altered by a chemical change.

Concepts Relating to the Position and Motion of Objects

The proper use of tools and machinery depends on an understanding of basic physics, which includes the study of motion and the interactions of *mass*, *force*, and *energy*. These terms are used every day, but their exact meanings are difficult to define. In fact, they're usually defined in terms of each other.

The matter in the universe (atoms and molecules) is characterized in terms of its *mass*, which is measured in kilograms in the *International System of Units (SI)*. The amount of mass that occupies a given volume of space is termed *density*.

Mass occupies space, but it's also a component that inversely relates to acceleration when a force is applied to it. This *force* is the application of *energy* to an object with the intent of changing its position (mainly its acceleration).

To understand *acceleration*, it's necessary to relate it to displacement and velocity. The *displacement* of an object is simply the distance it travels. The *velocity* of an object is the distance it travels in a unit of time, such as miles per hour or meters per second:

$$Velocity = \frac{Distance\ Traveled}{Time\ Required}$$

There's often confusion between the words "speed" and "velocity." Velocity includes speed *and* direction. For example, a car traveling east and another traveling west can have the same speed of 30 miles per hour (mph), but their velocities are different. If movement eastward is considered positive, then movement westward is negative. Thus, the eastbound car has a velocity of 30 mph while the westbound car has a velocity of -30 mph.

The fact that velocity has a *magnitude* (speed) and a direction makes it a vector quantity. A *vector* is an arrow pointing in the direction of motion, with its length proportional to its magnitude.

Vectors can be added geometrically as shown below. In this example, a boat is traveling east at 4 *knots* (nautical miles per hour) and there's a current of 3 knots (thus a slow boat and a very fast current). If the boat travels in the same direction as the current, it gets a "lift" from the current and its speed is 7 knots. If the boat heads *into* the current, it has a forward speed of only 1 knot (4 knots – 3 knots = 1 knot) and makes very little headway.

As shown in the figure below, the current is flowing north across the boat's path. Thus, for every 4 miles of progress the boat makes eastward, it drifts 3 miles to the north.

Working with Velocity Vectors

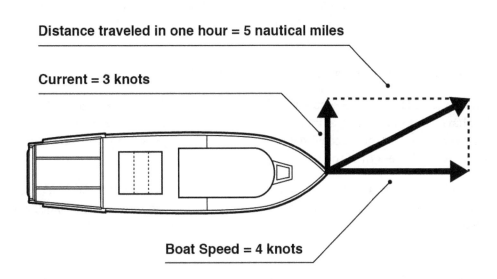

Distance traveled in one hour = 5 nautical miles

Current = 3 knots

Boat Speed = 4 knots

The total distance traveled is calculated using the *Pythagorean Theorem* for a right triangle, which should be memorized as follows:

$$a^2 + b^2 = c^2 \text{ or } c = \sqrt{a^2 + b^2}$$

Of course, the problem above was set up using a Pythagorean triple (3, 4, 5), which made the calculation easy.

Another example where velocity and speed are different is with a car traveling around a bend in the road. The speed is constant along the road, but the direction (and therefore the velocity) changes continuously.

The *acceleration* of an object is the change in its velocity in a given period of time:

$$Acceleration = \frac{Change\ in\ Velocity}{Time\ Required}$$

Newton's Laws

Isaac Newton's three laws of motion describe how the acceleration of an object is related to its mass and the forces acting on it. The three laws are:

- Unless acted on by a force, a body at rest tends to remain at rest; a body in motion tends to remain in motion with a constant velocity and direction.

- A force that acts on a body accelerates it in the direction of the force. The larger the force, the greater the acceleration; the larger the mass, the greater its inertia (resistance to movement and acceleration).

- Every force acting on a body is resisted by an equal and opposite force.

To understand Newton's laws, it's necessary to understand forces. These forces can push or pull on a mass, and they have a magnitude and a direction. Forces are represented by a vector, which is the arrow lined up along the direction of the force with its tip at the point of application. The magnitude of the force is represented by the length of the vector.

The figure below shows a mass acted on or "pushed" by two equal forces (shown here by vectors of the same length). Both vectors "push" along the same line through the center of the mass, but in opposite directions. What happens?

A Mass Acted on by Equal and Opposite Forces

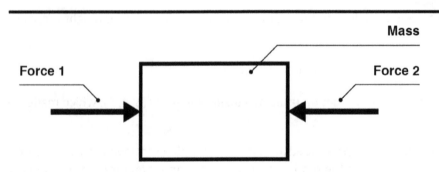

According to Newton's third law, every force on a body is resisted by an equal and opposite force. In the figure above, Force 1 acts on the left side of the mass. The mass pushes back. Force 2 acts on the right side, and the mass pushes back against this force too. The net force on the mass is zero, so according to Newton's first law, there's no change in the *momentum* (the mass times its velocity) of the mass. Therefore, if the mass is at rest before the forces are applied, it remains at rest. If the mass is in motion with a constant velocity, its momentum doesn't change. So, what happens when the net force on the mass isn't zero, as shown in the figure below?

A Mass Acted on by Unbalanced Forces

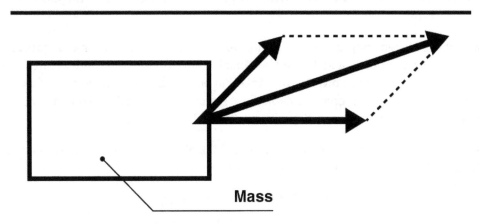

Mass

Notice that the forces are vector quantities and are added geometrically the same way that velocity vectors are manipulated.

Here in the figure above, the mass is pulled by two forces acting to the right, so the mass accelerates in the direction of the net force. This is described by Newton's second law:

Force = Mass x Acceleration

The force (measured in *newtons*) is equal to the product of the mass (measured in kilograms) and its acceleration (measured in meters per second squared or meters per second, per second). A better way to look at the equation is dividing through by the mass:

Acceleration = Force/Mass

This form of the equation makes it easier to see that the acceleration of an object varies directly with the net force applied and inversely with the mass. Thus, as the mass increases, the acceleration is reduced for a given force. To better understand, think of how a baseball accelerates when hit by a bat. Now imagine hitting a cannonball with the same bat and the same force. The cannonball is more massive than the baseball, so it won't accelerate very much when hit by the bat.

In addition to forces acting on a body by touching it, gravity acts as a force at a distance and causes all bodies in the universe to attract each other. The *force of gravity (F_g)* is proportional to the masses of the two objects (*m* and *M*) and inversely proportional to the square of the distance (r^2) between them (and *G* is the proportionality constant).

This is shown in the following equation:

$$F_g = G\frac{mM}{r^2}$$

The force of gravity is what causes an object to fall to Earth when dropped from an airplane. Understanding gravity helps explain the difference between mass and weight. Mass is a property of an object that remains the same while it's intact, no matter where it's located. A 10-kilogram cannonball has the same mass on Earth as it does on the moon. On Earth, it *weighs* 98.1 newtons because of the attractive force of gravity, so it accelerates at 9.81 m/s^2. However, on the moon, the same cannonball has a weight of only about 16 newtons. This is because the gravitational attraction on the moon is approximately one-sixth that on Earth. Although Earth still attracts the body on the moon, it's so far away that its force is negligible.

For Americans, there's often confusion when talking about mass because the United States still uses "pounds" as a measurement of weight. In the traditional system used in the United States, the unit of mass is called a *slug*. It's derived by dividing the weight in pounds by the acceleration of gravity (32 feet/s^2); however, it's rarely used today. To avoid future problems, test takers should continue using SI units and *remember to express mass in kilograms and weight in Newtons*.

Another way to understand Newton's second law is to think of it as an object's change in momentum, which is defined as the product of the object's mass and its velocity:

Momentum = Mass x Velocity

Which of the following has the greater momentum: a pitched baseball, a softball, or a bullet fired from a rifle?

A bullet with a mass of 5 grams (0.005 kilograms) is fired from a rifle with a muzzle velocity of 2200 mph. Its momentum is calculated as:

$$2200\frac{miles}{hour} \times \frac{5,280\ feet}{mile} \times \frac{m}{3.28\ feet} \times \frac{hour}{3600\ seconds} \times 0.005kg = 4.92\frac{kg.m}{seconds}$$

A softball has a mass between 177 grams and 198 grams and is thrown by a college pitcher at 50 miles per hour. Taking an average mass of 188 grams (0.188 kilograms), a softball's momentum is calculated as:

$$50\frac{miles}{hour} \times \frac{5280\ feet}{mile} \times \frac{m}{3.28\ ft} \times \frac{hour}{3600\ seconds} \times 0.188kg = 4.19\frac{kg.m}{seconds}$$

That's only slightly less than the momentum of the bullet. Although the speed of the softball is considerably less, its mass is much greater than the bullet's.

A professional baseball pitcher can throw a 145-gram baseball at 100 miles per hour. A similar calculation (try doing it!) shows that the pitched hardball has a momentum of about 6.48 kg.m/seconds. That's more momentum than a speeding bullet!

So why is the bullet more harmful than the hard ball? It's because the force that it applies acts on a much smaller area.

Instead of using acceleration, Newton's second law is expressed here as the change in momentum (with the delta symbol "Δ" meaning "change"):

$$Force = \frac{\Delta\ Momentum}{\Delta\ Time} = \frac{\Delta\ (Mass\ \times\ Velocity)}{\Delta\ Time} = Mass\ \times\ \frac{\Delta\ Velocity}{\Delta\ Time}$$

The rapid application of force is called *impulse*. Another way of stating Newton's second law is in terms of the impulse, which is the force multiplied by its time of application:

$$Impluse = Force\ \times\ \Delta\ Time = Mass\ \times\ \Delta\ Velocity$$

In the case of the rifle, the force created by the pressure of the charge's explosion in its shell pushes the bullet, accelerating it until it leaves the barrel of the gun with its *muzzle velocity* (the speed the bullet has when it leaves the muzzle). After leaving the gun, the bullet doesn't accelerate because the gas pressure is exhausted. The bullet travels with a constant velocity in the direction it's fired (ignoring the force exerted against the bullet by friction and drag).

Similarly, the pitcher applies a force to the ball by using their muscles when throwing. Once the ball leaves the pitcher's fingers, it doesn't accelerate and the ball travels toward the batter at a constant speed (again ignoring friction and drag). The speed is constant, but the velocity can change if the ball travels along a curve.

Projectile Motion

According to Newton's first law, if no additional forces act on the bullet or ball, it travels in a straight line. This is also true if the bullet is fired in outer space. However, here on Earth, the force of gravity continues to act so the motion of the bullet or ball is affected.

What happens when a bullet is fired from the top of a hill using a rifle held perfectly horizontal? Ignoring air resistance, its horizontal velocity remains constant at its muzzle velocity. Its vertical velocity (which is zero when it leaves the gun barrel) increases because of gravity's acceleration. Each passing second, the bullet traces out the same distance horizontally while increasing distance vertically (shown in the figure below). In the end, the projectile traces out a *parabolic curve*.

Projectile Path for a Bullet Fired Horizontally from a Hill (Ignoring Air Resistance)

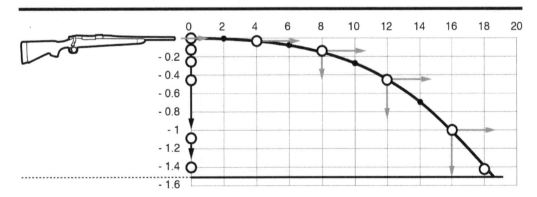

This vertical, downward acceleration is why a pitcher must put an arc on the ball when throwing across home plate. Otherwise the ball will fall at the batter's feet.

It's also interesting to note that if an artillery crew simultaneously drops one cannonball and fires another one horizontally, the two cannonballs will hit the ground at the same time since both balls are accelerating at the same rate and experience the same changes in vertical velocity.

What if air resistance is taken into account? This is best answered by looking at the horizontal and vertical motions separately.

The horizontal velocity is no longer constant because the initial velocity of the projectile is continually reduced by the resistance of the air. This is a complex problem in fluid mechanics, but it's sufficient to note that that the projectile doesn't fly as far before landing as predicted from the simple theory.

The vertical velocity is also reduced by air resistance. However, unlike the horizontal motion where the propelling force is zero after the cannonball is fired, the downward force of gravity acts continuously. The downward velocity increases every second due to the acceleration of gravity. As the velocity increases, the resisting force (called *drag*) increases with the square of the velocity. If the projectile is fired or dropped from a sufficient height, it reaches a terminal velocity such that the upward drag force equals the downward force of gravity. When that occurs, the projectile falls at a constant rate.

This is the same principle that's used for a parachute. Its drag (caused by its shape that scoops up air) is sufficient enough to slow down the fall of the parachutist to a safe velocity, thus avoiding a fatal crash on the ground.

So, what's the bottom line? If the vertical height isn't too great, a real projectile will fall short of the theoretical point of impact. However, if the height of the fall is significant and the drag of the object results in a small terminal fall velocity, then the projectile can go further than the theoretical point of impact.

What if the projectile is launched from a moving platform? In this case, the platform's velocity is added to the projectile's velocity. That's why an object dropped from the mast of a moving ship lands at the base of the mast rather than behind it. However, to an observer on the shore, the object traces out a parabolic arc.

Angular Momentum

In the previous examples, all forces acted through the center of the mass, but what happens if the forces aren't applied through the same line of action, like in the figure below?

A Mass Acted on by Forces Out of Line with Each Other

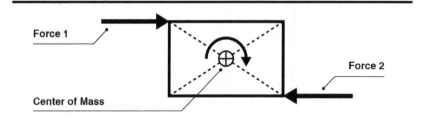

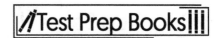

When this happens, the two forces create *torque* and the mass rotates around its center of gravity. In the figure above, the center of gravity is the center of the rectangle ("Center of Mass"), which is determined by the two, intersecting main diagonals. The center of an irregularly shaped object is found by hanging it from two different edges, and the center of gravity is at the intersection of the two "plumb lines."

Newton's second law still applies when the forces form a moment pair, but it must be expressed in terms of angular acceleration and the moment of inertia. The *moment of inertia* is a measure of the body's resistance to rotation, similar to the mass's resistance to linear acceleration. The more compact the body, the less the moment of inertia and the faster it rotates, much like how an ice skater spinning with outstretched arms will speed up as the arms are brought in close to the body.

The concept of torque is important in understanding the use of wrenches and is likely to be on the test. The concept of torque and moment/lever arm will be taken up again below, when the physics of simple machines is presented.

Conservation of Angular Momentum

An object moving in a circular motion also has momentum; for circular motion, it is called angular momentum. This is determined by rotational inertia and rotational velocity and the distance of the mass from the axis of rotation or center of rotation. When objects are exhibiting circular motion, they also demonstrate the conservation of angular momentum, meaning that the angular momentum of a system is always constant, regardless of the placement of the mass. Rotational inertia can be affected by how far the mass of the object is placed with respect to the center of rotation (axis of rotation). The larger the distance between the mass and the center of rotation, the slower the rotational velocity. Conversely, if the mass is closer to the center of rotation, the rotational velocity increases. A change in one affects the other, thus conserving the angular momentum. This holds true as long as no external forces act upon the system.

For example, an ice skater spinning on one ice skate extends their arms out for a slower rotational velocity. When the skater brings their arms in close to their body (or lessens the distance between the mass and the center of rotation), their rotational velocity increases, and they spin much faster. Some skaters extend their arms straight up above their head, which causes an extension of the axis of rotation, thus removing any distance between the mass and the center of rotation and maximizing their rotational velocity.

Another example is when a person selects a horse on a merry-go-round: the placement of their horse can affect their ride experience. All of the horses are traveling with the same rotational speed, but in order to travel along the same plane as the merry-go-round turns, a horse on the outside will have a greater linear speed, due to it being farther away from the axis of rotation. Another way to think of it is that an outside horse has to cover a lot more ground than a horse on the inside, in order to keep up with the rotational speed of the merry-go-round platform. Thrill seekers should always select an outer horse.

Principles of Light, Heat, Electricity, and Magnetism

The term *energy* typically refers to an object's ability to perform work. This can include a transfer of heat from one object to another, or from an object to its surroundings. Energy is usually measured in Joules. There are two main categories of energy: renewable and non-renewable.

- Renewable: energy produced from the exhaustion of a resource that can be replenished. Burning wood to produce heat, then replanting trees to replenish the resource is an example of using renewable energy.

- Non-renewable: energy produced from the exhaustion of a resource that cannot be replenished. Burning coal to produce heat would be an example of a non-renewable energy. Although coal is a natural resource found in/on the earth that is mined or harvested from the earth, it cannot be regrown or replenished. Other examples include oil and natural gas (fossil fuels).

Temperature is measured in degrees Celsius (C) or Kelvin (K). Temperature should not be confused with heat. Heat is a form of energy: a change in temperature or a transfer of heat can also be a measure of energy. The amount of energy measured by the change in temperature (or a transfer) is the measure of heat.

Heat energy (thermal energy) can be transferred through the following ways:

Conduction
Conduction is the heating of one object by another through the actual touching of molecules, in order to transfer heat across the objects involved. A spiral burner on an electric stovetop heats from one molecule touching another to transfer the heat via conduction.

Convection
Heat transfer due to the movement/flow of molecules from areas of high concentration to ones of low concentration. Warmer molecules tend to rise, while colder molecules tend to sink. The heat in a house will rise from the vents in the floor to the upper levels of the structure and circulate in that manner, rising and falling with the movement of the molecules. This molecular movement helps to heat or cool a house and is often called convection current.

Radiation
The sun warms the earth through radiation or radiant energy. Radiation does not need any medium for the heat to travel; therefore, the heat from the sun can radiate to the earth across space.

Greenhouse Effect
The sun transfers heat into the earth's atmosphere through radiation traveling in waves. The atmosphere helps protect the earth from extreme exposure to the sun, while reflecting some of the waves continuously within the atmosphere, creating habitable temperatures. The rest of the waves are meant to dissipate out through the atmosphere and back into space. However, humans have created pollutants and released an overabundance of certain gasses into the earth's atmosphere, causing a layer of blockage. So, the waves that should be leaving the atmosphere continue to bounce back upon the earth repeatedly, thus contributing to global warming. This is a negative effect from the extra re-radiation of the sun's energy and causes planetary overheating.

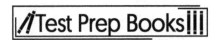

This additional warming is not something easily or quickly reversed. Because the rate of reflection within the atmosphere only multiplies the more a light wave is bounced around, it will take a concerted effort to undue past reflectance and stop future reflectance of the light waves in the earth's atmosphere. Once the re-reflectance occurs, it duplicates exponentially, along with the additional compounding of more waves. Each degree the atmospheric temperature increases has a profound effect on the delicate balance of our planet, including the melting of polar ice caps, the rise of tidal currents—which cause strong weather systems—and the depletion of specific ecosystems necessary to sustain certain species of animals or insects, to name a few.

Electrostatics

Electrostatics is the study of electric charges at rest. A charge comes from an atom having more or fewer electrons than protons. If an atom has more electrons than protons, it has a negative charge. If an atom has fewer electrons than protons, it has a positive charge. It is important to remember that opposite charges attract each other, while like charges repel each other. So, a negative attracts a positive, a negative repels a negative, and similarly, a positive repels a positive. Just as energy cannot be created or destroyed, neither can charge; charge is transferred. This transfer can be done through touch.

If a person wears socks and scuffs their feet across carpeting, they are transferring electrons to the carpeting through friction. If that person then goes to touch a light switch, they will receive a small shock, which is the electrons transferring from the switch to their hand. The person lost electrons to the carpet, which left them with a positive charge; therefore, the electrons from the switch attract to the person for the transfer. The shock is the electrons jumping from the switch to the person's finger.

Another method of charging an object is through induction. Induction is when a charged object is brought near, but not touched to, a neutral conducting object. The charged object will cause the electrons within the conductor to move. If the charged object is negative, the electrons will be induced away from the charged object and vice versa.

Yet another way to charge an object is through polarization. Polarization can be achieved by simply reconfiguring the electrons on an object. If a person were to rub a balloon on their hair, the balloon would then stick to a wall. This is because rubbing the balloon causes it to become negatively charged and when the balloon is held against a neutral wall, the negatively charged balloon repels all of the wall's electrons, causing a positively charged surface on the wall. This type of charge would be temporary, due to the massive size of the wall, and the charges would quickly redistribute.

Electric Current

Electrical current is the process by which electrons carry charge. In order to make the electrons move so that they can carry a charge, a change in voltage must be present. On a small scale, this is demonstrated through the electrons travelling from the light switch to a person's finger in the example where the person scuffed their socks on a carpet. The difference between the switch and the finger caused the electrons to move. On a larger and more sustained scale, this movement would need to be more controlled. This can be achieved through batteries/cells and generators. Batteries or cells have a chemical reaction that takes place inside, causing energy to be released and a charge to be able to move freely. Generators convert mechanical energy into electric energy.

If a wire is run from touching the end of a battery to the end of a light bulb, and then another is run from touching the base of the light bulb to the opposite end of the original battery, the light bulb will light up. This is due to a complete circuit being formed with the battery and the electrons being carried across the voltage drop (the two ends of the battery). The appearance of the light from the bulb is the visible heat caused by the friction of the electrons moving through the filament.

Electric Energy

Electric energy can be derived from a number of sources including coal, wind, sun, and nuclear reactions. Electricity has numerous applications, including being able to transfer into light, sound, heat, or magnetic forces.

Magnetic Forces

Magnetic forces can occur naturally in certain types of materials. If two straight rods are made from iron, they will naturally have a negative end (pole) and a positive end (pole). These charged poles react just like any charged item: opposite charges attract and like charges repel. They will attract each other when set up positive to negative, but if one rod is turned around, the two rods will repel each other due to the alignment of negative to negative and positive to positive.

These types of forces can also be created and amplified by using an electric current.

The relationship between magnetic forces and electrical forces can be explored by sending an electric current through a stretch of wire, which creates an electromagnetic force around the wire from the charge of the current, as long as the flow of electricity is sustained. This magnetic force can also attract and repel other items with magnetic properties. Depending upon the strength of the current in the wire, a smaller or larger magnetic force can be generated around this wire. As soon as the current is cut off, the magnetic force also stops.

Magnetic Energy

Magnetic energy can be harnessed, or controlled, from natural sources or from a generated source (a wire carrying electric current). Magnetic forces are used in many modern applications, including the creation of super-speed transportation. Super-magnets are used in rail systems and supply a cleaner form of energy than coal or gasoline.

Sound/Acoustic Energy

Just like light, sound travels in waves and both are forms of energy. The transmittance of a sound wave produced when plucking a guitar string sends vibrations at a specific frequency through the air, resulting in one's ear hearing a specific note or sets of notes that form a chord. If the same guitar is plugged into an electric amplifier, the strength of the wave is increased, producing what is perceived as a "louder" note. If a glass of water is set on the amplifier, the production of the sound wave can also be visually observed in the vibrations in the water. If the guitar were being plucked loudly enough and in great succession, the force created by the vibrations of the sound waves could even knock the glass off of the amplifier.

Waves can travel through different mediums. When they reach a different material (i.e., light traveling from air to water), they can bend around and through the new material. This is called refraction.

If one observes a straw in half a glass of water from above, the straw appears to be bent at the height of the water. The straw is still straight, but the observation of light passing from air to water (different materials) makes the straw seem as though it bends at the water line. This illusion occurs because the human eye can perceive the light travels differently through the two materials. The light might slow down in one material, or refract or reflect off of the material, causing differences in an object's appearance.

In another example, imagine a car driving straight along a paved road. If one or two of the tires hit the gravel along the side of the road, the entire car will pull in that direction, due to the tires in the gravel now traveling slower than the tires on the paved road. This is what happens when light travels from one medium to another: its path becomes warped, like the path of the car, rather than traveling in a straight line. This is why a straw appears to be bent when the light travels from water to air; the path is warped.

When waves encounter a barrier, like a closed door, parts of the wave may travel through tiny openings. Once a wave has moved through a narrow opening, the wave begins to spread out and may cause interference. This process is called diffraction.

Principles of Matter and Atomic Structure

Protons, Neutrons, and Electrons
The structure of an atom has two major components: the atomic nucleus and the atomic shells (also known as orbitals). The nucleus is found in the center of an atom. The three major subatomic particles are protons, neutrons, and electrons and are found in the atomic nucleus and shells.

Protons are found in the atomic nucleus and are positively charged particles. The addition or removal of protons from an atom's nucleus creates an entirely different element. *Neutrons* are also found in the atomic nucleus and are neutral particles, meaning they have no net electrical charge. The addition or removal of neutrons from an atom's nucleus does not create a different element but instead creates a lighter or heavier form of that element called an isotope. *Electrons* are found orbiting in the atomic shells around the nucleus and are negatively charged particles. A proton or a neutron has nearly 2,000 times the mass of an electron.

Electrons and Chemical Bonds
Electrons orbit the nucleus in atomic shells, or electron clouds, each of which can accommodate a certain number of electrons. For example, the first atomic shell can accommodate two electrons, the second atomic shell can hold a maximum of eight electrons, and the third atomic shell can house a maximum of eighteen electrons. The negatively charged electrons orbiting the nucleus are attracted to the positively charged protons in the nucleus via electromagnetic force. The attraction of opposite electrical charges gives rise to chemical bonds, which refers to the ways atoms are attached to each other.

Chemical bonding typically results in the formation of a new substance, called a compound. Only the electrons in the outermost atomic shell are able to form chemical bonds. These electrons are known as valence electrons, and they are what determines the chemical properties of an atom.

Chemical Bonds Between Atoms
Chemical bonds refer to the manner in which atoms are attached to one another. Atoms may be held together with three fundamental types of chemical bonds—ionic, covalent, or hydrogen.

Ionic Bonding
In an *ionic bond*, an atom loses one or more electrons to another who gains them. The atoms do this so that they can achieve a full outermost shell of electrons, which is the configuration that is most stable, and these are typically the strongest types of bonds. Ionic bonds typically occur in metals and nonmetals to form salts like $MgCl_2$ and BF_3. When an atom has a different number of electrons than its number of protons, it is termed an *ion*. A positively charged ion is referred to as a *cation*, while a negatively charged ion is referred to as an *anion*. For example, a sodium cation (Na^{+1}) can form an ionic bond with a

chlorine anion (Cl^{-1}), which forms sodium chloride ($NaCl$), which is also known as table salt. The majority of the molecules formed with ionic bonds such as this have a crystalline structure.

Covalent Bonding

In a *covalent bond*, two or more atoms share one or more electrons. Covalent bonds are the most plentiful type of bond making up the human body. They are typically found in molecules containing carbon. Only six elements typically form covalent bonds: carbon (C), nitrogen (N), phosphorus (P), oxygen (O), sulfur (S), and hydrogen (H). Covalent bonds are a crucial source of energy for living organisms. Covalent bonds are comparable in strength to ionic bonds, but stronger than hydrogen bonds, and are typically used to bind the basic macromolecules—carbohydrates, lipids, nucleic acids, and proteins—together.

Hydrogen Bonding

Hydrogen bonds are temporary and weak. They typically occur between two partial, opposite electrical charges. For example, hydrogen bonds form when a hydrogen (H) atom is in the vicinity of nitrogen (N), fluorine (F), or oxygen (O) atoms. These partial electrical charges are called *dipoles*, and are caused by the unequal sharing of electrons between covalent bonds. Water is the most prevalent molecule that forms hydrogen bonds.

Hydrogen bonds contribute to the adhesiveness and cohesiveness properties of molecules like water. Adhesiveness confers glue-like properties to molecules, which ensure they stick or connect more easily with other molecules—much like wetting a suction cup before sticking it to a surface. Cohesiveness refers to a molecule's ability to form hydrogen bonds with itself. For example, the cohesiveness of water is the reason why it has a high boiling point, which is a physical property.

Properties of Water

Water is the most abundant molecule on Earth. It is a compound composed of hydrogen and oxygen with the chemical formula H_2O. Water is also *polar*, which means it is negatively charged at one end and positively charged at the other end. The oxygen is more *electronegative* than the hydrogens, meaning that its protons pull in more of the electrons than do the hydrogens. This leaves the oxygen with a partial negative charge and the hydrogens with a partial positive charge.

Water is *amphoteric* and *self-ionizable* in that it tends to dissociate, or split, into hydrogen ions (H^+) and hydroxyl ions (OH^-) randomly. No other substance on Earth may be found naturally in all three states of matter—liquid, solid, and gas. Water is also unique due to its liquid state being more dense than its solid state (ice), which is why ice floats in liquid water. Water is colorless, odorless, and tasteless. It freezes at 32 °F and boils at 212 °F. Pure water has a pH of 7, which makes it neutral. Water is considered the universal solvent because it can dissolve many substances.

Many of the properties of water are due to its polarity and hydrogen bonding. Water has the following properties:

- Cohesiveness: Refers to the force of attraction between molecules of identical substances. Cohesiveness is mainly the result of hydrogen bonding in water.

- Adhesiveness: Refers to the force of attraction between molecules of different substances, like water and glass. Adhesiveness is mainly the result of hydrogen bonding in water.

- High specific heat: Refers to the amount of heat required to raise the temperature of water by one degree Celsius. Water's high specific heat is a consequence of hydrogen bonding. As a result, water can store a great deal of heat energy.

- High surface tension: The cohesiveness of water molecules is responsible for its high surface tension. The strong cohesion of its molecules makes water sticky and elastic.

- High heat of vaporization: Refers to the energy required to transform a given amount of water from a liquid to a gas at a given pressure. Water's high heat of vaporization is due to hydrogen bonding.

Measurable Properties of Atoms

All matter is made of atoms. Atoms are the most basic component of an element that still retains its properties. All of the elements known to man are catalogued in the periodic table, a chart of elements arranged by increasing atomic number. The atomic number refers to the number of protons in an atom's nucleus. It can be found either in the upper left-hand corner of the box or directly above an element's chemical symbol on the periodic table. For example, the atomic number for hydrogen (H) is 1. The term "atomic mass" refers to the sum of protons and neutrons in an atom's nucleus. The atomic mass can be found beneath an element's abbreviation on the periodic table. For example, the average atomic mass of hydrogen (H) is 1.008. Because protons have a positive charge and neutrons have a neutral charge, an atom's nucleus typically has a positive electrical charge. Electrons orbiting the nucleus have a negative charge. As a result, elements with equal numbers of protons and electrons have no net charge.

Atoms that have gained or lost electrons wind up having a net electrical charge and are termed ions. The following are the primary ions pertinent to human health:

Bicarbonate	HCO_3^-	A major buffer in blood. The lungs and kidneys regulate its concentration.
Chloride	Cl^-	Important in stomach acid and usually ingested as the salts sodium chloride (NaCl) and potassium chloride (KCl).
Calcium	Ca^{2+}	Important for muscle contraction and bone construction.
Copper	Cu	Specialized chemical reactions in the cell.
Iodine	I	Specialized chemical reactions in the cell.
Iron	Fe	Important in hemoglobin for transport of oxygen as well as part of the electron transport chain.
Magnesium	Mg^{2+}	Important in chlorophyll and animal energy production as well as a constituent of bone.
Phosphate	PO_4^{3-}	A minor intracellular pH buffer that is regulated by the kidneys and is an important factor in bone.
Potassium	K^+	The most plentiful mineral inside of cells and is important for nerve and muscle function.
Sodium	Na^+	The most common mineral outside of cells and is important for water and osmolarity regulation as well as nerve and muscle function.
Sulfate	SO_4^{2-}	A minor pH buffer for body fluids.

For the majority of light atoms, the number of protons is similar to the number of neutrons. An isotope is a variation of an element having the same number of protons, but a different number of neutrons. For example, all isotopes of carbon (C) have six protons. However, C-12 has six neutrons, C-13 has seven neutrons, and C-14 has eight neutrons.

Some isotopes are radioactive and result in nuclear decay. Not all radioactive isotopes are harmful, and some are even useful to scientists and physicians. For example, C-14 is radioactive and can be used in the process of radiocarbon dating, which can be used to determine the age of organic remains. A radioactive isotope of gold (Au-198) can be utilized to treat ovarian, prostate, and brain cancer.

Periodicity and the Periodic Table

Periodicity

Periodicity refers to the repeating patterns, or trends, in the properties of elements. The atomic number and atomic structure are the key determinants of the properties of elements. During the mid-1800s, the Russian chemist Dmitri Mendeleev utilized the principal of periodicity to arrange elements in a manner similar to the modern periodic table. Mendeleev's periodic table was arranged in rows according to increasing atomic mass and in columns according to similar chemical behavior. The modern periodic table is arranged in order of increasing atomic number, which is defined as the number of protons in an atom's nucleus. Elements near each other are more similar than elements that are distant on the periodic table.

Periodic Table

The periodic table catalogues all of the elements known to man, currently 118. It is one of the most important references in the science of chemistry. Information that can be gathered from the periodic table includes the element's atomic number, atomic mass, and chemical symbol. The first periodic table was rendered by Mendeleev in the mid-1800s and was ordered according to increasing atomic mass. The modern periodic table is arranged in order of increasing atomic number. It is also arranged in horizontal rows known as periods, and vertical columns known as families, or groups. The periodic table contains seven periods and eighteen families. Elements in the periodic table can also be classified into three major groups: metals, metalloids, and nonmetals. Metals are concentrated on the left side of the periodic table, while nonmetals are found on the right side. Metalloids occupy the area between the metals and nonmetals.

Due to the fact the periodic table is ordered by increasing atomic number, the electron configurations of the elements show periodicity. As the atomic number increases, electrons gradually fill the shells of an atom. In general, the start of a new period corresponds to the first time an electron inhabits a new shell. Other trends in the properties of elements in the periodic table are:

Atomic radius: One-half the distance between the nuclei of atoms of the same element.

Electronegativity: A measurement of the tendency of an atom to form a chemical bond.

Ionization energy: The amount of energy needed to remove an electron from a gas or ion.

Electron affinity: The ability of an atom to accept an electron.

Moving left to right in a period, trends reveal decreasing atomic radius, increasing electronegativity, increasing ionization energy, and increasing electron affinity. Moving from top to bottom in a group, trends reveal increasing atomic radius, decreasing electronegativity, and decreasing ionization energy. The trend of decreasing electron affinity is only seen in Group 1 of the periodic table.

States of Matter—Liquids, Gases, and Solids

There are three fundamental states of matter—liquid, gas, and solid. The molecules in a liquid are not in an orderly arrangement and can move past one another. Weak intermolecular forces contribute to a liquid having an indefinite shape, but definite volume. Lastly, a liquid conforms to the shape of its container, is not easily compressible, and flows quite easily.

The molecules in a gas have a large amount of space between them. A gas will diffuse indefinitely if unconfined, while it will assume the shape and volume of its container if enclosed. In other words, a gas has no definite shape or volume. Lastly, a gas is compressible and flows quite easily.

The molecules in a solid are closely packed together, which restricts their movement. Very strong intermolecular forces contribute to a solid having a definite shape and volume. Furthermore, a solid is not easily compressible and does not flow easily.

Vaporization, Evaporation, and Condensation

States of matter are able to undergo phase transitions. Vaporization refers to the transformation of a solid or liquid into a gas. There are two types of vaporization—evaporation and boiling. Evaporation is a surface phenomenon and involves the conversion of a liquid into a gas below the boiling temperature at a given pressure. Evaporation is also an important component of the water cycle. Boiling occurs below the surface and involves the conversion of liquid into a gas at or above the boiling temperature. Condensation represents the conversion of a gas into a liquid. It is the reverse of evaporation. Condensation is also most often synonymous with the water cycle. It is a crucial component of distillation.

Chemical Reactions

Types of Chemical Reactions

Chemical reactions are characterized by a chemical change in which the starting substances, or reactants, differ from the substances formed, or products. Chemical reactions may involve a change in color, the production of gas, the formation of a precipitate, or changes in heat content.

The following are the basic types of chemical reactions:

Reaction Type	Definition	Example
Decomposition	A compound is broken down into two or more smaller elements or compounds.	$2H_2O \rightarrow 2H_2 + O_2$
Synthesis	Two or more elements or compounds are joined together.	$2H_2 + O_2 \rightarrow 2H_2O$
Single Displacement	A single element or ion takes the place of another in a compound. Also known as a substitution reaction.	$Zn + 2HCl \rightarrow ZnCl_2 + H_2$
Double Displacement	Two elements or ions exchange a single atom each to form two different compounds, resulting in different combinations of cations and anions in the final compounds. Also known as a metathesis reaction.	$H_2SO_4 + 2NaOH \rightarrow Na_2So_4 + 2H_2O$
Oxidation-Reduction	Elements undergo a change in oxidation number. Also known as a redox reaction.	$2S_2O_3^{2-}(aq) + I_2(aq) \rightarrow S_4O_6^{2-}(aq) + 2I^-$ (aq)
Acid-Base	Involves a reaction between an acid and a base, which usually produces a salt and water	$HBr + NaOH \rightarrow NaBr + H_2O$
Combustion	A hydrocarbon (a compound composed of only hydrogen and carbon) reacts with oxygen to form carbon dioxide and water.	$CH_4 + 2O_2 \rightarrow CO_2 + 2H_2O$

Balancing Chemical Reactions

Chemical reactions are expressed using chemical equations. Chemical equations must be balanced with equivalent numbers of atoms for each type of element on each side of the equation. Antoine Lavoisier, a French chemist, was the first to propose the Law of Conservation of Mass for the purpose of balancing a chemical equation. The law states, "Matter is neither created nor destroyed during a chemical reaction."

The reactants are located on the left side of the arrow, while the products are located on the right side of the arrow. Coefficients are the numbers in front of the chemical formulas. Subscripts are the numbers to the lower right of chemical symbols in a formula. To tally atoms, one should multiply the formula's coefficient by the subscript of each chemical symbol. For example, the chemical equation $2H_2 + O_2 \rightarrow 2H_2O$ is balanced. For H, the coefficient of 2 multiplied by the subscript 2 = 4 hydrogen atoms. For O, the coefficient of 1 multiplied by the subscript 2 = 2 oxygen atoms. Coefficients and subscripts of 1 are

understood and never written. When known, the form of the substance is noted with (g)=gas, (s)=solid, (l)=liquid, or (aq)=aqueous.

Catalysts

Catalysts are substances that accelerate the speed of a chemical reaction. A catalyst remains unchanged throughout the course of a chemical reaction. In most cases, only small amounts of a catalyst are needed. Catalysts increase the rate of a chemical reaction by providing an alternate path requiring less activation energy. Activation energy refers to the amount of energy required for the initiation of a chemical reaction.

Catalysts can be homogeneous or heterogeneous. Catalysts in the same phase of matter as its reactants are homogeneous, while catalysts in a different phase than reactants are heterogeneous. It is important to remember catalysts are selective. They don't accelerate the speed of all chemical reactions, but catalysts do accelerate specific chemical reactions.

Enzymes

Enzymes are a class of catalysts instrumental in biochemical reactions, and in most, if not all, examples are proteins. Like all catalysts, enzymes increase the rate of a chemical reaction by providing an alternate path requiring less activation energy. Enzymes catalyze thousands of chemical reactions in the human body. Enzymes are proteins and possess an active site, which is the part of the molecule that binds the reacting molecule, or substrate. The "lock and key" analogy is used to describe the substrate key fitting precisely into the active site of the enzyme lock to form an enzyme-substrate complex.

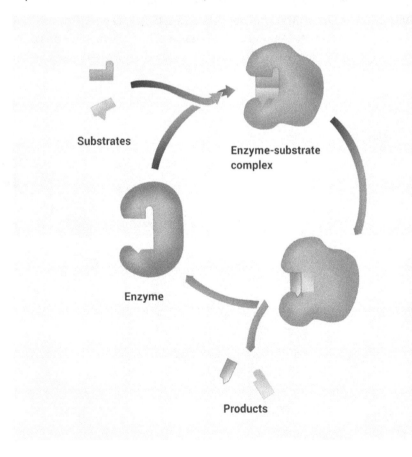

255

Many enzymes work in tandem with cofactors or coenzymes to catalyze chemical reactions. *Cofactors* can be either inorganic (not containing carbon) or organic (containing carbon). Organic cofactors can be either coenzymes or prosthetic groups tightly bound to an enzyme. *Coenzymes* transport chemical groups from one enzyme to another. Within a cell, coenzymes are continuously regenerating and their concentrations are held at a steady state.

Several factors including temperature, pH, and concentrations of the enzyme and substrate can affect the catalytic activity of an enzyme. For humans, the optimal temperature for peak enzyme activity is approximately body temperature at 98.6 °F, while the optimal pH for peak enzyme activity is approximately 7 to 8. Increasing the concentrations of either the enzyme or substrate will also increase the rate of reaction, up to a certain point.

The activity of enzymes can be regulated. One common type of enzyme regulation is termed *feedback inhibition*, which involves the product of the pathway inhibiting the catalytic activity of the enzyme involved in its manufacture.

pH, Acids, and Bases

pH refers to the power or potential of hydrogen atoms and is used as a scale for a substance's acidity. In chemistry, pH represents the hydrogen ion concentration (written as $[H^+]$) in an aqueous, or watery, solution. The hydrogen ion concentration, $[H^+]$, is measured in moles of H^+ per liter of solution.

The pH scale is a logarithmic scale used to quantify how acidic or basic a substance is. pH is the negative logarithm of the hydrogen ion concentration: $pH = -\log [H^+]$. A one-unit change in pH correlates with a ten-fold change in hydrogen ion concentration. The pH scale typically ranges from zero to 14, although it is possible to have pHs outside of this range. Pure water has a pH of 7, which is considered neutral. pH values less than 7 are considered acidic, while pH values greater than 7 are considered basic, or alkaline.

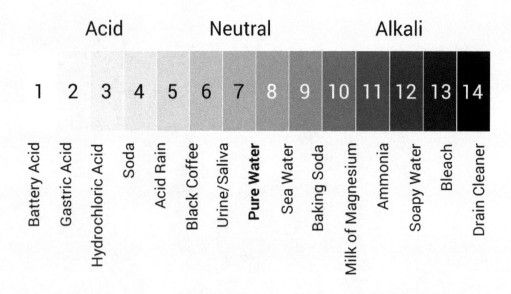

Generally speaking, an acid is a substance capable of donating hydrogen ions, while a base is a substance capable of accepting hydrogen ions. A buffer is a molecule that can act as either a hydrogen ion donor or acceptor. Buffers are crucial in the blood and body fluids, and prevent the body's pH from fluctuating into dangerous territory. pH can be measured using a pH meter, test paper, or indicator sticks.

Earth Science

Properties of Earth Materials

Geology is the study of the nature and composition of the rocks and materials that make up the Earth, how they were formed, and the physical and chemical processes that have changed Earth over time.

Earth can be imagined as a giant construction of billions of Lego blocks, and that these blocks represent different minerals. A *mineral* is a naturally occurring inorganic solid composed of certain chemical elements (or atoms) in a defined crystalline structure. When minerals are aggregated together with other minerals, organic compounds (carbon-containing remains of decomposed plant or animal matter), and/or mineraloids (minerals that lack a defined crystalline structure), rocks are formed. Rock types are classified based on their mechanism of formation and the materials of their compositions.

The three fundamental classifications of rocks include:

- Sedimentary
- Igneous
- Metamorphic

Sedimentary rocks form at the Earth's surface (on land and in bodies of water) through deposition and cementation of fragments of other rocks, organic matter, and minerals. These materials, called sediment, are deposited and accumulate in layers called strata, which get pressed into a solid over time when more sediment settles on top. Sedimentary rocks are further classified as either clastic/detrital, biochemical, chemical, or other. *Clastic* or *detrital rocks* are composed of other inorganic rocks or organic particles, respectively. *Biochemical rocks* have an organic component (like coal, which is composed of decayed plant matter). *Chemical rocks* form from a solution containing dissolved materials that became supersaturated, and minerals precipitate out of solution. Halite, or rock salt, is an example of a chemical sedimentary rock. Sedimentary rocks that do not fit into these types are categorized as "other." These rocks are formed from fragments formed by asteroid or comet impacts or from fragments of volcanic lava.

Igneous rocks are composed of molten material beneath the Earth's surface called *magma* and are classified based on where the magma cooled and solidified; they can be intrusive/plutonic, extrusive/volcanic, or hypabyssal. *Intrusive* or *plutonic rocks,* such as granite, form when magma cools slowly within or beneath the Earth's surface. Because they solidify slowly, these rocks tend to have a coarse grain, larger crystalline structure of their mineral constituents, and rough appearance. By contrast, *unbold the word "or"* form from rapid cooling as magma escapes the Earth's surface as lava and have a smooth or fine-grained appearance, with tiny crystals or ones that are too small to see. A common example of an extrusive igneous rock is glassy obsidian. *Hypabyssal rocks* are formed at levels between intrusive and extrusive (just below the surface); they aren't nearly as common.

Metamorphic rocks form from the transformation of other rocks via a process called metamorphism. This transformation happens when existing rocks—sedimentary, igneous, or other metamorphic rocks—are subjected to significant heat and pressure, which causes physical and/or chemical changes. Based on their appearance, metamorphic rocks are classified as either foliated or non-foliated. *Foliated rocks* are layered or folded, which means they form from compression in one direction and result in visible layers or banding within the rock. Examples include gneiss or slate. *Non-foliated rocks,* such as marble, receive equal pressure from all directions and thus have a homogenous appearance.

It should be noted that classification is not always completely clear and some rocks don't quite fit the criteria for one of these three categories, so they are sometimes lumped together in a category called "other rocks." A classic example is a fossil. A *fossil* is a rock formed from the remains or impression of dead plants or animals, but it doesn't fit into any biochemical class of rocks because fossils themselves are wholly composed of organic material and only formed under strict conditions, although they are commonly found within sedimentary rock.

Earth's Systems, Processes, Geologic Structures, and Time

Plate Tectonics

The theory of plate tectonics states that the Earth's superficial layer (the crust and upper mantle, together called the *lithosphere*) is a collection of variably-sized plates that move and interact with each other on top of the more molten *asthenosphere* in the mantle below. It is estimated that there are between 9 and 15 major plates and up to 40 minor plates. Some of these plates are *oceanic,* which means they contain an ocean basin, while others are *continental* and carry a landmass. The line formed by the meeting of two plates is called a *fault*. A well-known example is the San Andreas Fault, where the Pacific and North American plates meet. These faults are classified as *convergent* (plates colliding into each other) or *divergent* (plates moving away from each other). *Transform boundaries* occur when two plates slide past each other in opposite directions horizontally.

Major plates of the lithosphere

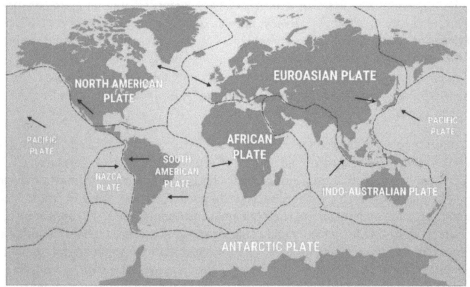

The larger plates are:

- North American plate
- Eurasian plate
- Pacific plate
- Australian plate
- Antarctic plate
- South American plate
- African plate
- Indian plate

For the most part, these plates contain the countries/continents for which they are named. Scientists theorize that these plates were once part of a supercontinent, called *Pangea,* that existed over 175 million years ago. Evidence for Pangea's existence comes from geological findings, corresponding shapes of adjacent but currently independent landmasses, and fossils of identical species found in areas that used to be joined but have since separated.

Planet Earth is not a perfect sphere; it's actually slightly ellipsoid or oblong (like a football), and it's divided into three main layers from surface to center: the crust, mantle, and core. The crust and core are solid, while the mantle possesses a more fluid quality. In plate tectonics, the plates are considered to be the Earth's *lithosphere* (the crust and uppermost solid portion of mantle). One can visualize the plates as flat planks of wood covering the surface of a pool bobbing up and down and bumping, sliding, or moving away from each other. These different plate interactions create Earth's landscape.

At convergent boundaries, the collision of two plates can give rise to mountain ranges. Another possibility is that one plate slides atop another and pushes it down, creating magma and volcanoes, a process called *subduction.* This frequently happens when an oceanic and continental plate converge, and the denser oceanic plate sinks below under the continental plate. Volcanic activity also occurs at divergent boundaries—when the two plates separate—and magma escapes to the surface and solidifies. This is how new land is formed. Faults are not the only places a volcano may appear; plates may contain areas called *hot spots*, and when these hot spots reach the surface, a volcano forms. Earthquakes are another consequence of plate tectonics, occurring as a result of sliding or colliding plates. Friction energy created between two plates can become an earthquake.

Weather, Atmosphere, and the Water Cycle

The study of the Earth's weather, atmosphere, and water cycle is called *meteorology*. The weather experienced in an area at a given time is a product of Earth's atmosphere above that location. The different atmospheric layers are defined by their temperature but are thought of by their distance above sea level (listed from lowest to highest):

- Troposphere: Sea level to 11 miles above sea level
- Stratosphere: 11 miles to 31 miles above sea level
- Mesosphere: 31 miles to 50 miles above sea level
- Ionosphere: 50 miles to 400 miles above sea level
- Exosphere: 400 miles to 800 miles above sea level

Above the exosphere is outer space. Together, the ionosphere and exosphere are considered the *thermosphere*. The ozone layer lies within the stratosphere, while the troposphere is where the conditions that lead to the observable weather on Earth originates. The manner in which the temperature changes is different in each of these layers. In the troposphere, it gets colder as the layer gets further from sea level, while the stratosphere gets warmer, the mesosphere gets colder, and the thermosphere gets warmer in this same direction.

The *atmosphere* is a layer of gas particles floating in space. The atmospheric levels are created by gravity and its pull on those particles. The Earth's atmosphere is mostly comprised of nitrogen and oxygen (78% and 21%, respectively) along with significantly lower amounts of other gases including 1% argon and 0.039% carbon dioxide. Gas particles have mass, and pressure is higher at the bottom (at the surface of Earth) because of all that mass in the layers above. Therefore, the farther away one gets from Earth's surface, the air gets thinner, as there are fewer air molecules compressing lower particles together. That's why breathing becomes more difficult when a person climbs to higher altitudes. To visualize this, one can imagine that when a person takes a breath, he or she is breathing in a cup of air. At higher altitudes, that cup is still the same size; it just holds fewer particles of gas.

Weather is a state of the atmosphere at a given place and time based on conditions such as air pressure, temperature, and moisture. Weather includes conditions like clouds, storms, temperature, tornadoes, hurricanes, and blizzards. In a given location on Earth, the weather experienced varies day-to-day, based on atmospheric conditions. The average weather for a particular area over a long period (usually over 30 years) is defined as that area's *climate*. The main force driving weather is the atmospheric variations between different areas on Earth. The major circles of latitude experience these differences because of the Sun. When there's a large difference in temperature between two areas, a jet stream forms. Pressure differences occur when there are temperature differences caused by surface variations (like mountains and valleys).

The *water cycle* is another factor that drives weather. It is the movement of water above, within, and on the surface of the Earth. During any phase of the cycle, water can exist in any of its three phases: liquid, ice, and vapor.

The processes that drive the cycle are:

- Precipitation: Rain, snow, hail, and sleet

- Canopy interception: Precipitation that falls on trees and doesn't hit the ground, eventually evaporating back into the atmosphere

- Snowmelt: Runoff from melting snow

- Runoff: Water moving across land that either seeps into the ground, evaporates, gets stored as lakes, or gets extracted by living organisms. It also includes surface and channel runoff.

- Infiltration: Water that moves from ground surface into the ground

- Subsurface flow: Moving water underground

- Transpiration: Release of vapor into the atmosphere from plants and soil

- Percolation: Water that flows down through soil and rocks

- Evaporation: Transformation from liquid water to gaseous vapor driven primarily by solar radiation

- Sublimation: Transformation of ice to gaseous vapor, never becoming a liquid

- Deposition: Transformation of vapor to solid ice

- Condensation: Transformation of vapor to liquid water

Water in the atmosphere exists as clouds, which are visible masses made of water droplets, tiny crystals of ice, dust, and various chemicals. The study of clouds, called *nephology*, is a subspecialty of meteorology. There are many types of clouds, and they can be classified based on the altitude at which they occur. It's important to note that clouds primarily occur in the troposphere. The classes of clouds are:

- High-Clouds: Occurring between 5,000 and 13,000 meters above sea level
- Cirrus: Thin and wispy "mare's tail" appearance
- Cirrocumulus: Rows of small puffs
- Cirrostratus: Thin sheets that cover the sky
- Middle clouds: Occurring between 2,000 and 7,000 meters above sea level
- Altocumulus: Gray and white, made up of water droplets
- Altostratus: Grayish or bluish gray clouds
- Low clouds: Occurring below 2,000 meters above sea level
- Stratus: Gray and cover the sky
- Stratocumulus: Gray and lumpy low-lying clouds
- Nimbostratus: Dark gray with uneven bases that occur with rain or snow

Here are examples of some different types of clouds:

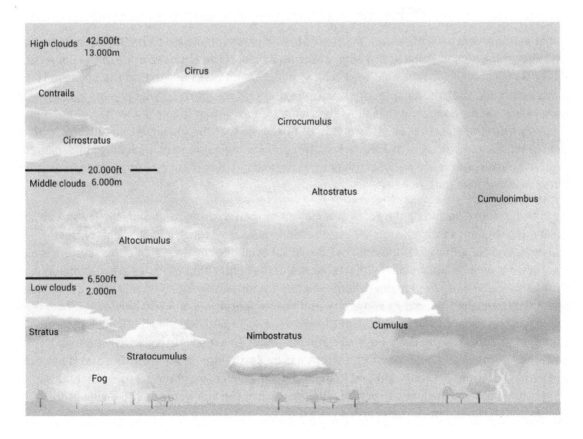

The vast majority of water on Earth is found in its oceans, which contain salt water, unlike lakes and other bodies, which contain freshwater. Salt water has a higher density than freshwater due to the dissolved salt. The total mass of all oceanic water is estimated to be 1.4×10^{24} grams, covering approximately 70% of the Earth's surface (361,254,000 km^2). The deepest part of the ocean is an area called the Challenger Deep in the Mariana Trench where the ocean floor is nearly 11,000 meters below sea level. The major oceans are the Pacific, Atlantic, and Indian Oceans.

The Earth is also divided into different *biomes* or *ecosystems*. The simplest classification dictates that there are five different ecosystems:

- Aquatic: Contains marine and freshwater systems and found at all latitudes
- Desert: Areas of low precipitation and usually windy; can be found at nearly all latitudes
- Forest: Tropical, temperate, or boreal; types of trees depend on latitude
- Grassland: Prairies, savannahs, steppes; can exist in tropical, temperate, and cold climates
- Tundra: Found far from the equator toward either pole; receives the least amount of total annual sunlight

Earth's Movements and Position in the Solar System

Astronomy is the study of celestial bodies, or objects in space, and how they interact with each other. It is one of the oldest sciences; people have always looked up at the night sky with curiosity and amazement. Earth is a celestial body; others include the Sun, moon, other planets, black holes, satellites, asteroids, meteors, comets, stars, and galaxies. Information about celestial bodies and processes is obtained from observation via visible light and radiation throughout the universe. These observations are used in a field called theoretical astronomy, where scientists create theoretical models to explain or predict celestial processes or events.

From what astronomers have observed, the size of the universe is believed to be 91 billion light years. That means that if a person stood at the edge of the on a light at the opposite edge of the universe, it would take 91 billion years for the light to reach the person. And it's believed that the universe is constantly enlarging. The prevailing theory is that the universe was born 13 billion years ago, after an explosion called the *Big Bang*, and that the debris from that explosion (planets and stars) has been floating away from the epicenter ever since. It is also believed that the universe is mainly made up of dark energy (about 73%) along with 23% dark matter and 4% regular matter, including stars, planets, and living organisms. The space between planets, stars, and galaxies is *interstellar space*, which is filled with *interstellar medium* (gas and dust).

Objects in the universe, including the Earth, are clumps of matter. Consistent with the laws of physics and gravity, these clumps form clusters. These clusters are solar systems, galaxies, galaxy clusters, superclusters, and something astronomers call the Great Wall of Galaxies. Earth's solar system is a planetary system on an ecliptic plane with a large sun in the center that provides gravitational pull. The Sun is primarily made of hydrogen and helium—metals comprise only 2% of its mass. It's 1.3 million kilometers wide, weighs 1.989×10^{30} kilograms, and has temperatures of 5,800 Kelvin (9980 degrees Fahrenheit) on the surface (also called the photosphere) and 15,600,000 Kelvin (28 million degrees Fahrenheit) in its core. It is the Sun's huge size and mass that give it so much gravity—enough to compress the hydrogen and helium (which exist as gas on Earth) into liquid form. The Sun is basically a series of giant explosions creating light and heat, and it's that huge gravitational force that pulls in those explosions and maintains the Sun's structure. The pressure at the Sun's core is 250 billion atmospheres, making it over 150 times denser than Earth's water!

Astronomers used to think there were nine total planets in the Solar System, but Pluto is now considered a dwarf planet, along with Ceres, Haumea, Makemake, and Eris. The eight planets can be divided into four inner (terrestrial) and four outer (Jovian) planets. Generally, the terrestrial planets are small, and the Jovian planets are larger and less dense with rings and moons.

Listed in order of closest to the Sun to furthest, the planets are:

- Mercury: The smallest planet in the system and the one that is closest to the Sun. Because it's so close, it only takes about 88 days for Mercury to completely orbit the Sun It has a large iron core and the surface has craters like Earth's moon. There's no atmosphere, and it doesn't have any orbiting moons/satellites. From Earth, Mercury looks bright.

- Venus: The second planet is bright and has about the same size, composition, and gravity as Earth. It orbits the Sun every 225 days Its atmosphere creates clouds of sulfuric acid, and there's even thunder and lightning.

- Earth: The third planet orbits the Sun every year (about 365 days). Scientists believe it's the only planet in this system that's capable of supporting life.

- Mars: The Red Planet looks red because there's iron oxide on the surface. It also takes around 687 days to complete its orbit around the Sun. Interestingly, a day on Mars (one rotation about its axis), is very similar to the 24-hour day on Earth. Mars has a thin atmosphere as well as the largest mountain, canyon, and crater that astronomers have ever been able to see. Volcanoes, valleys, deserts, and polar ice caps like those on Earth have also been seen on its surface.

- Jupiter: The largest planet in the solar system is comprised mainly of hydrogen and helium (helium makes up 25% of its mass). The atmosphere on Jupiter has band-like clouds made of ammonia crystals that create tremendous storms and turbulence on the surface. Winds blow at around 100 meters per second, or over 220 miles per hour!

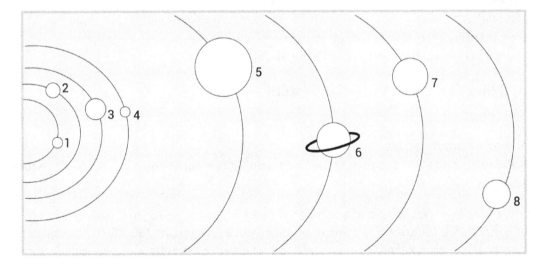

- Saturn: The second-largest planet is comprised mainly of hydrogen and helium along with other trace elements. The core is believed to be rock and ice. Saturn has a layer of metallic hydrogen. Winds are even stronger than those on Jupiter, reaching up to 1,100 miles per hour. Saturn has 61 moons, but the planet is most famous for its beautiful rings. There's no definitive explanation for how the rings formed, but two popular theories say they could be remnants from when Saturn itself formed, or they were moons that were destroyed in the past.

- Uranus: This planet has the coldest atmosphere of any in the Solar System, with a temperature that reaches -224.2 degrees Celsius (-371.56 °F). It's also mainly made of hydrogen and helium, but it has water, ammonia, methane, and even some hydrocarbons (the material humans are made of). A solid surface has yet to be observed through the thick layer of gas covering the planet. Uranus has 27 known moons.

- Neptune: The furthest planet is the third-largest by mass, and the second-coldest. It has 12 orbiting moons, an atmosphere like Uranus', a Great Dark Spot, and the strongest recorded winds of any planet in the system (reaching speeds of 2,100 kilometers per hour).

The nearest star to Earth's solar system is Proxima Centauri, which is about 270,000 Astronomical Units away. The closest object to Earth is the Moon, which is about 384,401 kilometers or 238,910 miles away. The Moon has two phases as it revolves around Earth—the waxing phase and the waning phase—which each last about two weeks. During the *waxing phase*, the Moon goes from new (black) to full moon; then it *wanes*, going from full to black. The Moon appears white because it's illuminated by the Sun. The edge of the shadow on the Moon is called the *lunar terminator*. The phases of waxing and waning in the Northern Hemisphere are:

- Waxing: The right side of the Moon is illuminated
- New moon (dark): The Moon rises and sets with the Sun
- Crescent: Tiny sliver illuminated on the right side
- First quarter: The right half is illuminated; its phase is due south on the meridian
- Gibbous: More than half is illuminated on the right side
- Full moon: Rises at sunset and sets at sunrise
- Waning: The left side of the Moon is illuminated
- Gibbous: More than half is illuminated on the left side
- Last quarter: Half-illuminated on the left side; rises at midnight and sets at noon
- Crescent: Tiny sliver illuminated on the left
- New moon (dark): Rises and sets with the Sun again

The Sun, Other Stars, and the Solar System

The Sun
Features and Characteristics
The Sun is the largest object in the solar system and is also its center. It contains 99.8% of the total mass of the solar system.

The majority of the Sun is composed of a number of gases including hydrogen (70%), helium (28%), and some light metals (2%). The core of the Sun is extreme with a pressure of 250 billion atmospheres and a temperature of 15.6 million Kelvin (K). Compared to Earth, the Sun is 109 times physically larger.

It has the following four distinct regions, or layers:

- Photosphere: The visible surface of the sun. The *photosphere* does not have a solid surface as it is composed of gases. It is approximately 400 km thick. Temperatures on the surface of the Sun average 5,800 K. It can contain sunspots, which are relatively cool regions averaging a

temperature of 3,800 K. As a result, *sunspots* are seen as dark areas in the photosphere. Sunspots are thought to be large areas of magnetic storms.

- Chromosphere: Located just above the photosphere, the *chromosphere* is 2,000 kilometers deep. Although it is farther from the center of the Sun, the chromosphere attains greater temperatures than the photosphere. The temperatures of the chromosphere vary between 4,500 K and 8,000 K. It has a reddish color and can only be seen during a total solar eclipse.

- Corona: Located above the chromosphere, the *corona* is a Latin term meaning *crown.* Temperatures in the corona are in excess of 1 million K. It stretches millions of miles into space. The solar corona can only be seen during a total solar eclipse as a white crown encircling the Sun.

- Magnetosphere: The *magnetosphere*, or *heliosphere*, is a magnetic sphere caused by solar winds of the Sun. It is often described as the bubble containing our solar system. The magnetosphere shields and protects the Earth from harmful cosmic radiation and extends beyond Pluto.

Energy

The Sun is the primary external source of energy on Earth and produces this energy through nuclear fusion, where two or more atomic nuclei fuse to make one larger nucleus. This process releases the energy to fuel the Sun and the stars. Theoretically, the amount of energy that nuclear fusion can possibly produce is limitless. In the core of the Sun, nuclear fusion converts hydrogen to helium.

Science Process

Interpreting and Applying

Interpreting Observed Data or Information

An important skill for scientific comprehension is the ability to interpret observed data and information. Scientific studies and research articles can be challenging to understand, but by familiarizing oneself with the format, language, and presentation of such information, fluency regarding the research process, results, and significance of such results can be improved. Most scientific research articles published in a journal follow the same basic format and contain the following sections:

- Abstract: A brief summary in paragraph form of each of the individual sections that are detailed below.

- Introduction or background: this section introduces the research question, states why the question is important, lays the groundwork for what is already known and what remains unknown, and states the hypothesis. Typically, this section includes a miniature literature review of previously published studies and details the results and or gaps in the research that these studies failed to achieve.

- Methods: this section details the experimental process that was conducted including information about how subjects were recruited, their demographic information, steps that were followed in the methodology, and what statistical analyses were performed on the data.

- Results: this section includes a narrative description of the results that were found as well as tables and graphs of the findings.

- Conclusion: this section interprets the results to put meaning to the numbers. Essentially, it answers the question, so what? It also discusses strengths, weaknesses, and shortcomings of the performed experiment as well as needed areas of future research.

As mentioned, the majority of scientific data is presented in tables, charts or graphs, and statistics. In order to interpret such information, one must be comfortable with these presentation methods.

In general, there are two types of data: qualitative and quantitative. Science passages may contain both, but simply put, quantitative data is reflected numerically and qualitative is not. Qualitative data is based on its qualities. In other words, qualitative data tends to present information more in subjective generalities (for example, relating to size or appearance). Quantitative data is based on numerical findings such as percentages. Quantitative data will be described in numerical terms. While both types of data are valid, the test taker will more likely be faced with having to interpret quantitative data through one or more graphic(s), and then be required to answer questions regarding the numerical data. The section of this study guide briefly addresses how data may be displayed in line graphs, bar charts, circle graphs, and scatter plots. A test taker should take the time to learn the skills it takes to interpret quantitative data. An example of a line graph is as follows:

Cell Phone Use in Kiteville, 2000-2006

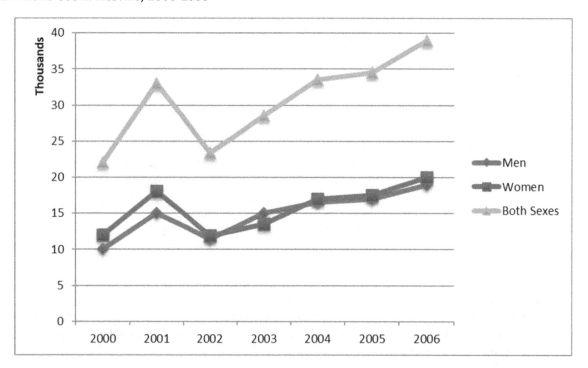

A line graph presents quantitative data on both horizontal (side to side) and vertical (up and down) axes. It requires the test taker to examine information across varying data points. When reading a line graph, a test taker should pay attention to any headings, as these indicate a title for the data it contains. In the above example, the test taker can anticipate the line graph contains numerical data regarding the use of cellphones during a certain time period. From there, a test taker should carefully read any outlying words or phrases that will help determine the meaning of data within the horizontal and vertical axes. In

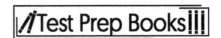

this example, the vertical axis displays the total number of people in increments of 5,000. Horizontally, the graph displays yearly markers, and the reader can assume the data presented accounts for a full seven-year period. In addition, the line graph also uses different shapes to mark its data points. Some data points represent the number of men. Some data points represent the number of women, and a third type of data point represents the number of both sexes combined.

A test taker may be asked to read and interpret the graph's data, then answer questions about it. For example, the test may ask, *In which year did men seem to decrease cellphone use?* then require the test taker to select the correct answer. Similarly, the test taker may encounter a question such as *Which year yielded the highest number of cellphone users overall?* The test taker should be able to identify the correct answer as 2006.

A **bar graph** presents quantitative data through the use of lines or rectangles. The height and length of these lines or rectangles corresponds to the magnitude of the numerical data for that particular category or attribute. The data presented may represent information over time, showing shaded data over time or over other defined parameters. A bar graph will also utilize horizontal and vertical axes. An example of a bar graph is as follows:

Population Growth in Major U.S. Cities

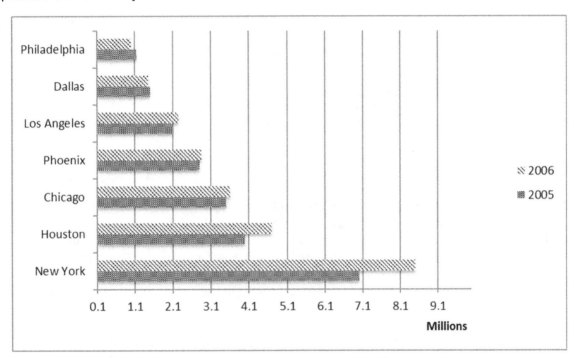

Reading the data in a bar graph is similar to the skills needed to read a line graph. The test taker should read and comprehend all heading information, as well as information provided along the horizontal and vertical axes. Note that the graph pertains to the population of some major U.S. cities. The "values" of these cities can be found along the left side of the graph, along the vertical axis. The population values can be found along the horizontal axes. Notice how the graph uses shaded bars to depict the change in population over time, as the heading indicates. Therefore, when the test taker is asked a question such as, *Which major U.S. city experienced the greatest amount of population growth during the depicted two-year cycle,* the reader should be able to determine a correct answer of New York. It is important to pay particular attention to color, length, data points, and both axes, as well as any outlying header information in order to be able to answer graph-like test questions.

A circle graph (also sometimes referred to as a pie chart) presents quantitative data in the form of a circle. The same principles apply: the test taker should look for numerical data within the confines of the circle itself but also note any outlying information that may be included in a header, footer, or to the side of the circle. A circle graph will not depict horizontal or vertical axis information, but it will instead rely on the reader's ability to visually take note of segmented circle pieces and apply information accordingly. An example of a circle graph is as follows:

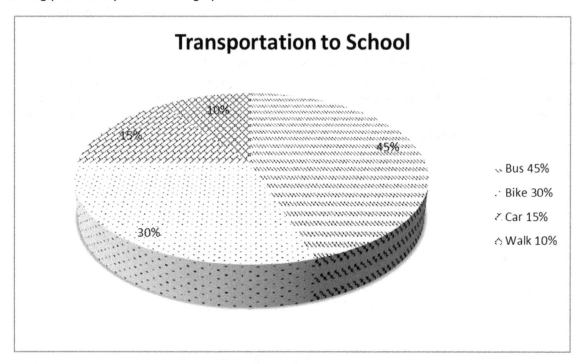

Notice the heading "Transportation to School." This should indicate to the test taker that the topic of the circle graph is how people traditionally get to school. To the right of the graph, the reader should comprehend that the data percentages contained within it directly correspond to the method of transportation. In this graph, the data is represented through the use shades and pattern. Each transportation method has its own shade. For example, if the test taker was then asked, *Which method of school transportation is most widely utilized,* the reader should be able to identify school bus as the correct answer.

Be wary of test questions that ask test takers to draw conclusions based on information that is not present. For example, it is not possible to determine, given the parameters of this circle graph, whether the population presented is of a particular gender or ethnic group. This graph does not represent data from a particular city or school district. It does not distinguish between student grade levels and, although the reader could infer that the typical student must be of driving age if cars are included, this is not necessarily the case. Elementary school students may rely on parents or others to drive them by personal methods. Therefore, do not read too much into data that is not presented. Only rely on the quantitative data that is presented in order to answer questions.

A scatter plot or scatter diagram is a graph that depicts quantitative data across plotted points. It will involve at least two sets of data. It will also involve horizontal and vertical axes.

An example of a scatter plot is as follows:

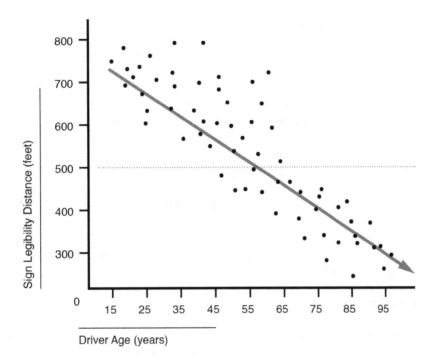

Driver Age (years)

The skills needed to address a scatter plot are essentially the same as in other graph examples. Note any topic headings, as well as horizontal or vertical axis information. In the sample above, the reader can determine the data addresses a driver's ability to correctly and legibly read road signs as related to their age. Again, note the information that is absent. The test taker is not given the data to assess a time period, location, or driver gender. It simply requires the reader to note an approximate age to the ability to correctly identify road signs from a distance measured in feet. Notice that the overall graph also displays a trend. In this case, the data indicates a negative one and possibly supports the hypothesis that as a driver ages, their ability to correctly read a road sign at over 500 feet tends to decline over time. If the test taker were to be asked, *At what approximation in feet does a sixteen-year-old driver correctly see and read a street sign,* the answer would be the option closest to 700 feet.

Reading and examining scientific data in excerpts involves all of a reader's contextual reading, data interpretation, drawing logical conclusions based only on the information presented, and their application of critical thinking skills across a set of interpretive questions. Thorough comprehension and attention to detail is necessary to achieve test success.

Applying Scientific Principles

The scientific method provides the framework for studying and learning about the world in a scientific fashion. The scientific method has been around since at least the 17th century and is a codified way to answer natural science questions. Due to objectivity, the scientific method is impartial and its results are highly repeatable; these are its greatest advantages. There is no consensus as to the number of steps involved in executing the scientific method, but the following six steps are needed to fulfill the criteria for correct usage of the scientific method:

- Ask a question: Most scientific investigations begin with a question about a specific problem.

- Make observations: Observations will help pinpoint research objectives on the quest to answer the question.

- Create or propose a hypothesis: The hypothesis represents a possible solution to the problem. It is a simple statement predicting the outcome of an experiment designed to investigate the research question.

- Formulate an experiment: The experiment tests the proposed hypothesis.

- Test the hypothesis: The outcome of the experiment to test the hypothesis is the most crucial step in the scientific method.

- Accept or reject the hypothesis: Using results from the experiment, a scientist can conclude to accept or reject the hypothesis.

Several key nuances of the scientific method include:

- The hypothesis must be verifiable and falsifiable. Falsifiable refers to the possibility of a negative solution to the hypothesis. The hypothesis should also have relevance, compatibility, testability, simplicity, and predictive power.

- Investigation must utilize both deductive and inductive reasoning. Deductive reasoning employs a logical process to arrive at a conclusion using premises considered true, while inductive reasoning employs an opposite approach. Inductive reasoning allows scientists to propose hypotheses in the scientific method, while deductive reasoning allows scientist to apply hypotheses to particular situations.

- An experiment should incorporate an independent, or changing, variable and a dependent, or non-changing, variable. It should also utilize both a control group and an experimental group. The experimental group will ultimately be compared against the control group.

A scientific explanation has three crucial components—a claim, evidence, and logical reasoning. A claim makes an assertion or conclusion focusing on the original question or problem. The evidence provides backing for the claim and is usually in the form of scientific data. The scientific data must be appropriate and sufficient. The scientific reasoning connects the claim and evidence and explains why the evidence supports the claim.

Scientific explanations must fit certain criteria and be supported by logic and evidence. The following represent scientific explanation criteria. The proposed explanation:

- Must be logically consistent
- Must abide by the rules of evidence
- Must report procedures and methods
- Must be open to questions and possible modification
- Must be based on historical and current scientific knowledge

The scientific method encourages the growth and communication of new information and procedures among scientists. Explanations of how the natural changes and works based on fiction, personal convictions, religious morals, mystical influences, superstitions, or authorities are not scientific; therefore, these explanations are irrelevant to the scientific community.

Scientific explanations have two fundamental characteristics. First, they should explain all scientific data and observations gleaned from experiments. Second, they should allow for predictions that can be verified with future experiments.

Analyzing

Discerning an Appropriate Research Question

Part of the process of scientific inquiry is researching a problem or question. Before an experiment can be designed, proper research should be conducted into the question. The initial question needs to be well formed and based in logical reasoning. A literature review should be conducted on existing material pertaining to the subject in question, and confirmation of any experimentation on the question that has been conducted prior should be made. If prior experimentation exists, what were the results obtained and were any conclusions drawn from those results? In addition, research should be done on all possible information regarding the initial question, the experiment, how to investigate the question, and what tools will be necessary to draw conclusions and explain any findings. Just as an experiment must be unbiased, so should any research regarding the experiment. All sources of information need to be proven reliable and accredited. For example, a person's account of their opinion on a situation does not constitute as a valid source for research. Sources should be free of opinion or speculation.

During experimentation, research should be conducted with appropriate mechanisms for observation and measurement. Knowing the proper tools and units for accurately measuring a volume or a mass is a fundamental skill of research. Researchers also need to be held to standards of ethics and honesty. The independent repetition of an experiment helps to ensure this level of accountability. Often, the most reliable resources are those of accredited experimenters, universities, and other research laboratories. In order for such sources to publish information, they should demonstrate strict adherence to scientific methods, precise measurements for observations, and specific mathematical reporting.

It is often common for different scientists in the same place, or even separate countries, to be conducting experiments to test the same hypothesis. This does not always lead to a race to see who finishes first, but it can lead to cooperative research and shared accolades if the results prove successful. Awards for research, discoveries, and scientific application are often used by the scientific community to show appreciation for advancements in science.

Scientific questions can be derived from a multitude of sources including observation, experience, or even just wondering how something is made or works. In order to answer these questions, experiments should be designed and conducted to try to achieve a solution. At the end of an experiment, there often is no clear solution and a new experiment must be designed to test the same question. If a sound, logical solution is reached through experimentation, then it must be repeatable, by the experimenter and any other person wishing to test this solution. This entire process is commonly referred to as the scientific method.

A question or situation exists, a hypothesis (or a well-educated guess) is formulated, an experiment is designed to test this guess, a prediction is made as to what the outcome might be based upon research, and a conclusion is formed (either the guess was correct or not). This simple method is repeated over and over, as much as necessary for each question, idea, or proposed investigation.

An experiment must be carefully designed to include concerns for safety, use of proper instrumentation for measurement, systematic methods of documentation or data collection, appropriate mathematics for analysis of data and for the interpretation to draw valid conclusions. These conclusions must be explainable and verifiable by an outside source.

The importance of having an independent party test a solution is one of the critical parts of scientific inquiry. This ensures an experiment is free from bias, truly repeatable, and documentable to multiple sources. Without this confirmation, people could make erroneous claims and cause disastrous results. There would be no order to the inquiry of science.

In scientific experimentation, safety, respect for living things, and the effect on an environment must be acknowledged and protected, as necessary. There exist universal rules for research in order to preserve these underlying tenets. Most researchers or facilities that demonstrate an adherence to these rules garner the most support from others in the scientific community when accepting ideas.

Identifying Reasons for a Procedure and Analyzing Limitations

Valid experiments must start with a valid hypothesis. There must be one independent variable for any scientific question and a measurable dependent variable that is used to investigate the question. The hypothesis is a statement that explains how changing the independent variable would affect the dependent variable, and is often stated as an *if (independent variable plus verb), then (dependent variable plus verb)* sentence.

Dependent variables must be measurable. Common dependent variables include mass (measured with a balance or scale), length (measured with a ruler), and volume (measured with a graduated cylinder).

It is also important to develop an independent variable, which has at least two known conditions. The normal condition is called the control group, and any conditions different from the control group are called experimental groups.

For example, a simple experiment could be investigating how fertilizer affects plant growth.

Here are some examples of valid hypotheses that collect quantitative data:

- If fertilizer is added to the soil, then plant height will increase.
- If fertilizer is added to the soil, then plant leaf number will increase.

Here are some examples of invalid hypotheses:

- If fertilizer is added to the soil, then the plant will grow better.
- If the plant soil changes, then the plant will grow better.

Notice in the valid hypothesis that there is a clear independent variable: the addition of fertilizer to soil. Also, both dependent variable options are quantitative and can be measured. Height can be measured with a ruler, and number of leaves can be measured by counting.

The invalid hypotheses contain immeasurable changes. "Growing better" is not specific enough and is not something that can be measured with a scientific instrument. "Soil changes" is not specific enough

and could involve many different changes including amount of soil, type of fertilizer, or amount of fertilizer.

Variable development

After a valid hypothesis is developed with specific changes and measurements in mind, the details of the experiment must be confirmed before developing a procedure.

Defining constants, a control group, experimental groups, as well as making a decision on a measurement device should be included.

With the proposed hypothesis in mind (*If fertilizer is added to the soil, then plant height will increase*), several constants should be defined. The type of plant is important because some grow faster than others, so they all should be the same. The brand and amount of soil should be the same, the type of fertilizer should be the same, the shape and type of pot should be the same, and the amount of watering and light exposure should be consistent between all test groups. If any one of these factors is different, it will be impossible to tell if any observable change was due to that factor as opposed to the independent variable. The independent variable should be the *only* changed factor, and in this case, the only change should be the addition of fertilizer.

A control group is also important to include in all experiments. In this case, it would be soil with no fertilizer added. This is important because there would be no way to know if the fertilizer was actually helpful unless the unaffected condition was included. Including at least two experimental variables is important to identify varying amounts of change; good options in this case would be to include soil with 5% fertilizer and 10% fertilizer.

The dependent variable should be something easy to measure. In this case, a good dependent variable is length; the measurement should consist of stretching out the plant and measuring its length with a string, then using the ruler to measure the length of the straightened string. This method of measurement will account for possible drooping and will be more accurate.

Finally, it is important to include multiple trials of each test group, in this case multiple seeds in each fertilizer condition. This helps ensure confidence in the results; some experiments inevitably go wrong through the forces of nature (possibly a dud seed), so multiple trials are a critical part of any experiment.

Identifying the reason for a procedure in published studies involves correctly identifying the research question or what the research group was attempting to investigate. The hypothesis must be picked out as well; different research questions and hypotheses will necessitate different experimental procedures. For example, if, instead of wondering if fertilizer would increase plant growth in height, researchers were interested in determining if fertilizer increased plant longevity, the procedure would be altered. Instead of measuring plant length with a string, researchers would count the number of days that different conditions of the plant survived.

Limitations of a procedure should be critically examined. This skill is important in developing a sounder future experiment and reducing sources of error for more reliable results. The list of possible limitations is innumerable but common issues stem from problems with the sample population, such as a small study group, too heterogeneous or homogenous groups, or results from the group not being translatable to the population at large, due to unique characteristics of the group. Issues with materials or instruments, calibration, and accuracy of equipment can induce measurement error. There may be limitations to the procedure itself as well. For example, there may be variables that are hard to isolate, confounding the results with the study variables and outcomes. In the plant growth experiment, perhaps the lab is closed on weekends so growth cannot be measured two days a week and instead gets lumped or averaged on Mondays.

Selecting the Best Procedure

Developing a procedure takes critical thinking and brainstorming about any foreseeable problems and possible solutions. For the example of the fertilizer experiment, several parameters should be kept in mind.

- The experiment should last for at least a week so that the plants have a chance to grow.
- Water and light should be a critical part of plant care.
- The string/ruler measurement system is important as well.

Beyond that, the sky is the limit. A valid procedure must be specific.

Example procedure:

Day 1

1. Place 2 cups of soil of the no fertilizer mix (0% fertilizer) in the pot.
2. Add five seeds in the pattern of the picture below.
3. Pour ½ cup tap water over seeds.
4. Sprinkle ¼ cup soil over watered seeds.
5. Place group name on pot.
6. Repeat steps #1-5 with:
7. Low fertilizer mix (5% fertilizer)
8. High fertilizer mix (10% fertilizer mix)
9. Put the pots in the light box

Day 2 through 14

10. Water plants with 1/3 cup water every school day.
11. Day 14: Carefully pull each plant out of the soil. Carefully stretch it out and measure its length with a piece of string. Then measure the piece of string against a ruler and tell your teacher the value.

Data collection and results

A table should be generated to collect the experimental data. For example:

	Plant Height (cm)					
Amount of Fertilizer	Group 1	Group 2	Group 3	Group 4	Group 5	Average
None	5	6.4	4.7	5.5	6	
Low	12	10	12	13	13	
High	19	22	18.2	18.5	18	

After data collection is complete, a graphical representation of the results should be generated. By convention, the independent variable should be the x-axis and the dependent variable should be the y-axis. Appropriate statistical analyses should be performed after data collection is complete, as part of the presentation of results. Description statistics include the calculations of mean, median, mode, range, etc. Significance level is typically set at $p < 0.05$ or $p < 0.001$. Other statistical analyses can be conducted; common tests are independent and dependent t-tests, which are used to identify differences among groups.

Once all of the results have been calculated and visually represented, the importance and interpretation of such results are explored and presented in a discussion section. Any necessary procedural modifications for future experiments and weaknesses and strengths of the current procedure are addressed, and areas of further interest are identified.

In general, selecting the best procedure involves evaluating which method will be the best for isolating the variables of interest, selecting the appropriate study group, forming a like control group, and investigating the most accurate—yet accessible—instruments and equipment to measure outcomes.

Evaluating and Generalizing

Distinguishing Among Hypotheses, Assumptions, Data, and Conclusions

In the scientific method, scientists must always take care to recognize and analyze alternative scientific explanations and models. A scientific argument is valid only if alternative explanations and models have been examined, tested, and proved wrong. During scientific investigations, scientists often propose several alternative explanations and models which introduces creativity into the scientific method. A scientific team should and must consider as many plausible alternative explanations and models, with the ultimate aim being the elimination of all but one. Other scientists may be a source of competing alternative scientific explanations and models.

In the early stages of scientific investigations, several alternative explanations and models may be circulating to explain a certain observation or phenomenon. Over the years, some of the alternative explanations and models are eliminated as a result of testing disproving the hypotheses. Eventually, one single explanation or model dominates the scientific landscape. Often, aspects of eliminated explanations and models are incorporated into the dominant model.

A scientific argument makes a science-oriented claim supported by evidence and historical or current scientific knowledge. Opinions and simple ideas are not arguments because they lack the aforementioned characteristics.

The goal of a scientific argument is very different than that of a confrontational argument. The goal of a confrontational argument is for one person's point of view to win over that of another's. The goal of a scientific argument is to generate, verify, communicate, debate, and modify explanations in an effort to better understand a particular scientific topic. The following are six key parts of a scientific argument:

- Question: The question is often derived from an event, observation, or dispute sparking the interest of a scientist.

- Assumption: An assumption utilizes prior knowledge to entertain a possible solution to the question being researched.

- Claim: A claim makes an assertion derived from the scientific data gathered during the scientific investigation that makes an effort to definitively answer the question being researched. A strong claim answers the question being researched, stands apart, and explains matters utilizing a cause and effect relationship.

- Evidence: Evidence consists of measurements, observations, or calculations included in the scientific data that support the validity of the claim. It is a pattern that emerges during the analysis of scientific data. Strong evidence is pattern-based, accurate, reliable, and sufficient.

- Explanation: An explanation is a summary, oral or written, based on the claim that provides justification of the evidence.

- Rebuttal: An opponent to the original claim, including its supportive evidence, typically provides a rebuttal. Opponents may also provide contrary evidence to change, modify, or disprove the original claim.

Judging the Basis of Information for a Given Conclusion

Sound scientific conclusions should be formed only after appropriate experimental procedures, research, and analyses have been conducted. Conclusions may be inaccurate if errors occurred or misinformation was obtained anywhere along the scientific process continuum. For example, without adequate background research, experiments may be improperly contrived and designed. A reader may come to a wrongful conclusion if information is obtained from unreliable sources. With all scientific research, education, and investigation, the credibility of the source is of utmost importance. Moreover, if experimental protocols are not objective and scientifically sound, results can be confounded and thus, conclusions drawn from them will be misinformed. For example, if researchers are examining how fertilization effects plant growth, they must ensure that they isolate the independent variable, which, is this case, is the amount of fertilizer received. Researchers must also include a control group of plants receiving only water, to ensure that any differences in growth are due to only the effects of the fertilizer. If, for example, the researchers put the unfertilized control group in the sun but keep the fertilized plants in a dark cabinet, differences in growth may be attributable to more than just the application of fertilizer. In fact, researchers may incorrectly conclude that the addition of fertilizer did not increase growth (because any benefits could have been negated by the lack of sun, stunting growth). In this way, it is imperative to develop an ability to critically critique the process through which conclusions were drawn as well as evaluate the credibility of the sources supporting the conclusion.

Determining the Relevance for Answering a Question

A *red herring* is a logical fallacy in which irrelevant information is introduced to alter an argument's trajectory. Red herrings are the irrelevant information used to fallaciously and slyly divert the argument

into an unrelated topic. This fallacy is common in thriller movies or television shows in which the audience is led to believe that a character is the villain or mastermind, while the true villain remains a secret. Red herrings can be used to distract the reader with irrelevant information in either the question or answer choice. A red herring is sometimes referred to as a *straw man*, since this fallacy attacks a different argument than the one presented. Consider the following example of a red herring:

> The government must immediately issue tax cuts to strengthen the economy. A strong middle class is the backbone of a fully functional economy, and tax cuts will increase the discretionary spending necessary to support the middle class. After all, it's extremely important for our society to be open-minded and to limit racial discord.

On its face, the argument looks fine. The conclusion is obvious—the government needs to pass tax cuts to strengthen the economy. This is based on the premise that the cuts will increase discretionary spending, which will strengthen the middle class, and the economy will be stronger. However, the last sentence is a red herring. The argument does not address how a society should be open-minded and avoid racism; it holds no connection with the main thrust of the argument. In this scenario, look out for answer choices that address the red herring rather than the essential argument.

Red herrings will be extremely common in the answer choices. The test makers know that test takers are aware of sound logic and flawed reasoning, but they also know that test takers are working quickly to identify key words and phrases. As a result, the test makers will include many red herrings in the answer choices.

The test makers commonly write appealing answer choices, but take the language to such an unjustified extreme that it is rendered incorrect. The extreme language usually will take the argument too far.

An example of an argument appealing to the extreme is:

> Weight lifting breaks down muscles and rebuilds them. If one just kept exercising and never stopped, his or her body would deteriorate and eventually fall apart.

This argument is clearly illogical. The author correctly describes what weight lifting does to the body, but then takes the argument to an unjustified extreme. If someone continually lifted weights, his or her body would not deteriorate and fall apart due to the weights breaking down muscle. The weightlifter may eventually die from a heart attack or dehydration, but it would not be because of how weight lifting rebuilds muscle.

Judging the Reliability of Sources

In all domains, but particularly with science, the credibility and reliability of sources must be assessed prior to accepting any of the source's claims. Incorrectly assuming that the information in a given scientific text or report is accurate can lead to a chain of issues in the scientific process. If a text presents misinformation yet the reader accepts it as factual, the reader may go on to design a new experiment based on these fallacies or incorporate such errors into an academic report or project. Unfortunately, these issues are increasingly common, with the greater reliance on Internet sources for research. The Internet is riddled with websites that contain unsubstantiated evidence and inaccurate scientific claims. If uninformed readers peruse one of these sites, they may encounter incorrect information and introduce errors into their own work.

The most credible sources in the scientific space come from peer-reviewed scientific journals, which publish experiments and literature reviews that have gone through a high level of scrutiny. Other valid sources include current published academic textbooks and educational or government-sponsored websites or publications such as Nova, the Smithsonian, or PBS. Websites ending in .gov or .edu are typically more reliable, having undergone a certain level of review and fact-checking. When using sources that cite other texts or resources, students are advised to also review these original sources directly as well. This helps ensure that source is factually sound and that the interpretation of its information is indeed correct in its incorporation in the current resource.

Practice Questions

1. Which statement about white blood cells is true?
    a. B cells are responsible for antibody production.
    b. White blood cells are made in the white/yellow cartilage before they enter the bloodstream.
    c. Platelets, a special class of white blood cell, function to clot blood and stop bleeding.
    d. The majority of white blood cells only activate during the age of puberty, which explains why children and the elderly are particularly susceptible to disease.

2. Which locations in the digestive system are sites of chemical digestion?
    1. Mouth
    2. Stomach
    3. Small Intestine

    a. II only
    b. III only
    c. II and III only
    d. I, II, and III

3. What is the theory that certain physical and behavioral survival traits give a species an evolutionary advantage?
    a. Gradualism
    b. Evolutionary advantage
    c. Punctuated equilibrium
    d. Natural selection

4. Which of the following structures is unique to eukaryotic cells?
    a. Cell walls
    b. Nuclei
    c. Cell membranes
    d. Organelles

5. Which is the cellular organelle used for digestion to recycle materials?
    a. The Golgi apparatus
    b. The lysosome
    c. The centrioles
    d. The mitochondria

6. Which of the following leads to diversity in meiotic division but not mitotic division?
    a. Tetrad formation
    b. Disassembly of the mitotic spindle
    c. Extra/fewer chromosomes due to nondisjunction
    d. Fertilization by multiple sperm

7. Why do arteries have valves?
    a. They have valves to maintain high blood pressure so that capillaries diffuse nutrients properly.
    b. Their valves are designed to prevent backflow due to their low blood pressure.
    c. The valves have no known purpose and thus appear to be unnecessary.
    d. They do not have valves, but veins do.

8. If the pressure in the pulmonary artery is increased above normal, which chamber of the heart will be affected first?
    a. The right atrium
    b. The left atrium
    c. The right ventricle
    d. The left ventricle

9. What is the purpose of sodium bicarbonate when released into the lumen of the small intestine?
    a. It works to chemically digest fats in the chyme.
    b. It decreases the pH of the chyme so as to prevent harm to the intestine.
    c. It works to chemically digest proteins in the chyme.
    d. It increases the pH of the chyme so as to prevent harm to the intestine.

10. Which of the following describes a reflex arc?
    a. The storage and recall of memory
    b. The maintenance of visual and auditory acuity
    c. The autoregulation of heart rate and blood pressure
    d. A stimulus and response controlled by the spinal cord

11. Describe the synthesis of the lagging strand of DNA.
    a. DNA polymerases synthesize DNA continuously after initially attaching to a primase.
    b. DNA polymerases synthesize DNA discontinuously in pieces called Okazaki fragments after initially attaching to primases.
    c. DNA polymerases synthesize DNA discontinuously in pieces called Okazaki fragments after initially attaching to RNA primers.
    d. DNA polymerases synthesize DNA discontinuously in pieces called Okazaki fragments which are joined together in the end by a DNA helicase.

12. Which of the following are chief factors that are associated with increased birth and fertility rates?
    I. Public education
    II. Low infant mortality
    III. Low urbanization
    IV. Increased cholesterol intake
    V. No access to contraception

    a. I and II only
    b. II and V only
    c. I, II, and IV only
    d. III and V only

13. Which of the following characterizes developed countries in Stage 4 of the Demographic Transition Model?
   a. Low birth rate and high immigration rate.
   b. Low birth rate and low mortality rate.
   c. High birth rate and rapidly decreasing death rate.
   d. High birth rate and high emigration rate.

Use the image below to answer question #14:

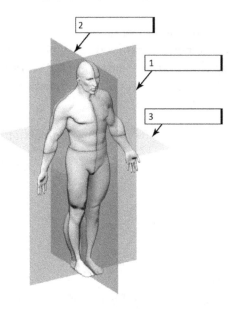

14. Identify the correct sequence of the 3 primary body planes as numbered 1, 2, and 3 in the above image.
   a. Plane 1 is coronal, plane 2 is sagittal, and plane 3 is transverse.
   b. Plane 1 is sagittal, plane 2 is coronal, and plane 3 is medial.
   c. Plane 1 is coronal, plane 2 is sagittal, and plane 3 is medial.
   d. Plane 1 is sagittal, plane 2 is coronal, and plane 3 is transverse.

15. Which of the following is NOT a major function of the respiratory system in humans?
   a. It provides a large surface area for gas exchange of oxygen and carbon dioxide.
   b. It helps regulate the blood's pH.
   c. It helps cushion the heart against jarring motions.
   d. It is responsible for vocalization.

16. Which of the following is NOT a function of the forebrain?
   a. To regulate blood pressure and heart rate
   b. To perceive and interpret emotional responses like fear and anger
   c. To perceive and interpret visual input from the eyes
   d. To integrate voluntary movement

17. What is the major difference between somatic and germline mutations?
    a. Somatic mutations usually benefit the individual while germline mutations usually harm them.
    b. Since germline mutations only affect one cell, they are less noticeable than the rapidly dividing somatic cells.
    c. Somatic mutations are not expressed for several generations, but germline mutations are expressed immediately.
    d. Germline mutations are usually inherited while somatic mutations will affect only the individual.

18. The sun is a major external source of energy. Which of the following is the best demonstration of this?
    a. Flowers tend to bloom in the morning, after dawn.
    b. Large animals like bears do not need to eat food when hibernating.
    c. Deserts can reach scorching temperatures in daylight but subzero temperatures at night.
    d. The tides of the ocean are highly dependent on the movement of the Moon, the celestial body that is highly reflective to sunlight.

19. What coefficients are needed to balance the following combustion equation?

$$\_ C_2H_{10} + \_ O_2 \rightarrow \_ H_2O + \_ CO_2$$

    a. 1:5:5:2
    b. 1:9:5:2
    c. 2:9:10:4
    d. 2:5:10:4

20. What is the purpose of a catalyst?
    a. To increase a reaction's rate by increasing the activation energy
    b. To increase a reaction's rate by increasing the temperature
    c. To increase a reaction's rate by decreasing the activation energy
    d. To increase a reaction's rate by decreasing the temperature

21. Most catalysts found in biological systems are
    a. Special lipids called cofactors.
    b. Special proteins called enzymes.
    c. Special lipids called enzymes.
    d. Special proteins called cofactors.

22. Which statement is true about the pH of a solution?
    a. A solution cannot have a pH less than 1.
    b. The more hydroxide ions in the solution, the higher the pH.
    c. If an acid has a pH of greater than 2, it is considered a weak base.
    d. A solution with a pH of 2 has ten times the amount of hydrogen ions than a solution with a power of 1.

23. Salts like sodium iodide (NaI) and potassium chloride (KCl) use what type of bond?
    a. Ionic bonds
    b. Disulfide bridges
    c. Covalent bonds
    d. London dispersion forces

24. Which of the following is unique to covalent bonds?
    a. Most covalent bonds are formed between the elements H, F, N, and O.
    b. Covalent bonds are dependent on forming dipoles.
    c. Bonding electrons are shared between two or more atoms.
    d. Molecules with covalent bonds tend to have a crystalline solid structure.

25. Which of the following describes a typical gas?
    a. Indefinite shape and indefinite volume
    b. Indefinite shape and definite volume
    c. Definite shape and definite volume
    d. Definite shape and indefinite volume

26. Of the following, which is the closest biologically to the extinct dodo (R. cucullatus)?
    a. The spotted sandgrouse (P. senegallus) of the same subclass (Columbimorphae).
    b. The ostrich (S. camelus) of the same class (Aves).
    c. The hummingbird hawk-moth (M. stellatarum) of the same kingdom (Animalia).
    d. The rock dove (C. livia) of the same family (Columbidae).

27. When describing photosynthetic autotrophs, which statement is correct?
    a. These organisms use aerobic respiration to break down glucose obtained through eating other organisms.
    b. All of these organisms use chloroplasts to perform photosynthesis.
    c. These organisms form the base of most food webs.
    d. These organisms, unlike heterotrophs, do not use glycolysis or the Krebs cycle.

28. Which of the following is an example of differentiation?
    a. A white blood cell ruptures after being infected by a virus.
    b. A sperm fuses with an egg and forms a zygote.
    c. A stem cell develops to become better suited for the production of insulin.
    d. A liver cell's DNA becomes damaged, so it enters the G0 stage.

29. Which human cellular metabolic pathway is more energy efficient and why?
    a. Anaerobic, because oxygen isn't depleted.
    b. Aerobic, because all of the energy produced is used for ATP.
    c. Anaerobic, because the process takes less time.
    d. Aerobic, because more ATP is produced.

30. Which of the following is directly transcribed from DNA and represents the first step in protein building?
    a. siRNA
    b. rRNA
    c. mRNA
    d. tRNA

31. What information does a genotype give that a phenotype does not?
    a. The genotype necessarily includes the proteins coded for by its alleles.
    b. The genotype will always show an organism's recessive alleles.
    c. The genotype must include the organism's physical characteristics.
    d. The genotype shows what an organism's parents looked like.

Use the image below to answer question #32:

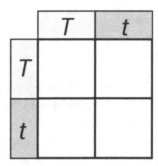

32. Which statement is supported by the Punnett square above, if "T" = Tall and "t" = short?
    a. Both parents are homozygous tall.
    b. 100% of the offspring will be tall because both parents are tall.
    c. There is a 25% chance that an offspring will be short.
    d. The short allele will soon die out.

33. Which of the following is a chief difference between evaporation and boiling?
    a. Liquids boil only at the surface while they evaporate equally throughout the liquid.
    b. Evaporating substances change from gas to liquid while boiling substances change from liquid to gas.
    c. Evaporation happens in nature while boiling is a manmade phenomenon.
    d. Evaporation can happen below a liquid's boiling point.

34. Which of the following CANNOT be found in a human cell's genes?
    a. Sequences of amino acids to be transcribed into mRNA
    b. Lethal recessive traits like sickle cell anemia
    c. Mutated DNA
    d. DNA that codes for proteins the cell doesn't use

35. Which of the following is a special property of water?
    a. Water easily flows through phospholipid bilayers.
    b. A water molecule's oxygen atom allows fish to breathe.
    c. Water is highly cohesive which explains its high melting point.
    d. Water can self-hydrolyze and decompose into hydrogen and oxygen.

36. What is an isotope? For any given element, it is an atom with
    a. a different atomic number.
    b. a different number of protons.
    c. a different number of electrons.
    d. a different mass number.

37. What is the electrical charge of the nucleus?
    a. A nucleus always has a positive charge.
    b. A stable nucleus has a positive charge, but a radioactive nucleus may have no charge and instead be neutral.
    c. A nucleus always has no charge and is instead neutral.
    d. A stable nucleus has no charge and is instead neutral, but a radioactive nucleus may have a charge.

38. A student believes that there is an inverse relationship between sugar consumption and test scores. To test this hypothesis, he recruits several people to eat sugar, wait one hour, and take a short aptitude test afterwards. The student will compile the participants' sugar intake levels and test scores. How should the student conduct the experiment?
    a. One round of testing, where each participant consumes a different level of sugar.
    b. Two rounds of testing: The first, where each participant consumes a different level of sugar, and the second, where each participant consumes the same level as they did in Round 1.
    c. Two rounds of testing: The first, where each participant consumes the same level of sugar as each other, and the second, where each participant consumes the same level of sugar as each other but at higher levels than in Round 1.
    d. One round of testing, where each participant consumes the same level of sugar.

39. Which of the following is the most effective method to create a scatter plot?
    a. Using graph paper to plot points
    b. Inputting information into a spreadsheet to generate a scatter plot
    c. Searching a statistics textbook for a similar scatter plot and using that
    d. Using an electronic tablet to draw one

40. A researcher is exploring factors that contribute to the GPA of college students. While the sample is small, the researcher is trying to determine what the data shows. What can be reasoned from the table below?

Student	Maintains a Calendar?	Takes Notes?	GPA
A	sometimes	often	3.1
B	never	always	3.9
C	never	never	2.0
D	sometimes	often	2.7

    a. No college students consistently maintain a calendar of events.
    b. There is an inverse correlation between maintaining a calendar and GPA, and there is a positive correlation between taking notes and GPA.
    c. There is a positive correlation between maintaining a calendar and GPA, and there is no correlation between taking notes and GPA.
    d. There is no correlation between maintaining a calendar and GPA, and there is a positive correlation between taking notes and GPA.

41. Four different groups of the same species of peas are grown and exposed to differing levels of sunlight, water, and fertilizer as documented in the table below. The data in the water and fertilizer columns indicates how many times the peas are watered or fertilized per week, respectively. Group 2 is the only group that withered. What is a reasonable explanation for this occurrence?

Group	Sunlight	Water	Fertilizer
1	partial sun	4 mL/hr	1
2	full sun	7 mL/hr	1
3	no sun	14 mL/hr	2
4	partial sun	3 mL/hr	2

    a. Insects gnawed away the stem of the plant.
    b. The roots rotted due to poor drainage.
    c. The soil type had nutrition deficiencies.
    d. This species of peas does not thrive in full sunlight.

42. Which of the following functions corresponds to the parasympathetic nervous system?
    a. It stimulates the fight-or-flight response.
    b. It increases heart rate.
    c. It stimulates digestion.
    d. It increases bronchiole dilation.

43. According to the periodic table, which of the following elements is the least reactive?
    a. Fluorine
    b. Silicon
    c. Neon
    d. Gallium

44. The Human Genome Project is a worldwide research project launched in 1990 to map the entire human genome. Although the Project was faced with the monumental challenge of analyzing tons and tons of data, its objective was completed in 2003 and ahead of its deadline by two years. Which of the following inventions likely had the greatest impact on this project?
    a. The sonogram
    b. X-ray diffraction
    c. The microprocessor
    d. Magnetic Resonance Imaging (MRI)

45. Which of the following inventions likely had the greatest improvement on the ability to combat nutrition deficiencies in developing countries?
    a. Food products fortified with dietary vitamins and minerals
    b. Integrated statistical models of fish populations
    c. Advances so that microscopes can use thicker tissue samples
    d. Refrigerated train cars for transportation of food

46. Which of the following depicts a form of potential energy?
    a. The light given off by a lamp
    b. The gravitational pull of a black hole
    c. The heat from a microwaved burrito
    d. The motion of a pendulum

47. Anya was paid by Company X to analyze dwindling honeybee populations of the Southwest. After measuring hive populations over several months, she noticed no patterns in the geographic distributions of the deaths after comparisons with local maps of interest. This supported her hypothesis, so she took samples of the honey and the bees from the hives and performed dozens of dissections to confirm her suspicions. Which of the following is the most likely hypothesis upon which this research was performed?

    a. Honeybees are being killed off and their hives destroyed by other extremely aggressive species of bees from the South.

    b. Honeybees are contracting parasites in large droves.

    c. Honeybees are so sensitive to certain pesticides that they die on contact.

    d. Honeybees die in larger numbers around cell phone towers.

48. Explain the Law of Conservation of Mass as it applies to this reaction: $2 H_2 + O_2 \rightarrow 2 H_2O$.

    a. Electrons are lost.

    b. The hydrogen loses mass.

    c. New oxygen atoms are formed.

    d. There is no decrease or increase of matter.

49. At what point in its swing does a pendulum have the most mechanical energy?

    a. At the top of its swing, just before going into motion

    b. At the bottom of its swing, in full motion

    c. Halfway between the top of its swing and the bottom of its swing

    d. It has the same amount of mechanical energy throughout its path

50. The energy of motion is also referred to as what?

    a. Potential energy

    b. Kinetic energy

    c. Solar energy

    d. Heat energy

51. Which of the following are functions of the urinary system?

    I. Synthesizing calcitriol and secreting erythropoietin

    II. Regulating the concentrations of sodium, potassium, chloride, calcium, and other ions

    III. Reabsorbing or secreting hydrogen ions and bicarbonate

    IV. Detecting reductions in blood volume and pressure

    a. I, II, and III

    b. II and III

    c. II, III, and IV

    d. All of the above

52. Using anatomical terms, what is the relationship of the sternum relative to the deltoid?

    a. Medial

    b. Lateral

    c. Superficial

    d. Posterior

53. Which of the following is a standard or series of standards to which the results from an experiment are compared?
    a. A control
    b. A variable
    c. A constant
    d. Collected data

54. What is the LAST phase of mitosis?
    a. Prophase
    b. Telophase
    c. Anaphase
    d. Metaphase

55. Which of the following is a type of boundary between two tectonic plates?
    a. Continental
    b. Oceanic
    c. Convergent
    d. Fault

56. Volcanic activity can occur in which of the following?
    a. Convergent boundaries
    b. Divergent boundaries
    c. The middle of a tectonic plate
    d. All of the above

57. Where is most of the Earth's weather generated?
    a. The troposphere
    b. The ionosphere
    c. The thermosphere
    d. The stratosphere

58. What type of cloud is seen when looking at the sky during a heavy rainstorm?
    a. High-Clouds
    b. Altocumulus
    c. Stratus
    d. Nimbostratus

59. What is the largest planet in our solar system and what is it mostly made of?
    a. Saturn, rocks
    b. Jupiter, ammonia
    c. Jupiter, hydrogen
    d. Saturn, helium

60. Viruses belong to which of the following classifications?
    a. Domain Archaea
    b. Kingdom Monera
    c. Kingdom Protista
    d. None of the above

Answer Explanations

1. A: When activated, B cells create antibodies against specific antigens. White blood cells are generated in red and yellow bone marrow, not cartilage. Platelets are not a type of white blood cell and are typically cell fragments produced by megakaryocytes. White blood cells are active throughout nearly all of one's life and have not been shown to specially activate or deactivate because of life events like puberty or menopause.

2. D: Mechanical digestion is physical digestion of food and tearing it into smaller pieces using force. This occurs in the stomach and mouth. Chemical digestion involves chemically changing the food and breaking it down into small organic compounds that can be utilized by the cell to build molecules. The salivary glands in the mouth secrete amylase that breaks down starch, which begins chemical digestion. The stomach contains enzymes such as pepsinogen/pepsin and gastric lipase, which chemically digest protein and fats, respectively. The small intestine continues to digest protein using the enzymes trypsin and chymotrypsin. It also digests fats with the help of bile from the liver and lipase from the pancreas. These organs act as exocrine glands because they secrete substances through a duct. Carbohydrates are digested in the small intestine with the help of pancreatic amylase, gut bacterial flora and fauna, and brush border enzymes like lactose. Brush border enzymes are contained in the towel-like microvilli in the small intestine that soak up nutrients.

3. D: The theory that certain physical and behavioral traits give a species an evolutionary advantage is called natural selection. Charles Darwin developed the theory of natural selection that explains the evolutionary process. He postulated that heritable genetic differences could aid an organism's chance of survival in its environment. The organisms with favorable traits pass genes to their offspring, and because they have more reproductive success than those that do not contain the adaptation, the favorable gene spreads throughout the population. Those that do not contain the adaptation often extinguish, thus their genes are not passed on. In this way, nature "selects" for the organisms that have more fitness in their environment. Birds with bright colored feathers and cacti with spines are examples of "fit" organisms.

4. B: The structure exclusively found in eukaryotic cells is the nucleus. Animal, plant, fungi, and protist cells are all eukaryotic. DNA is contained within the nucleus of eukaryotic cells, and they also have membrane-bound organelles that perform complex intracellular metabolic activities. Prokaryotic cells (archae and bacteria) do not have a nucleus or other membrane-bound organelles and are less complex than eukaryotic cells.

5. B: The cell structure responsible for cellular storage, digestion and waste removal is the lysosome. Lysosomes are like recycle bins. They are filled with digestive enzymes that facilitate catabolic reactions to regenerate monomers. The Golgi apparatus is designed to tag, package, and ship out proteins destined for other cells or locations. The centrioles typically play a large role only in cell division when they ratchet the chromosomes from the mitotic plate to the poles of the cell. The mitochondria are involved in energy production and are the powerhouses of the cell.

6. A: Crossing over, or genetic recombination, is the rearrangement of chromosomal sections in tetrads during meiosis, and it results in each gamete having a different combination of alleles than other gametes. The disassembly of the mitotic spindle happens only after telophase and is not related to diversity. While nondisjunction does cause diversity in division and is highly noticeable in gametes formed through meiosis, it can also happen through mitotic division in somatic cells. Although an egg being fertilized by multiple sperm would lead to interesting diversity in the offspring (and possibly fraternal twins), this is not strictly a byproduct of meiotic division.

7. D: Veins have valves, but arteries do not. Valves in veins are designed to prevent backflow, since they are the furthest blood vessels from the pumping action of the heart and steadily increase in volume (which decreases the available pressure). Capillaries diffuse nutrients properly because of their thin walls and high surface area and are not particularly dependent on positive pressure.

8. C: The blood leaves the right ventricle through a semi-lunar valve and goes through the pulmonary artery to the lungs. Any increase in pressure in the artery will eventually affect the contractibility of the right ventricle. Blood enters the right atrium from the superior and inferior venae cava veins, and blood leaves the right atrium through the tricuspid valve to the right ventricle. Blood enters the left atrium from the pulmonary veins carrying oxygenated blood from the lungs. Blood flows from the left atrium to the left ventricle through the mitral valve and leaves the left ventricle through a semi-lunar valve to enter the aorta.

9. D: Sodium bicarbonate, a very effective base, has the chief function to increase the pH of the chyme. Chyme leaving the stomach has a very low pH, due to the high amounts of acid that are used to digest and break down food. If this is not neutralized, the walls of the small intestine will be damaged and may form ulcers. Sodium bicarbonate is produced by the pancreas and released in response to pyloric stimulation so that it can neutralize the acid. It has little to no digestive effect.

10. D: A reflex arc is a simple nerve pathway involving a stimulus, a synapse, and a response that is controlled by the spinal cord—not the brain. The knee-jerk reflex is an example of a reflex arc. The stimulus is the hammer touching the tendon, reaching the synapse in the spinal cord by an afferent pathway. The response is the resulting muscle contraction reaching the muscle by an efferent pathway. None of the remaining processes is a simple reflex. Memories are processed and stored in the hippocampus in the limbic system. The visual center is located in the occipital lobe, while auditory processing occurs in the temporal lobe. The sympathetic and parasympathetic divisions of the autonomic nervous system control heart and blood pressure.

11. C: The lagging strand of DNA falls behind the leading strand because of its discontinuous synthesis. DNA helicase unzips the DNA helices so that synthesis can take place, and RNA primers are created by the RNA primase for the polymerases to attach to and build from. The lagging strand is synthesizing DNA in a direction that is hard for the polymerase to build, so multiple primers are laid down so that the entire length of DNA can be synthesized simultaneously, piecemeal. These short pieces of DNA being synthesized are known as Okazaki fragments and are joined together by DNA ligase.

12. D: Birth and fertility rates are typically higher in developing countries than in developed countries, and there are a wide variety of reasons for this. Some are rational or obvious, like how access to contraception decreases the number of pregnancies a woman is susceptible to, while others are justifiable, like how high infant mortality would cause women to want to have more children so that more of them make it to adulthood. Other correlations are evident, but not so easily explainable: Public education and urbanization rates are inversely correlated with national birth rates, so that if one goes up, the other goes down. Cholesterol intake has no proven association with birth rates.

13. B: In the Demographic Transition Model, developed countries tend to have low death or mortality rates because of improved healthcare and low birth rates due to access to contraception and myriad other factors. A high birth rate combined with a rapidly decreasing death rate indicates a country that is in the Early Expanding Stage 2 of the DTM, and neither emigration nor immigration plays a major role in any of the DTM stages.

14. A: The three primary body planes are coronal, sagittal, and transverse. The coronal or frontal plane, named for the plane in which a corona or halo might appear in old paintings, divides the body vertically into front and back sections. The sagittal plane, named for the path an arrow might take when shot at the body, divides the body vertically into right and left sections. The transverse plane divides the body horizontally into upper or superior and lower or inferior sections. There is no medial plane, per se. The anatomical direction medial simply references a location close or closer to the center of the body than another location.

15. C: Although the lungs may provide some cushioning for the heart when the body is violently struck, this is not a major function of the respiratory system. Its most notable function is that of gas exchange for oxygen and carbon dioxide, but it also plays a vital role in the regulation of blood pH. The aqueous form of carbon dioxide, carbonic acid, is a major pH buffer of the blood, and the respiratory system directly controls how much carbon dioxide stays and is released from the blood through respiration. The respiratory system also enables vocalization and forms the basis for the mode of speech and language used by most humans.

16. A: The forebrain contains the cerebrum, the thalamus, the hypothalamus, and the limbic system. The limbic system is chiefly responsible for the perception of emotions through the amygdala, while the cerebrum interprets sensory input and generates movement. Specifically, the occipital lobe receives visual input, and the primary motor cortex in the frontal lobe is the controller of voluntary movement. The hindbrain, specifically the medulla oblongata and brain stem, control and regulate blood pressure and heart rate.

17. D: Germline mutations in eggs and sperm are permanent, can be on the chromosomal level, and will be inherited by offspring. Somatic mutations cannot affect eggs and sperm, and therefore are not inherited by offspring. Mutations of either kind are rarely beneficial to the individual, but do not necessarily harm them. Germline cells divide much more rapidly than do somatic cells, and a mutation in a sex cell would promulgate and affect many thousands of its daughter cells.

18. C: Deserts' temperatures are extremely hot in the day and cold at night because of the warming effects of the sun's solar rays, so this is the best example of the sun's energy. Although some flowers do tend to bloom after dawn, this is probably due to day/night cycles regulated by the presence of light rather than intense amounts of energy. Hibernating animals tend to use large repositories of stored nutrients as energy sources rather than relying on the sun's energy, and they may in fact be in caves or hidden underground to shelter them from the sun or weather. The tides are more dependent on the moon due to its gravity rather than any effects its albedo moonlight may have.

19. C: 2:9:10:4. These are the coefficients that follow the law of conservation of matter. The coefficient times the subscript of each element should be the same on both sides of the equation.

20. C: A catalyst functions to increase reaction rates by decreasing the activation energy required for a reaction to take place. Inhibitors would increase the activation energy or otherwise stop the reactants from reacting. Although increasing the temperature usually increases a reaction's rate, this is not true in all cases, and most catalysts do not function in this manner.

21. B: Biological catalysts are termed enzymes, which are proteins with conformations that specifically manipulate reactants into positions which decrease the reaction's activation energy. Lipids do not usually affect reactions, and cofactors, while they can aid or be necessary to the proper functioning of enzymes, do not make up the majority of biological catalysts.

22. B: Substances with higher amounts of hydrogen ions will have lower pHs, while substances with higher amounts of hydroxide ions will have higher pHs. Choice *A* is incorrect because it is possible to have an extremely strong acid with a pH less than 1, as long as its molarity of hydrogen ions is greater than 1. Choice *C* is false because a weak base is determined by having a pH lower than some value, not higher. Substances with pHs greater than 2 include anything from neutral water to extremely caustic lye. Choice *D* is false because a solution with a pH of 2 has ten times fewer hydrogen ions than a solution of pH 1.

23. A: Salts are formed from compounds that use ionic bonds. Disulfide bridges are special bonds in protein synthesis which hold the protein in their secondary and tertiary structures. Covalent bonds are strong bonds formed through the sharing of electrons between atoms and are typically found in organic molecules like carbohydrates and lipids. London dispersion forces are fleeting, momentary bonds which occur between atoms that have instantaneous dipoles but quickly disintegrate.

24. C: As in the last question, covalent bonds are special because they share electrons between multiple atoms. Most covalent bonds are formed between the elements H, F, N, O, S, and C, while hydrogen bonds are formed nearly exclusively between H and either O, N, or F of other molecules. Covalent bonds may inadvertently form dipoles, but this does not necessarily happen. With similarly electronegative atoms like carbon and hydrogen, dipoles do not form, for example. Crystal solids are typically formed by substances with ionic bonds like the salts sodium iodide and potassium chloride.

25. A: Gases like air will move and expand to fill their container, so they are considered to have an indefinite shape and indefinite volume. Liquids like water will move and flow freely, so their shapes change constantly, but do not change volume or density on their own. Solids change neither shape nor volume without external forces acting on them, so they have definite shapes and volumes.

26. D: As they belong to the same family as the dodo, rock doves are the dodo's closest relatives of those given. Even though the ostrich is also flightless and the sandgrouse is also a bird, they are only in the same class as the dodo and are no closer. The common phylogenetic classification goes thus: Domain, Kingdom, Phylum, Class, Order, Family, Genus, Species. The further down the list, the more specific the genetic description is and the closer two organisms are related.

27. C: Photosynthetic autotrophs are able to synthesize their own energy by converting water, carbon dioxide, and solar energy into food. They do this by fixing carbon dioxide into glucose and performing aerobic respiration. Depending on which sugars they use in respiration, these organisms typically do indeed utilize glycolysis and the Krebs cycle. Autotrophs by definition do not consume other organisms for the purposes of obtaining glucose. Some autotrophs are bacteria without chloroplasts, and they can perform photosynthesis on their own.

28. C: When cells differentiate, they specialize from a general cell into a more focused cell with certain jobs and duties. The best example of this is when pluripotent cells like stem cells differentiate during development to become a certain kind of cell that makes a certain kind of protein. Choice *A* references cell lysis, Choice *B* references fertilization, and Choice *D* references cell senescence.

29. D: Aerobic respiration produces 36 molecules of ATP, and anaerobic respiration, through fermentation, produces only 4 molecules per molecule of glucose. Oxygen is not present in the anaerobic pathway, so it can't be depleted. The aerobic pathway has far fewer steps than the anaerobic pathway. Only about half of the energy produced by the aerobic pathway is used for ATP. The remaining energy is used for heat.

30. C: mRNA is directly transcribed from DNA before being taken to the cytoplasm and translated by rRNA into a protein. tRNA transfers amino acids from the cytoplasm to the rRNA for use in building these proteins. siRNA is a special type of RNA which interferes with other strands of mRNA typically by causing them to get degraded by the cell rather than translated into protein.

31. B: Since the genotype is a depiction of the specific alleles that an organism's genes code for, it includes recessive genes that may or may not be otherwise expressed. The genotype does not have to name the proteins that its alleles code for; indeed, some of them may be unknown. The phenotype is the physical, visual manifestations of a gene, not the genotype. The genotype does not necessarily include any information about the organism's physical characters. Although some information about an organism's parents can be obtained from its genotype, its genotype does not actually show the parents' phenotypes.

32. C: One in four offspring (or 25%) will be short, so all four offspring cannot be tall. Although both of the parents are tall, they are hybrid or heterozygous tall, not homozygous. The mother's phenotype is for tall, not short. A Punnett square cannot determine if a short allele will die out. Although it may seem intuitive that the short allele will be expressed by lower numbers of the population than the tall allele, it still appears in 75% of the offspring (although its effects are masked in 2/3 of those). Besides, conditions could favor the recessive allele and kill off the tall offspring.

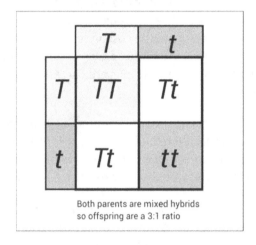

Both parents are mixed hybrids
so offspring are a 3:1 ratio

33. D: Evaporation takes place at the surface of a fluid while boiling takes place throughout the fluid. The liquid will boil when it reaches its boiling or vaporization temperature, but evaporation can happen due to a liquid's volatility. Volatile substances often coexist as a liquid and as a gas, depending on the pressure forced on them. The phase change from gas to liquid is condensation, and both evaporation and boiling take place in nature.

34. A: Human genes are strictly DNA and do not include proteins or amino acids. A human's genome and collection of genes will include even their recessive traits, mutations, and unused DNA.

35. C: Water's polarity lends it to be extremely cohesive and adhesive; this cohesion keeps its atoms very close together. Because of this, it takes a large amount of energy to melt and boil its solid and liquid forms. Phospholipid bilayers are made of nonpolar lipids and water, a polar liquid, cannot easily flow through it. Cell membranes use proteins called aquaporins to solve this issue and let water flow in and out. Fish breathe by capturing dissolved oxygen through their gills. Water can self-ionize, wherein it decomposes into a hydrogen ion (H+) and a hydroxide ion (OH-), but it cannot self-hydrolyze.

36. D: An isotope of an element has an atomic number equal to its number of protons, but a different mass number because of the additional neutrons. Even though there are differences in the nucleus, the behavior and properties of isotopes of a given element are identical. Atoms with different atomic numbers also have different numbers of protons and are different elements, so they cannot be isotopes.

37. A: The neutrons and protons make up the nucleus of the atom. The nucleus is positively charged due to the presence of the protons. The negatively charged electrons are attracted to the positively charged nucleus by the electrostatic or Coulomb force; however, the electrons are not contained in the nucleus. The positively charged protons create the positive charge in the nucleus, and the neutrons are electrically neutral, so they have no effect. Radioactivity does not directly have a bearing on the charge of the nucleus.

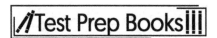

38. C: To gather accurate data, the student must be able compare a participant's test score from round 1 with their test score from round 2. The differing levels of intellect among the participants means that comparing participants' test scores to those of other participants would be inaccurate. This requirement excludes Choices *A* and *D*, which involve only one round of testing. The experiment must also involve different levels of sugar consumption from round 1 to round 2. In this way, the effects of different levels of sugar consumption can be seen on the same subjects. Thus, Choice *B* is incorrect because the experiment provides for no variation of sugar consumption. Choice *C* is the correct answer because it allows the student to compare each participant's test score from round 1 with their test score from round 2 after different levels of sugar consumption.

39. B: This question tests the test taker's ability to understand the role of technology to improve scientific investigations and communications. Choice *B* is the correct answer choice because, of the options listed, a spreadsheet-generated scatter plot is the most efficient, accurate way to create a scatter plot. Choice *A* can be accurate, but it is generally less efficient than using a spreadsheet. A person would be unlikely to find a scatter plot in a textbook that had the exact same point as the one they need to create. This method is also inefficient, so Choice *C* is incorrect. Choice *D* could work, but it would likely take longer and be less accurate than Choice *B*.

40. D: Based on this table, it can be reasoned that there is not a correlation between maintaining a calendar and GPA, since Student B never maintains a calendar but has the highest GPA of the cohort. Furthermore, it can be reasoned that there is a positive correlation between taking notes and GPA since the more notes a student takes, the higher the GPA they have. Thus, Choice *D* is the correct answer. Choice *A* offers an absolute that cannot be proven based on this study; thus, it is incorrect. Choices *B* and *C* are incorrect because they have at least one incorrect correlation.

41. D: *D* is the correct answer because excess sunlight is a common cause of plant wilting. Choices *A*, *B*, and *C* are all possible but unlikely to be a cause for wilting. Given that the test question asks for a reasonable explanation, sunlight is by far the most reasonable answer.

42. C: The parasympathetic nervous system is related to calm, peaceful times without stress that require no immediate decisions. It relaxes the fight-or-flight response, slows heart rate to a comfortable pace, and decreases bronchiole dilation to a normal size. The sympathetic nervous system, on the other hand, is in charge of the fight-or-flight response and works to increase blood pressure and oxygen absorption.

43. C: Neon, one of the noble gases, is chemically inert or not reactive because it contains eight valence electrons in the outermost shell. The atomic number is 10, with a 2.8 electron arrangement meaning that there are 2 electrons in the inner shell and the remaining 8 electrons in the outer shell. This is extremely stable for the atom, so it will not want to add or subtract any of its electrons and will not react under typical circumstances.

44. C: Because of the vast amounts of data that needed to be processed and analyzed, technological breakthroughs like innovations to the microprocessor were directly responsible for the ease of computing handled by the Human Genome Project. Although the sonogram and MRI technology are helpful to the healthcare industry in general, they would not have provided a great deal of help for sequencing and comprehending DNA data, in general. X-ray diffraction is a technique that helps visualize the structures of crystallized proteins, but cannot determine DNA bases with enough precision to help sequence DNA.

45. A: Many foods from developed countries are grown from plants which have been processed or bioengineered to include increased amounts of nutrients like vitamins and minerals that otherwise would be lost during manufacturing or are uncommon to the human diet. White rice, for example, is typically enriched with niacin, iron, and folic acid, while salt has been fortified with iodine for nearly a century. These help to prevent nutrition deficiencies. While it can be useful for fisheries to maintain models of fish populations so that they don't overfish their stock, this is not as immediately important to nutrition as are fortified and enriched foods. Although innovations to microscopes could lead to improved healthcare, this also has no direct effect on nutrition deficiency. Refrigerated train carts were historically a crucial invention around Civil War times and were used to transport meat and dairy products long distances without spoiling, but dietary deficiencies could be more easily remedied by supplying people with fortified foods containing those nutrients rather than spoilable meats.

46. B: In broad terms, energy is divided into kinetic and potential energy. Kinetic energy refers to an object in motion. It is the product of mass and velocity ($KE = \frac{1}{2}mv^2$). Potential energy refers to the capacity for doing work. Its gravitational configuration is the product of mass, acceleration due to gravity, and height ($PE = mgh$). Examples of kinetic energy include heat (which is the thermal energy from atoms and molecules moving around), waves like light, and physical motion. Potential energy examples include gravitational energy and chemical energy stored in bonds.

47. B: The most likely hypothesis that Anya is testing has something to with the pathophysiology of the bees, as she performed dissections on some of her samples. These dissections would be unnecessary if the bees were being killed off by another species, as the destruction of the hives would be obvious. As the deaths did not seem linked in any particular way geographically, it is also safe to assume that there was no correlation to cell phone towers, as maps of cell phone coverage would be readily available to her. If the pesticides were so toxic that the bees died on contact, then they wouldn't make it back to the hive to be available for her dissections or to drop off chemicals in the honey. The most likely of the choices is that parasites are killing off the bees, which would be easily communicable.

48. D: The law states that matter cannot be created or destroyed in a closed system. In this equation, there are the same number of molecules of each element on either side of the equation. Matter is not gained or lost, although a new compound is formed. As there are no ions on either side of the equation, no electrons are lost. The law prevents the hydrogen from losing mass and prevents oxygen atoms from being spontaneously spawned.

49. D: It has the same amount of mechanical energy throughout its path. Mechanical Energy is the total amount of energy in the situation; the sum of the potential energy and the kinetic energy. The amount of potential and kinetic energy both vary by the position of an object, but the mechanical energy remains constant.

50. B: Kinetic Energy. Kinetic energy is an energy an object has while moving; potential energy is energy an object has based on its position or height.

51. D: The urinary system has many functions, the primary of which is removing waste products and balancing water and electrolyte concentrations in the blood. It also plays a key role in regulating ion concentrations, such as sodium, potassium, chloride, and calcium, in the filtrate. The urinary system helps maintain blood pH by reabsorbing or secreting hydrogen ions and bicarbonate as necessary. Certain kidney cells can detect reductions in blood volume and pressure and then can secrete renin to activate a hormone that causes increased reabsorption of sodium ions and water. This serves to raise blood volume and pressure. Kidney cells secrete erythropoietin under hypoxic conditions to stimulate red blood cell production. They also synthesize calcitriol, a hormone derivative of vitamin D3, which aids in calcium ion absorption by the intestinal epithelium.

52. A: The sternum is medial to the deltoid because it is much closer (typically right on) the midline of the body, while the deltoid is lateral at the shoulder cap. Superficial means that a structure is closer to the body surface and posterior means that it falls behind something else. For example, skin is superficial to bone and the kidneys are posterior to the rectus abdominis.

53. A: A control is the component or group of the experimental design that isn't manipulated—it's the standard against which the resultant findings are compared, so Choice *A* is correct. A variable is an element of the experiment that is able to be manipulated, making Choice *B* false. A constant is a condition of the experiment outside of the hypothesis that remains unchanged in order to isolate the changes in the variables; therefore, Choice *C* is incorrect. Choice *D* is false because collected data are simply recordings of the observed phenomena that result from the experiment.

54. B: During telophase, two nuclei form at each end of the cell and nuclear envelopes begin to form around each nucleus. The nucleoli reappear, and the chromosomes become less compact. The microtubules are broken down by the cell, and mitosis is complete. The process begins with prophase as the mitotic spindles begin to form from centrosomes. Prometaphase follows, with the breakdown of the nuclear envelope and the further condensing of the chromosomes. Next, metaphase occurs when the microtubules are stretched across the cell and the chromosomes align at the metaphase plate. Finally, in the last step before telophase, anaphase occurs as the sister chromatids break apart and form chromosomes.

55. C: Convergent plate boundaries occur where two tectonic plates collide together. The denser oceanic plate will drop below the continental plate in a process called subduction.

56. D: Volcanic activity can occur at both fault lines and within the area of a tectonic plate at areas called hot spots. Volcanic activity is more common at fault lines because of cracks that allow the mantle's magma to more easily escape to the surface.

57. A: Technically, the troposphere is a layer of the atmosphere where the majority of the activity that creates weather conditions experienced on Earth occurs. The ozone layer is in the stratosphere; this is also where airplanes fly.

58. D: Stratus clouds are also grey, but nimbostratus clouds are the low clouds that appear during stormy weather. The other choices are usually seen on fair-weather days.

59. C: Jupiter is the largest planet in the solar system, and it is primarily composed of hydrogen and helium. Ammonia is in much lower quantity and usually found as a cloud within Jupiter's atmosphere.

60. D: Viruses are not classified as living organisms. They are neither prokaryotic or eukaryotic; therefore, they don't belong to any of the answer choices.

Social Studies

History

Analyzing Historical Sources and Recognizing Perspectives

Often history is interpreted or taught as a mere timeline or a series of bland facts about the past, but history should also be understood as a "lived experience." All humans are *a part of* history—they are the historical actors and personas who make positive (or negative) changes in the world. Human beings are constantly interacting with the super-structural forces of history. History occurs in interlocking webs of mutual reciprocity.

History happens in a context of local and global events. The people recording that history (in whatever format) are part of that context and therefore shaped by it. No one merely records statistics, facts, and figures. Each thing recorded is done so because it is important for some reason to the one recording it. Those who study history must do their best to understand the people and places they study as well as understand themselves in their own historical context.

Good historians ask questions prior to reading or studying what has been left for them by prior generations. What was important to the person who left this record? Were they rich or poor? Where they weak or powerful? What was their particular view of the world? What was their view of themselves and the group(s) they belonged to and their perceived place in history? These and other questions are critical to better understanding what was recorded and why it was considered important. It also helps provide a context for understanding the record left for posterity.

The historian must also understand their own biases, worldview, preconceptions, and context so that they can be aware of who they are and what they believe, because it influences the way they read, interpret, and understand the historical record.

If one were to analyze current events, they will get a clearer view of how this works. The same event can be recorded by two different people and sound like two different events because of the way the information is reported. For example, two people might report on an event during a time of war. The first might be a pacifist and therefore would be opposed to the war and that bias would be seen in how they reported on the conflict. Someone else might speak of the same events and make them seem heroic because they are very much in favor of their country's involvement in the conflict. The same historical event is being recorded but with two very different intents, understandings, and interpretations.

The historian who comes to this information (or the modern reader in the current events case) needs to also be aware of their personal views and how that affects their understanding of what they are reading. They may read sympathetically if they share the bias of the original author. They may also react in great opposition to what was recorded if their own view varies sharply from that of the original recorder.

Awareness of the times, backgrounds, purposes, and influences on both the original recorder and the one examining the record must be taken into account when analyzing historical sources.

Interconnections Among the Past, Present, and Future

Every time someone studies history, it is very much a collision of past, present, and future. Historians are concerned for the past, rooted in the present, and thinking about the future. Historical analysis is, therefore, a process infusing the present in the past in hopes of predicting (or deterring) certain social interactions in the future.

When examining the historical narratives of events, it is important to understand the relationship between *causes* and *effects*. A *cause* can be defined as something, whether an event, social change, or other factor, that contributes to the occurrence of certain events; the results of causes are called *effects*. Those terms may seem simple enough, but they have drastic implications on how one explores history. Events such as the American Revolution or the Civil Rights Movement may appear to occur spontaneously, but a closer examination will reveal that these events depended on earlier phenomena and patterns that influenced the course of history.

For example, although the battles at Concord and Lexington may seem to be instantaneous eruptions of violence during the American Revolution, they stemmed from a variety of factors. The most obvious influences behind those two battles were the assortment of taxes and policies imposed on the Thirteen Colonies following the French and Indian War from 1754 to 1763. Taxation without direct representation, combined with the deployment of British soldiers to enforce these policies, greatly increased American resistance. Earlier events, such as the Boston Massacre and the Boston Tea Party, similarly stemmed from conflicts between British soldiers and local colonists over perceived tyranny and rebelliousness. Therefore, the start of the American Revolution progressed from earlier developments.

Furthermore, there can be multiple causes and effects for any situation. The existence of multiple causes can be seen through the settling of the American West. Many historians have emphasized the role of *manifest destiny*—the national vision of expanding across the continent—as a driving force behind the growth of the United States. Yet there were many different influences behind the expansion westward. Northern abolitionists and southern planters saw the frontier as a way to either extend or limit slavery. Economic opportunities in the West also encouraged travel westward, as did the gradual pacification, relocation, or eradication of Native American tribes. In fact, manifest destiny as well as economic and political reasons played significant roles in justifying the pacification, relocation, or eradication of the Native American tribal nations.

Even an individual cause can be subdivided into smaller factors or stretched out in a gradual process. Although there were numerous issues that led to the Civil War, slavery was the primary cause. However, that topic stretched back to the very founding of the nation, and the existence of slavery was a controversial topic during the creation of the Declaration of Independence and the Constitution. The abolition movement as a whole did not start until the 1830s, but nevertheless, slavery is a cause that gradually grew more important over the following decades. In addition, opponents of slavery were divided by different motivations—some believed that it stifled the economy, while others focused on moral issues.

On the other end of the spectrum, a single event can have numerous results. The rise of the telegraph, for example, had several effects on American history. The telegraph allowed news to travel much quicker and turned events into immediate national news, such as the sinking of the USS *Maine,* which sparked the Spanish-American War. In addition, the telegraph helped make railroads run more efficiently by improving the links between stations. The faster speed of both travel and communications led to a shift in time itself, and localized times were replaced by standardized *time zones* across the nation.

By looking at different examples of cause and effect closely, it becomes clear that no event occurs without one—if not multiple—causes behind it, and that each historical event can have a variety of direct and indirect consequences.

One of the most critical elements of cause-and-effect relationships is how they are relevant not only in studying history but also in contemporary events. People must realize that events and developments today will likely have a number of consequences later on. Therefore, the study of cause and effect remains vital in understanding the past, the present, and the future.

Specific Eras in World and U.S. History

Classical Civilizations

There were a number of powerful civilizations during the classical period. Mesopotamia was home to one of the earliest civilizations between the Euphrates and the Tigris rivers in the Near East. The rivers provided water and vegetation for early humans, but they were surrounded by desert. This led to the beginning of irrigation efforts to expand water and agriculture across the region, which resulted in the area being known as the Fertile Crescent.

The organization necessary to initiate canals and other projects led to the formation of cities and hierarchies, which would have considerable influence on the structure of later civilizations. For example, the new hierarchies established different classes within the societies, such as kings, priests, artisans, and workers. Over time, these city-states expanded to encompass outside territories, and the city of Akkad became the world's first empire in 2350 B.C. In addition, Mesopotamian scribes developed systemized drawings called pictograms, which were the first system of writing in the world; furthermore, the creation of wedge-shaped cuneiform tablets preserved written records for multiple generations.

Later, Mesopotamian kingdoms made further advancements. For example, Babylon established a sophisticated mathematical system based on numbers from one to sixty; this not only influenced modern concepts, such as the number of minutes in each hour, but also created the framework for math equations and theories. In addition, the Babylonian king Hammurabi established a complex set of laws, known as the Code of Hammurabi, which would set a precedent for future legal systems.

Meanwhile, another major civilization began to form around the Nile River in Africa. The Nile's relatively predictable nature allowed farmers to use the river's water and the silt from floods to grow many crops along its banks, which led to further advancements in irrigation. Egyptian rulers mobilized the kingdom's population for incredible construction projects, including the famous pyramids. Egyptians also improved pictographic writing with their more complex system of hieroglyphs, which allowed for more diverse styles of writing. The advancements in writing can be seen through the Egyptians' complex system of religion, with documents such as the *Book of the Dead* outlining not only systems of worship and pantheons of deities but also a deeper, more philosophical concept of the afterlife.

While civilizations in Egypt and Mesopotamia helped to establish class systems and empires, other forms of government emerged in Greece. Despite common ties between different cities, such as the Olympic Games, each settlement, known as a polis, had its own unique culture. Many of the cities were oligarchies, in which a council of distinguished leaders monopolized the government; others were dictatorships ruled by tyrants. Athens was a notable exception by practicing an early form of democracy

in which free, landholding men could participate, but it offered more freedom of thought than other systems.

Taking advantage of their proximity to the Mediterranean Sea, Greek cities sent expeditions to establish colonies abroad that developed their own local traditions. In the process, Greek merchants interacted with Phoenician traders, who had developed an alphabetic writing system built on sounds instead of pictures. This diverse network of exchanges made Greece a vibrant center of art, science, and philosophy. For example, the Greek doctor Hippocrates established a system of ethics for doctors called the Hippocratic Oath, which continues to guide the modern medical profession. Complex forms of literature were created, including the epic poem "The Iliad," and theatrical productions were also developed. Athens in particular sought to spread its vision of democratic freedom throughout the world, which led to the devastating Peloponnesian War between allies of Athens and those of oligarchic Sparta from 431 to 404 B.C.

Alexander the Great helped disseminate Greek culture to new regions. Alexander was in fact an heir to the throne of Macedon, which was a warrior kingdom to the north of Greece. After finishing his father's work of unifying Greece under Macedonian control, Alexander successfully conquered Mesopotamia, which had been part of the Persian Empire. The spread of Greek institutions throughout the Mediterranean and Near East led to a period of Hellenization, during which various civilizations assimilated Greek culture; this allowed Greek traditions, such as architecture and philosophy, to endure into the present day.

Greek ideas were later assimilated, along with many other concepts, into the Roman Empire. Located west of Greece on the Italian peninsula, the city of Rome gradually conquered its neighbors and expanded its territories abroad; by 44 B.C., Rome had conquered much of Western Europe, northern Africa, and the Near East. Romans were very creative, and they adapted new ideas and innovated new technologies to strengthen their power. For example, Romans built on the engineering knowledge of Greeks to create arched pathways, known as aqueducts, to transport water for long distances and devise advanced plumbing systems.

One of Rome's greatest legacies was its system of government. Early Rome was a republic, a democratic system in which leaders are elected by the people. Although the process still heavily favored wealthy elites, the republican system was a key inspiration for later institutions such as the United States. Octavian "Augustus" Caesar later made Rome into an empire, and the senate had only a symbolic role in the government. The new imperial system built on the examples of earlier empires to establish a vibrant dynasty that used a sophisticated legal code and a well-trained military to enforce order across vast regions. Even after Rome itself fell to barbarian invaders in fifth century A.D., the eastern half of the empire survived as the Byzantine Empire until 1453 A.D. Furthermore, the Roman Empire's institutions continued to influence and inspire later medieval kingdoms, including the Holy Roman Empire; even rulers in the twentieth century called themselves Kaiser and Tsar, titles which stem from the word "Caesar."

In addition, the Roman Empire was host to the spread of new religious ideas. In the region of Israel, the religion of Judaism presented a new approach to worship via monotheism, which is the belief in the existence of a single deity. An offshoot of Judaism called Christianity spread across the Roman Empire and gained popularity. While Rome initially suppressed the religion, it later backed Christianity and allowed the religious system to endure as a powerful force in medieval times.

Twentieth-Century Development and Transformations in World History

At the turn of the twentieth century, imperialism had led to powers, such as France, the United States, and Japan, to establish spheres of influence throughout the world. The combination of imperial competition and military rivalries led to the outbreak of World War I when Archduke Ferdinand of Austria was assassinated in 1914. The war pitted the Allies, including England, France, and Russia, against the Central Powers of Austria-Hungary, Germany, and the Ottoman Empire—a large Islamic realm that encompassed Turkey, Palestine, Saudi Arabia, and Iraq. The rapid advances in military technology turned the war into a prolonged bloodbath that took its toll on all sides. By the end of the war in 1918, the Ottoman Empire had collapsed, the Austrian-Hungarian Empire was split into multiple countries, and Russia had descended into a civil war that would lead to the rise of the Soviet Union and Communism.

The Treaty of Versailles ended the war, but the triumphant Allies also levied heavy fines on Germany, which led to resentment that would be accentuated by the Great Depression of the 1930s. The Great Depression destabilized the global economy and led to the rise of fascism, a militarized and dictatorial system of government, in nations such as Germany and Italy. The rapid expansion of the Axis Powers of Germany, Italy, and Japan led to the outbreak of World War II. The war was even more global than the previous conflicts, with battles occurring in Europe, Africa, and Asia. World War II encouraged the development of new technologies, such as advanced radar and nuclear weapons, that would continue to influence the course of future wars.

In the aftermath of World War II, the United Nations was formed as a step toward promoting international cooperation. Based on the preceding League of Nations, the United Nations included countries from around the world and gave them a voice in world policies. The formation of the United Nations coincided with the independence of formerly colonized states in Africa and Asia, and those countries joined the world body. A primary goal of the United Nations was to limit the extent of future wars and prevent a third world war; while the United Nations could not prevent the outbreak of wars, it nevertheless tried to peacefully resolve them. In addition to promoting world peace, the United Nations also helped protect human rights.

Even so, the primary leadership in the early United Nations was held by the United States and its allies, which contributed to tensions with the Soviet Union. The United States and the Soviet Union, while never declaring war on each other, fueled a number of proxy wars and coups across the world in what would be known as the Cold War. Cold War divisions were especially noticeable in Europe, where communist regimes ruled the eastern region and democratic governments controlled the western portion. These indirect struggles often involved interference with foreign politics, and sometimes local people began to resent Soviet or American attempts to influence their countries. For example, American and Soviet interventions in Iran and Afghanistan contributed to fundamentalist Islamic movements. The Cold War ended when the Soviet Union collapsed in 1991, but the conflict affected nations across the globe and continues to influence current issues.

Another key development during the twentieth century, as noted earlier with the United Nations, was that most colonized nations broke free from imperial control and asserted their independence. Although these nations achieved autonomy and recognition in the United Nations, they still suffered from the legacies of imperialism. The borders of many countries in Africa and Asia were arbitrarily determined by colonists with little regard to the arrangement of native populations. Therefore, many former colonies have suffered conflicts between different ethnic groups; this was also the case with the British colony in India, which became independent in 1947. Violence occurred when it split into India and Pakistan because the borders were largely based on religious differences. In addition, former colonial powers

continue to assert economic control that inhibits the growth of native economies. On the other hand, the end of direct imperialism has helped a number of nations, such as India and Iran, rise as world powers that have significant influence on the world as a whole.

Additionally, there were considerable environmental reforms worldwide during the twentieth century. In reaction to the growing effects of industrialization, organizations around the world protested policies that damaged the environment. Many of these movements were locally based, but others expanded to address various environmental threats across the globe. The United Nations helped carry these environmental reforms forward by making them part of international policies. For example, in 1997, many members of the United Nations signed a treaty, known as the *Kyoto Protocol*, that tried to reduce global carbon dioxide emissions.

Most significantly, the twentieth century marked increasing globalization. The process had already been under way in the nineteenth century as technological improvements and imperial expansions connected different parts of the world, but the late twentieth century brought globalization to a new level. Trade became international, and local customs from different lands also gained prominence worldwide. Cultural exchanges occur on a frequent basis, and many people have begun to ponder the consequences of such rapid exchanges. One example of globalization was the 1993 establishment of the European Union—an economic and political alliance between several European nations.

European Exploration and Colonization in the U.S.

When examining how Europeans explored what would become the United States of America, one must first examine why Europeans came to explore the New World as a whole. In the fifteenth century, tensions increased between the Eastern and Mediterranean nations of Europe and the expanding Ottoman Empire to the east. As war and piracy spread across the Mediterranean, the once-prosperous trade routes across Asia's Silk Road began to decline, and nations across Europe began to explore alternative routes for trade.

Italian explorer Christopher Columbus proposed a westward route. Contrary to popular lore, the main challenge that Columbus faced in finding backers was not proving that the world was round. Much of Europe's educated elite knew that the world was round; the real issue was that they rightly believed that a westward route to Asia, assuming a lack of obstacles, would be too long to be practical. Nevertheless, Columbus set sail in 1492 after obtaining support from Spain and arrived in the West Indies three months later.

Spain launched further expeditions to the new continents and established *New Spain*. The colony consisted not only of Central America and Mexico, but also the American Southwest and Florida. France claimed much of what would become Canada, along with the Mississippi River region and the Midwest. In addition, the Dutch established colonies that covered New Jersey, New York, and Connecticut. Each nation managed its colonies differently, and thus influenced how they would assimilate into the United States. For example, Spain strove to establish a system of Christian missions throughout its territory, while France focused on trading networks and had limited infrastructure in regions such as the Midwest.

Even in cases of limited colonial growth, the land of America was hardly vacant, because a diverse array of Native American nations and groups were already present. Throughout much of colonial history, European settlers commonly misperceived native peoples as a singular, static entity. In reality, Native Americans had a variety of traditions depending on their history and environment, and their culture continued to change through the course of interactions with European settlers; for example, tribes such as the Cheyenne and Comanche used horses, which were introduced by white settlers, to become powerful warrior nations. However, a few generalizations can be made: many, but not all, tribes were matrilineal, which gave women a fair degree of power, and land was commonly seen as belonging to everyone. These differences, particularly European settlers' continual focus on land ownership, contributed to increasing prejudice and violence.

Situated on the Atlantic Coast, the Thirteen Colonies that would become the United States of America constituted only a small portion of North America. Even those colonies had significant differences that stemmed from their different origins. For example, the Virginia colony under John Smith in 1607 started with male bachelors seeking gold, whereas families of Puritans settled Massachusetts. As a result, the Thirteen Colonies—Virginia, Massachusetts, Connecticut, Maryland, New York, New Jersey, Pennsylvania, Delaware, Rhode Island, New Hampshire, Georgia, North Carolina, and South Carolina— had different structures and customs that would each influence the United States.

Competition among several imperial powers in eastern areas of North America led to conflicts that would later bring about the independence of the United States. The French and Indian War from 1754 to 1763, which was a subsidiary war of the Seven Years' War, ended with Great Britain claiming France's Canadian territories as well as the Ohio Valley. The same war was costly for all the powers involved, which led to increased taxes on the Thirteen Colonies. In addition, the new lands to the west of the colonies attracted new settlers, and they came into conflict with Native Americans and British troops that were trying to maintain the traditional boundaries. These growing tensions with Great Britain, as well as other issues, eventually led to the American Revolution, which ended with Britain relinquishing its control of the colonies.

Britain continued to hold onto its other colonies, such as Canada and the West Indies, which reflects the continued power of multiple nations across North America, even as the United States began to expand across the continent. Many Americans advocated expansion regardless of the land's current inhabitants, but the results were often mixed. Still, events both abroad and within North America contributed to the growth of the United States. For example, the rising tumult in France during the French Revolution and the rise of Napoleon led France to sell the Louisiana Purchase, a large chunk of land consisting not only of Louisiana but also much of the Midwest, to the United States in 1803. Meanwhile, as Spanish power declined, Mexico claimed independence in 1821, but the new nation became increasingly vulnerable to foreign pressure. In the Mexican-American War from 1846 to 1848, Mexico surrendered territory to the United States that eventually became California, Nevada, Utah, and New Mexico, as well as parts of Arizona, Colorado, and Wyoming.

Even as the United States sought new inland territory, American interests were also expanding overseas via trade. As early as 1784, the ship *Empress of China* traveled to China to establish trading connections. American interests had international dimensions throughout the nation's history. For example, during the presidency of Andrew Jackson, the ship *Potomac* was dispatched to the Pacific island of Sumatra in 1832 to avenge the deaths of American sailors. This incident exemplifies how U.S. foreign trade connected with imperial expansion.

This combination of continental and seaward growth adds a deeper layer to American development, because it was not purely focused on western expansion. For example, take the 1849 Gold Rush; a large number of Americans and other immigrants traveled to California by ship and settled western territories before more eastern areas, such as Nevada and Idaho. Therefore, the United States' early history of colonization and expansion is a complex network of diverse cultures.

The American Revolution and the Founding of the Nation in United States History

The American Revolution largely occurred as a result of changing values in the Thirteen Colonies that broke from their traditional relationship with England. Early on in the colonization of North America, the colonial social structure tried to mirror the stratified order of Great Britain. In England, the landed elites were seen as intellectually and morally superior to the common man, which led to a paternalistic relationship. This style of governance was similarly applied to the colonial system; government was left to the property-owning upper class, and the colonies as a whole could be seen as a child dutifully serving "Mother England."

However, the colonies' distance from England meant that actual, hereditary aristocrats from Britain only formed a small percentage of the overall population and did not even fill all the positions of power. By the mid-eighteenth century, much of the American upper class consisted of local families who acquired status through business rather than lineage. Despite this, representatives from Britain were appointed to govern the colonies. As a result, a rift began to form between the colonists and British officials.

Tensions began to rise in the aftermath of the French and Indian War of 1754 to 1763. To recover the financial costs of the long conflict, Great Britain drew upon its colonies to provide the desired resources. Since the American colonists did not fully subscribe to the paternal connection, taxation to increase British revenue, such as the Stamp Act of 1765, was met with increasing resistance. Britain sent soldiers to the colonies and enacted the 1765 Quartering Act to require colonists to house the troops. In 1773, the new Tea Act, which created a monopoly, led some colonists to raid a ship and destroy its contents in the Boston Tea Party.

Uncertain about whether they should remain loyal to Britain, representatives from twelve colonies formed the First Continental Congress in 1774 to discuss what they should do next. When Patriot militiamen at Lexington and Concord fought British soldiers in April 1775, the Revolutionary War began. While the rebel forces worked to present the struggle as a united, patriotic effort, the colonies remained divided throughout the war. Thousands of colonists, known as Loyalists or Tories, supported Britain. Even the revolutionaries proved to be significantly fragmented, and many militias only served in their home states. The Continental Congress was also divided over whether to reconcile with Britain or push for full separation. These issues hindered the ability of the revolutionary armies to resist the British, who had superior training and resources at their disposal.

Even so, the Continental Army, under General George Washington, gradually built up a force that utilized Prussian military training and backwoods guerrilla tactics to make up for their limited resources. Although the British forces continued to win significant battles, the Continental Army gradually reduced Britain's will to fight as the years passed. Furthermore, Americans appealed to the rivalry that other European nations had with the British Empire. The support was initially limited to indirect assistance, but aid gradually increased. After the American victory at the Battle of Saratoga in 1777, France and other nations began to actively support the American cause by providing much-needed troops and equipment.

In 1781, the primary British army under General Cornwallis was defeated by an American and French coalition at Virginia, which paved the way for negotiations. The Treaty of Paris in 1783 ended the war, recognized the former colonies' independence from Great Britain, and gave America control over territory between the Appalachian Mountains and Mississippi River. However, the state of the new nation was still uncertain. The new nation's government initially stemmed from the state-based structure of the Continental Congress and was incorporated into the Articles of Confederation in 1777.

The Articles of Confederation emphasized the ideals of the American Revolution, particularly the concept of freedom from unjust government. Unfortunately, the resulting limitations on the national government left most policies—even ones with national ramifications—up to individual states. For example, states sometimes simply decided to not pay taxes. Many representatives did not see much value in the National Congress and simply did not attend the meetings. Some progress was still made during the period, such as the Northwest Ordinance of 1787, which organized the western territories into new states; nevertheless, the disjointed links in the state-oriented government inhibited significant progress.

Although many citizens felt satisfied with this decentralized system of government, key intellectuals and leaders in America became increasingly disturbed by the lack of unity. An especially potent fear among them was the potential that, despite achieving official independence, other powers could threaten America's autonomy. In 1786, poor farmers in Massachusetts launched an insurrection, known as Shays' Rebellion, which sparked fears of additional uprisings and led to the creation of the *Constitutional Convention* in 1787.

While the convention initially intended to correct issues within the Articles of Confederation, speakers, such as James Madison, compellingly argued for the delegates to devise a new system of government that was more centralized than its predecessor. The Constitution was not fully supported by all citizens, and there was much debate about whether or not to support the new government. Even so, in 1788, the Constitution was ratified. Later additions, such as the Bill of Rights, would help protect individual liberty by giving specific rights to citizens. In 1789, George Washington became the first president of the newly created executive branch of the government, and America entered a new stage of history.

Major Events and Developments in U.S. History from Founding to Present

One early development was the growth of political parties—something that Washington tried and failed to stop from forming. Federalists, such as Alexander Hamilton, wanted to expand the national government's power, while Democratic-Republicans, such as Thomas Jefferson, favored states' rights. The United States suffered multiple defeats by Britain in the War of 1812, but individual American victories, such as the Battle of New Orleans, still strengthened nationalistic pride.

In the aftermath of the war, the Federalists were absorbed into the Democratic-Republicans, which began the Era of Good Feelings. However, two new parties eventually emerged. The Democrats, whose leader Andrew Jackson became president in 1828, favored "Jacksonian" democracy, which emphasized mass participation in elections. However, Jackson's policies largely favored white male landowners and suppressed opposing views. The Whigs supported Federalist policies but also drew on democratic principles, particularly with marginalized groups such as African Americans and women.

At the same time, settlers continued to expand west in search of new land and fortune. The Louisiana Purchase of 1803 opened up large amounts of land west of the Mississippi River, and adventurers pushed past even those boundaries toward the western coast. The vision of westward growth into the frontier is a key part of American popular culture, but the expansion was often erratic and depended on a combination of incentives and assurances of relative security. Hence, some areas, such as California and Oregon, were settled more quickly than other areas to the east. Some historians have pointed to the growth of the frontier as a means through which American democracy expanded.

However, the matter of western lands became an increasingly volatile issue as the controversy over slavery heightened. Not all northerners supported abolition, but many saw the practice as outdated and did not want it to expand. Abolitionists formed the Republican Party, and their candidate, Abraham

Lincoln, was elected as president in 1860. In response, southern states seceded and formed the Confederate States of America. The ensuing Civil War lasted from 1861 to 1865 and had significant consequences. Slavery was abolished in the United States, and the power of individual states was drastically curtailed. After being reunified, southern states worked to retain control over freed slaves, and the Reconstruction period was followed by Jim Crow segregation. As a result, blacks were barred from public education, unable to vote, and forced to accept their status as second-class citizens.

After the Civil War, the United States increasingly industrialized and became part of the larger Industrial Revolution, which took place throughout the western world. Steps toward industrialization had already begun as early as Jackson's presidency, but the full development of American industry took place in the second half of the nineteenth century. Railroads helped link cities like Chicago to locations across the West, which allowed for rapid transfer of materials. New technologies, such as electricity, allowed leisure time for those with enough wealth. Even so, the Gilded Age was also a period of disparities, and wealthy entrepreneurs rose while impoverished workers struggled to make their voices heard.

The late nineteenth and early twentieth century not only marked U.S. expansion within North America but also internationally. For example, after the Spanish-American War in 1898, the United States claimed control over Guam, Puerto Rico, and the Philippines. Rivalries in Europe culminated in World War I, in which great powers ranging from France to Russia vied for control in a bloody struggle. Americans did not enter the war until 1917, but we had a critical role in the final phase of the war. During the peace treaty process, President Woodrow Wilson sought to establish a League of Nations in order to promote global harmony, but his efforts only achieved limited success.

After World War I, the United States largely stayed out of international politics for the next two decades. Still, American businesses continued overseas ventures and strengthened the economy in the 1920s. However, massive speculation in the stock market in 1929 triggered the Great Depression—a financial crisis that spread worldwide as nations withdrew from the global economy. The crisis shepherded in the presidency of Franklin D. Roosevelt, who reformed the Democratic Party and implemented new federal programs known as the New Deal.

The Great Depression had ramifications worldwide and encouraged the rise of fascist governments in Italy and Germany. Highly dictatorial, fascism emphasized nationalism and militarism. World War II began when the Axis powers of Germany, Italy, and Japan built up their military forces and launched invasions against neighboring nations in 1939. As part of the Allies, which also included Britain, France, and the Soviet Union, America defeated the Axis powers in 1945 and asserted itself as a global force.

The Union of Soviet Socialist Republics had emerged through the Bolshevik Revolution in 1917 in Russia and militantly supported Communism—a socialist system of government that called for the overthrow of capitalism. Although the Soviet Union formed an alliance with the United States during World War II, relations chilled, and the Cold War began in 1947. Although no true war was declared between the two nations, both the Union of Soviet Socialist Republics and the United States engaged in indirect conflict by supporting and overthrowing foreign governments.

Meanwhile, the Civil Rights Movement began to grow as marginalized groups objected to racial segregation and abuse by whites across the nation. Civil rights leaders, such as Martin Luther King Jr., argued for nonviolent resistance, but others, such as Malcolm X, advocated more radical approaches. Civil rights groups became increasingly discontented during the Vietnam War because they felt they were being drafted for a foreign war that ignored domestic problems. Even so, significant reforms, such as the Voting Rights Act of 1965, opened up new opportunities for freedom and equality in America.

In 1991, the Soviet Union collapsed, leaving the United States as the dominant global power. However, as the United States struggled to fill the void left by the Soviet Union, questions arose about America's role in the world. Terrorist acts, such as the 9/11 attack on the World Trade Center in 2001, have shed doubt on the United States' ability to enforce its authority on an international scale.

Twentieth-Century Developments and Transformations in the United States

Although the United States began industrializing in the second half of the nineteenth century, American technology continued to develop in new directions throughout the course of the twentieth century. A key example was the invention of the modern assembly line. Assembly lines and conveyor belts had already become a prominent part of industrial work, but Henry Ford combined conveyor belts with the system of assembly workers in 1913 in order to produce Model T automobiles. This streamlined production system, in which multiple parts were assembled by different teams along the conveyors, allowed industries in the United States to grow ever larger.

Ford's assembly lines also promoted the growth of the automobile as a means of transportation. Early cars were an expensive and impractical novelty and were primarily the toys of the rich. The Model T, on the other hand, was relatively affordable, which made the car available to a wider array of consumers. Many of the automobiles' early issues, such as radiator leaks and fragile tires, were gradually corrected, and this made the car more appealing than horses. With the support of President Eisenhower, the Federal Aid Highway Act of 1956 paved the way for a network of interstates and highways across the nation.

At the same time, a revolutionary approach to transportation was emerging: flight. Blimps and balloons were already gaining popularity by the turn of the twentieth century, but aviators struggled to create an airplane. The first critical success was by the Wright Brothers in 1903, and they demonstrated that aircrafts did not need to be lighter than air. In time, airplanes surpassed the popularity of balloons and blimps, which tended to be more volatile. Aircraft also added a new dimension to warfare, and aircraft carriers became an integral piece of the American navy during World War II.

Furthermore, by demonstrating that heavier-than-air vehicles could actually carry passengers upward, the stage was set for the space race in the second half of the twentieth century. In 1958, the U.S. government created the National Aeronautics and Space Administration (NASA) to head the budding initiative to extend American power into space. After the Soviet Union successfully launched the Sputnik satellite into Earth's orbit in 1957 and sent the first human in space in 1961, the United States intensified its own space program through the Apollo missions. Apollo 11 successfully landed on the moon in 1969 with Buzz Aldrin and Neil Armstrong. Later ventures into space would focus on space shuttles and satellites, and the latter significantly enhanced communications worldwide.

Indeed, the twentieth century also made considerable advancements in communications and media. Inventions such as the radio greatly boosted communication across the nation and world, such that news could be reported immediately rather than take days. Furthermore, motion pictures evolved from black-and-white movies at theaters to full-color television sets in households. From animation to live films, television matured into a compelling art form in popular culture. Live-action footage gave a new layer to news broadcasts and proved instrumental in the public's reaction to events, such as the Civil Rights Movement and the Vietnam War. With the success of the space program, satellites became a fundamental piece of Earth's communications network by transmitting signals across the planet instantaneously.

Further communications advancements resulted from the development of computer technology. The early computers in the twentieth century were enormous behemoths that were too bulky and expensive for anything but government institutions. However, computers gradually became smaller while still storing large amounts of data. A turning point came with the 1976 release of the Apple computer by entrepreneurs Steve Wozniak and Steve Jobs. The computer had a simplistic design that made it marketable for a mass consumer audience, and computers eventually became household items. Similarly, the networks that would become the Internet originated as government systems, but in time they were extended to commercial avenues that became a vibrant element of modern communications.

However, other advancements in American science during the twentieth century were aimed toward more lethal purposes. In response to the multiple wars throughout the century, the United States built up a powerful military force, and new technologies were devised for that purpose. One of the deadliest creations was the atomic bomb, which split molecular atoms to produce powerful explosions; in addition to the sheer force of the bombs, the aftereffects included toxic radiation and electronic shutdowns. Developed and used in the last days of World War II, the nuclear bomb was the United States' most powerful weapon during the Cold War.

On the other hand, the twentieth century also marked new approaches to the natural environments in America. In reaction to the depletion of natural habitats by industrialization and overhunting, President Theodore Roosevelt helped preserve areas for what would become the National Parks in 1916. Laws, such as the Clean Water Act of 1972, helped improve the health of ecosystems, which benefitted not only wildlife but people across the nation. This also led to the development of alternative energy sources such as wind and solar power.

America continues to change and grow into the twenty-first century by building on preexisting ideas but also pioneering new concepts. As globalization becomes an increasingly prominent phenomenon, American businesses strive to adapt their products to consumers worldwide while also funneling in new ideas from other nations. Yet many of the current developments in American enterprises stem in part from earlier events in American history. For example, the environmental movement has expanded to address new issues such as global warming. NASA continues its space exploration endeavors, but entrepreneurs hope one day to travel to Mars. Therefore, the history of technology within the United States remains an engaging and relevant subject in the present.

Civics/Government

The Role of the Citizen in a Democratic Society

Citizens express their political beliefs and public opinion through participation in politics. The conventional ways citizens can participate in politics in a democratic state include:

- Obeying laws
- Voting in elections
- Running for public office
- Staying interested and informed of current events
- Learning U.S. history
- Attending public hearings to be informed and to express their opinions on issues, especially on the local level
- Forming interest groups to promote their common goals

- Forming political action committees (PACs) that raise money to influence policy decisions
- Petitioning government to create awareness of issues
- Campaigning for a candidate
- Contributing to campaigns
- Using mass media to express political ideas, opinions, and grievances

Obeying Laws

Citizens living in a democracy have several rights and responsibilities to uphold. The first duty is that they uphold the established laws of the government. In a democracy, a system of nationwide laws is necessary to ensure that there is some degree of order. Therefore, citizens must try to obey the laws and also help enforce them because a law that is inadequately enforced, such as early civil rights laws in the South, is almost useless. Optimally, a democratic society's laws will be accepted and followed by the community as a whole.

However, conflict can occur when an unjust law is passed. For example, much of the civil rights movement centered around Jim Crow laws in the South that supported segregation between black and whites. Yet these practices were encoded in state laws, which created a dilemma for African Americans who wanted equality but also wanted to respect the law. Fortunately, a democracy offers a degree of protection from such laws by creating a system in which government leaders and policies are constantly open to change in accordance with the will of citizens. Citizens can influence the laws that are passed by voting for and electing members of the legislative and executive branches to represent them at the local, state, and national levels.

Voting

In a democratic state, the most common way to participate in politics is by voting for candidates in an election. Voting allows the citizens of a state to influence policy by selecting the candidates who share their views and make policy decisions that best suit their interests, or candidates who they believe are most capable of leading the country. In the United States, all citizens—regardless of gender, race, or religion—are allowed to vote unless they have lost their right to vote through due process, such as felons.

Since the Progressive movement and the increased social activism of the 1890s to the 1920s that sought to eliminate corruption in government, direct participation in politics through voting has increased. Citizens can participate by voting in the following types of elections:

- Direct primaries: Citizens can nominate candidates for public office.

- National, state, and municipal elections: Citizens elect their representatives in government.

- Recall elections: Citizens can petition the government to vote an official out of office before their term ends.

- Referendums: Citizens can vote directly on proposed laws or amendments to the state constitution.

- Voter initiatives: Citizens can petition their local or state government to propose laws that will be approved or rejected by voters.

Running for Public Office

An extension of citizens' voting rights is their ability to run as elected officials. By becoming leaders in the government, citizens can demonstrate their engagement and help determine government policy. The involvement of citizens as a whole in the selection of leaders is vital in a democracy because it helps to prevent the formation of an elite cadre that does not answer to the public. Without the engagement of citizens who run for office, voters are limited in their ability to select candidates that appeal to them. In this case, voting options would become stagnant, which inhibits the ability of the nation to grow and change over time. As long as citizens are willing to take a stand for their vision of America, America's government will remain dynamic and diverse.

Citizen Interest

These features of a democracy give it the potential to reshape itself continually in response to new developments in society. In order for a democracy to function, it is of the utmost importance that citizens care about the course of politics and be aware of current issues. Apathy among citizens is a constant problem that threatens the endurance of democracies. Citizens should have a desire to take part in the political process, or else they simply accept the status quo and fail to fulfill their role as citizens. Moreover, they must have acute knowledge of the political processes and the issues that they can address as citizens. A fear among the Founding Fathers was the prevalence of mob rule, in which the common people did not take interest in politics except to vote for their patrons; this was the usual course of politics in the colonial era, as the common people left the decisions to the established elites. Without understanding the world around them, citizens may not fully grasp the significance of political actions and thereby fail to make wise decisions in that regard. Therefore, citizens must stay informed about current affairs, ranging from local to national or global matters, so that they can properly address them as voters or elected leaders.

Historical Knowledge

Furthermore, knowledge of the nation's history is essential for healthy citizenship. History continues to have an influence on present political decisions. For example, Supreme Court rulings often take into account previous legal precedents and verdicts, so it is important to know about those past events and how they affect the current processes. It is especially critical that citizens are aware of the context in which laws were established because it helps clarify the purpose of those laws. For example, an understanding of the problems with the Articles of Confederation allows people to comprehend some of the reasons behind the framework of the Constitution. In addition, history as a whole shapes the course of societies and the world; therefore, citizens should draw on this knowledge of the past to realize the full consequences of current actions. Issues such as climate change, conflict in the Middle East, and civil rights struggles are rooted in events and cultural developments that reach back centuries and should be addressed.

Therefore, education is a high priority in democracies because it has the potential to instill generations of citizens with the right mind-set and knowledge required to do their part in shaping the nation. Optimally, education should cover a variety of different subjects, ranging from mathematics to biology, so that individuals can explore whatever paths they wish to take in life. Even so, social studies are especially important because students should understand how democracies function and understand the history of the nation and world. Historical studies should cover national and local events as well because they help provide the basis for the understanding of contemporary politics. Social studies courses should also address the histories of foreign nations because contemporary politics increasingly has global consequences. In addition, history lessons should remain open to multiple perspectives, even those that might criticize a nation's past actions, because citizens should be exposed to diverse perspectives that they can apply as voters and leaders.

The Structure and Functions of Different Levels of Government in the United States

A *political institution* is an organization created by the government to enact and enforce laws, act as a mediator during conflict, create economic policy, establish social systems, and carry out some power. These institutions maintain a rigid structure of internal rules and oversight, especially if the power is delegated, like agencies under the executive branch.

The Constitution established a federal government divided into three branches: legislative, executive, and judicial.

The Three Branches of the U.S. Government

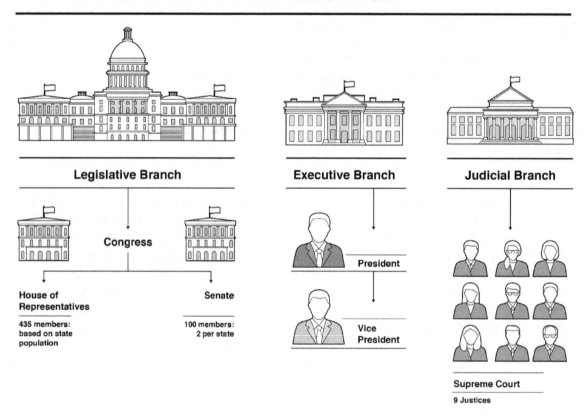

Legislative Branch

Congress

House of Representatives

435 members: based on state population

Senate

100 members: 2 per state

Executive Branch

President

Vice President

Judicial Branch

Supreme Court

9 Justices

Executive Branch

The executive branch is responsible for enforcing the laws. The executive branch consists of the president, the vice president, the president's cabinet, and federal agencies created by Congress to execute some delegated.

The president of the United States:

- Serves a four-year term and is limited to two terms in office
- Is the chief executive officer of the United States and commander-in-chief of the armed forces
- Is elected by the Electoral College
- Appoints cabinet members, federal judges, and the heads of federal agencies
- Vetoes or signs bills into law
- Handles foreign affairs, including appointing diplomats and negotiating treaties
- Must be thirty-five years old, a natural-born U.S. citizen, and have lived in the United States for at least fourteen years

The vice president:

- Serves four-year terms alongside and at the will of the president
- Acts as president of the Senate
- Assumes the presidency if the president is incapacitated
- Assumes any additional duties assigned by the president

The cabinet members:

- Are appointed by the president
- Act as heads for the fifteen executive departments
- Advise the president in matters relating to their departments and carry out delegated power

Note that the president can only sign and veto laws and cannot initiate them himself. As head of the executive branch, it is the responsibility of the president to execute and enforce the laws passed by the legislative branch.

Although Congress delegates their legislative authority to agencies in an enabling statute, they are located in the executive branch because they are tasked with executing their delegating authority. The president enjoys the power of appointment and removal over all federal agency workers, except those tasked with quasi-legislative or quasi-judicial powers.

Legislative Branch

The legislative branch is responsible for enacting federal laws. This branch possesses the power to declare war, regulate interstate commerce, approve or reject presidential appointments, and investigate the other branches. The legislative branch is *bicameral*, meaning it consists of two houses: the lower house, called the House of Representatives, and the upper house, known as the Senate. Both houses are elected by popular vote.

Members of both houses are intended to represent the interests of the constituents in their home states and to bring their concerns to a national level. Ideas for laws, called bills, are proposed in one chamber and then are voted upon according to the body's rules; should the bill pass the first round of voting, the other legislative chamber must approve it before it can be sent to the president.

The two houses (or chambers) are similar though they differ on some procedures such as how debates on bills take place.

House of Representatives

The House of Representatives is responsible for enacting bills relating to revenue, impeaching federal officers including the president and Supreme Court justices, and electing the president in the case of no candidate reaching a majority in the Electoral College.

In the House of Representatives:

- Each state's representation in the House of Representatives is determined proportionally by population, with the total number of voting seats limited to 435.

- There are six nonvoting members from Washington, D.C., Puerto Rico, American Samoa, Guam, Northern Mariana Islands, and the U.S. Virgin Islands.

- The Speaker of the House is elected by the other representatives and is responsible for presiding over the House. In the event that the president and vice president are unable to fulfill their duties, the Speaker of the House will succeed to the presidency.

- The representatives of the House serve two-year terms.

- The requirements for eligibility in the House include:

    o Must be twenty-five years of age
    o Must have been a U.S. citizen for at least seven years
    o Must be a resident of the state they are representing by the time of the election

Senate

The Senate has the exclusive powers to confirm or reject all presidential appointments, ratify treaties, and try impeachment cases initiated by the House of Representatives.

In the Senate:

- The number of representatives is one hundred, with two representatives from each state.
- The vice president presides over the Senate and breaks the tie, if necessary.
- The representatives serve six-year terms.
- The requirements for eligibility in the Senate include:
    o Must be thirty years of age
    o Must have been a U.S. citizen for the past nine years
    o Must be a resident of the state they are representing at the time of their election

Legislative Process

Although all members of the houses make the final voting, the senators and representatives serve on committees and subcommittees dedicated to specific areas of policy. These committees are responsible for debating the merit of bills, revising bills, and passing or killing bills that are assigned to their committee. If it passes, they then present the bill to the entire Senate or House of Representatives (depending on which they are a part of). In most cases, a bill can be introduced in either the Senate or the House, but a majority vote of both houses is required to approve a new bill before the president may sign the bill into law.

Judicial Branch

The *judicial branch*, though it cannot pass laws itself, is tasked with interpreting the law and ensuring citizens receive due process under the law. The judicial branch consists of the Supreme Court, the highest court in the country, overseeing all federal and state courts. Lower federal courts are the district courts and court of appeals.

The Supreme Court:

- Judges are appointed by the president and confirmed by the Senate.
- Judges serve until retirement, death, or impeachment.
- Judges possess sole power to judge the constitutionality of a law.
- Judges set precedents for lower courts based on their decisions.
- Judges try appeals that have proceeded from the lower district courts.

Checks and Balances

Notice that a system of checks and balances between the branches exists. This is to ensure that no branch oversteps its authority. They include:

- Checks on the Legislative Branch:
  - The president can veto bills passed by Congress.
  - The president can call special sessions of Congress.
  - The judicial branch can rule legislation unconstitutional.
- Checks on the Executive Branch:
  - Congress has the power to override presidential vetoes by a two-thirds majority vote.
  - Congress can impeach or remove a president, and the chief justice of the Supreme Court presides over impeachment proceedings.
  - Congress can refuse to approve presidential appointments or ratify treaties.
- Checks on the Judicial Branch:
  - The president appoints justices to the Supreme Court, as well as district court and court of appeals judges.
  - The president can pardon federal prisoners.
  - The executive branch can refuse to enforce court decisions.
  - Congress can create federal courts below the Supreme Court.
  - Congress can determine the number of Supreme Court justices.
  - Congress can set the salaries of federal judges.
  - Congress can refuse to approve presidential appointments of judges.
  - Congress can impeach and convict federal judges.

The three branches of government operate separately, but they must rely on each other to create, enforce, and interpret the laws of the United States.

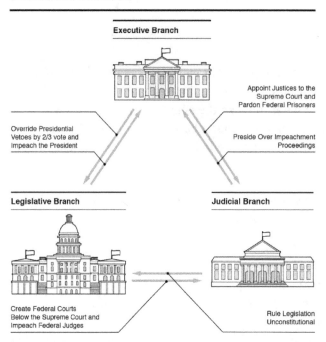

How Laws are Enacted and Enforced

To enact a new law:

- The bill is introduced to Congress.
- The bill is sent to the appropriate committee for review and revision.
- The approved bill is sent to the Speaker of the House and the majority party leader of the Senate, who places the bill on the calendar for review.
- The houses debate the merits of the bill and recommend amendments.

  - In the House of Representatives, those who wish to debate about a bill are allowed only a few minutes to speak, and amendments to the bill are limited.

  - In the Senate, debates and amendments are unlimited, and those who wish to postpone a vote may do so by filibuster, refusing to stop speaking.

- The approved bill is revised in both houses to ensure identical wording in both bills.
- The revised bill is returned to both houses for final approval.
- The bill is sent to the president, who may

  - Sign the bill into law

  - Veto the bill

  - Take no action, resulting in the bill becoming law if Congress remains in session for ten days or dying if Congress adjourns before ten days have passed

The Role of State Government

While the federal government manages the nation as a whole, state governments address issues pertaining to their specific territory. In the past, states claimed the right, known as nullification, to refuse to enforce federal laws that they considered unconstitutional. However, conflicts between state and federal authority, particularly in the South in regard to first, slavery, and later, discrimination, have led to increased federal power, and states cannot defy federal laws. Even so, the Tenth Amendment limits federal power to those powers specifically granted in the Constitution, and the rest of the powers are retained by the states and citizens. Therefore, individual state governments are left in charge of decisions with immediate effects on their citizens, such as state laws and taxes.

In this way, the powers of government are separated both horizontally between the three branches of government (executive, legislative, and judicial) and vertically between the levels of government (federal, state, and local).

Like the federal government, state governments consist of executive, judicial, and legislative branches, but the exact configuration of those branches varies between states. For example, while most states follow the bicameral structure of Congress, Nebraska has only a single legislative chamber. Additionally, requirements to run for office, length of terms, and other details vary from state to state. State governments have considerable authority within their states, but they cannot impose their power on other states.

Separation of Powers

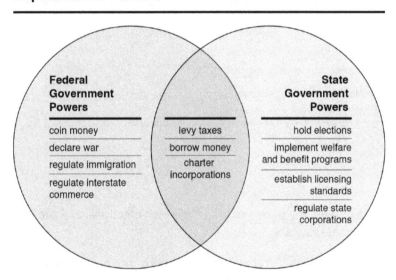

The Role of Local Government

Local governments, which include town governments, county boards, library districts, and other agencies, are especially variable in their composition. They often reflect the overall views of their state governments but also have their own values, rules, and structures. Generally, local governments function in a democratic fashion, although the exact form of government depends on its role. Depending on the location within the state, local government may have considerable or minimal authority based on the population and prosperity of the area; some counties may have strong influence in the state, while others may have a limited impact.

Native American Tribes

Native American tribes are treated as dependent nations that answer to the federal government but may be immune to state jurisdiction. As with local governments, the exact form of governance is left up to the tribes, which ranges from small councils to complex systems of government. Other U.S. territories, including the District of Columbia (site of Washington, D.C.) and acquired islands, such as Guam and Puerto Rico, have representation within Congress, but their legislators cannot vote on bills.

Election System

As members of a Constitutional Republic with certain aspects of a *democracy*, U.S. citizens are empowered to elect most government leaders, but the process varies between branch and level of government. Presidential elections at the national level use the *Electoral College* system. Rather than electing the president directly, citizens cast their ballots to select *electors* that represent each state in the college.

Legislative branches at the federal and state level are also determined by elections. In some areas, judges are elected, but in other states judges are appointed by elected officials. The U.S. has a *two-party system*, meaning that most government control is under two major parties: the Republican Party and the Democratic Party. It should be noted that the two-party system was not designed by the Constitution but gradually emerged over time.

Electoral Process

During the *electoral process*, the citizens of a state decide who will represent them at the local, state, and federal level. Different political officials that citizens elect through popular vote include but are not limited to:

- City mayor
- City council members
- State representative
- State governor
- State senator
- House member
- U.S. Senator
- President

The Constitution grants the states the power to hold their own elections, and the voting process often varies from city to city and state to state.

While a popular vote decides nearly all local and state elections, the president of the United States is elected by the *Electoral College*, rather than by popular vote. Presidential elections occur every four years on the first Tuesday after the first Monday in November.

The electoral process for the president of the United States includes:

Primary Elections and Caucuses

In a presidential election, *nominees* from the two major parties, as well as some third parties, run against each other. To determine who will win the nomination from each party, the states hold *primary elections* or *caucuses*.

During the primary elections, the states vote for who they want to win their party's nomination. In some states, primary elections are closed, meaning voters may only vote for candidates from their registered party, but other states hold *open primaries* in which voters may vote in either party's primary.

Some states hold *caucuses* in which the members of a political party meet in small groups, and the decisions of those groups determine the party's candidate.

Each state holds a number of delegates proportional to its population, and the candidate with the most delegate votes receives the domination. Some states give all of their delegates (*winner-take-all*) to the primary or caucus winner, while some others split the votes more proportionally.

Conventions

The two major parties hold national conventions to determine who will be the nominee to run for president from each party. The *delegates* each candidate won in the primary elections or caucuses are the voters who represent their states at the national conventions. The candidate who wins the most delegate votes is given the nomination. Political parties establish their own internal requirements and procedures for how a nominee is nominated.

Conventions are typically spread across several days, and leaders of the party give speeches, culminating with the candidate accepting the nomination at the end.

Campaigning

Once the nominees are selected from each party, they continue campaigning into the national election. Prior to the mid-1800s, candidates did not actively campaign for themselves, considering it dishonorable to the office, but campaigning is now rampant. Modern campaigning includes, but is not limited to:

- Raising money
- Meeting with citizens and public officials around the country
- Giving speeches
- Issuing policy proposals
- Running internal polls to determine strategy
- Organizing strategic voter outreach in important districts
- Participating in debates organized by a third-party private debate commission
- Advertising on television, through mail, or on the Internet

General Election

On the first Tuesday after the first Monday in November of an election year, every four years, the people cast their votes by secret ballot for president in a *general election*. Voters may vote for any candidate, regardless of their party affiliation. The outcome of the popular vote does not decide the election; instead, the winner is determined by the Electoral College.

Electoral College

When the people cast their votes for president in the general election, they are casting their votes for the *electors* from the *Electoral College* who will elect the president. In order to win the presidential election, a nominee must win 270 of the 538 electoral votes. The number of electors is equal to the total number of senators and representatives from each state plus three electoral votes for Washington D.C. which does not have any voting members in the legislative branch.

The electors typically vote based on the popular vote from their states. Although the Constitution does not require electors to vote for the popular vote winner of their state, no elector voting against the popular vote of their state has ever changed the outcome of an election. Due to the Electoral College, a nominee may win the popular vote and still lose the election.

For example, let's imagine that there only two states, Wyoming and Nebraska, in a presidential election. Wyoming has three electoral votes and awards them all to the winner of the election by majority vote. Nebraska has five electoral votes and also awards them all to the winner of the election by majority vote. If 500,000 people in Wyoming vote and the Republican candidate wins by a vote of 300,000 to 200,000, the Republican candidate will win the three electoral votes for the state. If the same number of people vote in Nebraska, but the Republican candidate loses the state by a vote of 249,000 to 251,000, the Democratic candidate wins the five electoral votes from that state. This means the Republican candidate will have received 549,000 popular votes but only three electoral votes, while the Democratic candidate will have received 451,000 popular votes but will have won five electoral votes. Thus, the Republican won the popular vote by a considerable margin, but the Democratic candidate will have been awarded more electoral votes, which are the only ones that matter.

	Wyoming	Nebraska	Total # of Votes
Republican Votes	300,000	249,000	**549,000**
Democratic Votes	200,000	251,000	**451,000**
Republican Electoral Votes	3	0	**3**
Democratic Electoral Votes	0	5	**5**

If no one wins the majority of electoral votes in the presidential election, the House of Representatives decides the presidency, as required by the Twelfth Amendment. They may only vote for the top three candidates, and each state delegation votes as a single bloc. Twenty-six votes, a simple majority, are required to elect the president. The House has only elected the president twice, in 1801 and 1825.

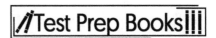

Here how many electoral votes each state and the District of Columbia have:

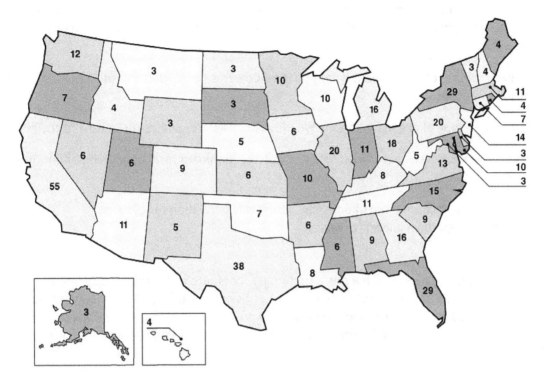

Purposes and Characteristics of Various Governance Systems

Government is the physical manifestation of the political entity or ruling body of a state. It includes the formal institutions that operate to manage and maintain a society. The form of government does not determine the state's *economic system*, though these concepts are often closely tied. Many forms of government are based on a society's economic system. However, while the form of government refers to the methods by which a society is managed, the term *economy* refers to the management of resources in a society. Many forms of government exist, often as hybrids of two or more forms of government or economic systems. Forms of government can be distinguished based on protection of civil liberties, protection of rights, distribution of power, power of government, and principles of Federalism.

Regime is the term used to describe the political conditions under which the citizens live under the ruling body. A regime is defined by the amount of power the government exerts over the people and the number of people who comprise the ruling body. It is closely related to the form of government because the form of government largely creates the political conditions. Regimes are governmental bodies that control both the form and the limit of term of their office. For example, authoritarianism is an example of a form of government and type of regime. A regime is considered to be ongoing until the culture, priorities, and values of the government are altered, ranging from the peaceful transitions of power between democratic political parties to the violent overthrow of the current regime.

The forms of government operated by regimes of government include:

Aristocracy

An *aristocracy* is a form of government composed of a small group of wealthy rulers, either holding hereditary titles of nobility or membership in a higher class. Variations of aristocratic governments include:

- Oligarchy: form of government where political power is consolidated in the hands of a small group of people

- Plutocracy: type of oligarchy where a wealthy elite class dominates the state and society

Though no aristocratic governments exist today, it was the dominant form of government during ancient times, including the:

- Vassals and lords during the Middle Ages, especially in relation to feudalism
- City-state of Sparta in ancient Greece

Authoritarian

An authoritarian state is one in which a single party rules indefinitely. The ruling body operates with unrivaled control and complete power to make policy decisions, including the restriction of denying civil liberties such as freedom of speech, press, religion, and protest. Forms of authoritarian governments include *autocracy*, *dictatorship*, and *totalitarian*—states or societies ruled by a single person with complete power over society.

Examples of states with authoritarian governments:

- Soviet Union
- Nazi Germany
- Modern-day North Korea

Democracy

Democracy is a form of government in which the people act as the ruling body by electing representatives to voice their views. Forms of democratic governments include:

- Direct democracy: democratic government in which the people make direct decisions on specific policies by majority vote of all eligible voters, like in ancient Athens

- Representative democracy: democratic government in which the people elect representatives to vote in a legislative body. This form of soft government providing for the election of representatives is also known as representative republic or indirect democracy. Representative democracy is currently the most popular form of government in the world.

The presidential and parliamentary systems are the most common forms of representative democracy. In the presidential system, the executive operates in its own distinct branch. Although the executive and legislative branches might enjoy powers checking each other, as in the American presidential system, the two functions are clearly separated. In addition, the president is typically both the head of state and head of government. Examples of presidential systems include Brazil, Nigeria, and the United States.

In the parliamentary system, the prime minister serves as the head of the government. The legislative branch, typically a parliament, elects the prime minister; thus, unlike in the presidential system, the parliament can replace the prime minister with a vote of no confidence. This practically means that the

parliament has considerable influence over the office of prime minister. Parliamentary systems often include a president as the head of state, but the office is mostly ceremonial, functioning like a figurehead. Examples of parliamentary systems include Germany, Australia, and Pakistan.

The presidential system is a form of government better designed to distribute power between separate branches of government. This theoretically provides more stability. The president serves for a limited term of years, while prime ministers serve until replaced after receiving a vote of no confidence.

In the parliamentary system, the interdependence and interconnectedness between the parliament and prime minister facilitates efficient and timely governance, capable of adjusting to developing and fluid situations. In contrast, the presidential system is more prone to political gridlock because there is no direct connection between the legislative and executive branches. The legislature in a presidential system cannot replace the executive, like in the parliamentary system. The separation of powers in a presidential system can lead to disagreement between the executive and legislature, causing gridlock and other delays in governance.

Federalism is a set of principles that divides power between a central government and regional governments. Sovereign states often combine into a federation, and to do so, they cede some degree of sovereignty to establish a functional central government to handle broad national policies. The United States Constitution structures the central government according to principles of Federalism. Canada is another example of a form of government with a Federalist structure.

Monarchy

Monarchy is a form of government in which the state is ruled by a *monarch*, typically a hereditary ruler. Monarchs have often justified their power due to some divine right to rule. Types of monarchies include:

- Absolute monarchy: a monarchy in which the monarch has complete power over the people and the state

- Constitutional monarchy: a type of monarchy in which the citizens of the state are protected by a constitution, and a separate branch, typically a parliament, makes legislative decisions. The monarch and legislature share power.

- Crowned republic: a type of monarchy in which the monarch holds only a ceremonial position and the people hold sovereignty over the state. It is defined by the monarch's lack of executive power.

Examples of monarchies:

- Kingdom of Saudi Arabia is an absolute monarchy.
- Australia is a crowned republic.

Economics

Fundamental Economic Concepts

Economics is the study of human behavior in response to the production, consumption, and distribution of assets or wealth. Economics can help individuals or societies make decisions or plans for themselves or communities, dependent upon their needs, wants, and resources. Economics is divided into two subgroups: microeconomics and macroeconomics.

Microeconomics is the study of individual or small group behaviors and patterns in relationship to such things as earning and spending money. It focuses on particular markets within the economy, and looks at single factors that could potentially affect individuals or small groups. For example, the use of coupons in a grocery store can affect an individual's product choice, quantity purchased, and overall savings that a person may later roll into a different purchase. Microeconomics is the study of scarcity, choice, opportunity costs, economics systems, factors of production, supply and demand, market efficiency, the role of government, distribution of income, and product markets.

Macroeconomics examines a much larger scale of the economy. It focuses on how a society or nation's goods, services, spending habits, and other factors affect the people of that entity. It focuses on aggregate factors such as demand and output. For example, if a national company moves its production overseas to save on costs, how will production, labor, and capital be affected? Macroeconomics analyzes all aggregate indicators and the microeconomic factors that influence the economy. Government and corporations use macroeconomic models to help formulate economic policies and strategies.

Microeconomics

Scarcity

People have different needs and wants, and the question arises, are the resources available to supply those needs and wants? Limited resources and high demand create scarcity. When a product is scarce, there is a short supply of it. For example, when the newest version of a cellphone is released, people line up to buy the phone or put their name on a wait list if the phone is not immediately available. The product, the new cellphone, may become a scarce commodity. In turn, because of the scarcity, companies may raise the cost of the commodity, knowing that if it is immediately available, people may pay more for the instant gratification—and vice versa. If a competing company lowers the cost of the phone but has contingencies, such as extended contracts or hidden fees, the buyer will still have the opportunity to purchase the scarce product. Limited resources and extremely high demand create scarcity and, in turn, cause companies to acquire opportunity costs.

Factors of Production

There are four factors of production:

- Land: both renewable and nonrenewable resources
- Labor: effort put forth by people to produce goods and services
- Capital: the tools used to create goods and services
- Entrepreneurship: persons who combine land, labor, and capital to create new goods and services

The four factors of production are used to create goods and services to make economic profit. All four factors strongly impact one another.

Supply and Demand

Supply and demand is the most important concept of economics in a market economy. Supply is the amount of a product that a market can offer. Demand is the quantity of a product needed or desired by buyers. The price of a product is directly related to supply and demand. The correlation between the price of a product and the demand necessary to distribute resources to the market go hand in hand in a market economy. For example, when there are a variety of treats at a bakery, certain treats are in higher demand than others. The bakery can raise the cost of the more demanded items as supplies get limited.

Conversely, the bakery can sell the less desirable treats by lowering the cost of those items as an incentive for buyers to purchase them.

Product Markets

Product markets are marketplaces where goods and services are bought and sold. Product markets provide sellers a place to offer goods and services to consumers, and for consumers to purchase those goods and services. The annual value of goods and services exchanged throughout the year is measured by the Gross Domestic Product (GDP), a monetary measure of goods and services made either quarterly or annually. Department stores, gas stations, grocery stores, and other retail stores are all examples of product markets. However, product markets do not include any raw, scarce, or trade materials.

Theory of the Firm

The behavior of firms is composed of several theories varying between short- and long-term goals. There are four basic firm behaviors: perfect competition, profit maximization, short run, and long run. Each firm follows a pattern, depending on its desired outcome. Theory of the Firm posits that firms, after conducting market research, make decisions that will maximize their profits since they are for-profit entities.

- Perfect competition:
- In perfect competition, several businesses are selling the same product at the same time.
- There are so many businesses and consumers that none will directly impact the market.
- Each business and consumer is aware of the competing businesses and markets.
- Profit maximization:
- Firms decide the quantity of a product that needs to be produced in order to receive maximum profit gains. Profit is the total amount of revenue made after subtracting costs.
- Short run:
- A short amount of time where fixed prices cannot be adjusted
- The quantity of the product depends on the varying amount of labor. Less labor means less product.
- Long run:
- An amount of time where fixed prices can be adjusted
- Firms try to maximize production while minimizing labor costs.

Overall, microeconomics operates on a small scale, focusing on how individuals or small groups use and assign resources.

Macroeconomics

Macroeconomics analyzes the economy as a whole. It studies unemployment, interest rates, price levels, and national income, which are all factors that can affect the nation as a whole, and not just individual households. Macroeconomics studies all large factors to determine how, or if, they will affect future trend patterns of production, consumption, and economic growth.

Measures of Economic Performance

It is important to measure economic performance to determine if an economy is growing, stagnant, or deteriorating. To measure the growth and sustainability of an economy, several indicators can be used. Economic indicators provide data that economists can use to determine if there are faulty processes or if some form of intervention is needed.

One of the main indicators to measure economic performance is the growth of the country's Gross Domestic Product (GDP). GDP growth provides important information that can be used to determine fiscal or financial policies. The GDP does not measure income distribution, quality of life, or losses due to natural disasters. For example, if a community lost everything to a hurricane, it would take a long time to rebuild the community and stabilize its economy. That is why there is a need to take into account more balanced performance measures when factoring overall economic performance.

Other indicators used to measure economic performance are unemployment or employment rates, inflation, savings, investments, surpluses and deficits, debt, labor, trade terms, the HDI (Human Development Index), and the HPI (Human Poverty Index).

Unemployment

Unemployment occurs when an individual does not have a job, is actively trying to find employment, and is not getting paid. Official unemployment rates do not factor in the number of people who have stopped looking for work, unlike true unemployment rates that do, causing them to be higher.

There are three types of unemployment: cyclical, frictional, and structural.

Cyclical
The product of a business cycle. This usually occurs during a recession.
Frictional
The difficulty of matching qualified workers for specific jobs. An example would be a person changing careers.
Structural
When a person no longer qualifies for a specific job, or failing out of a retraining course for a job.

Given the nature of a market economy and the fluctuations of the labor market, a 100 percent employment rate is impossible to reach.

Inflation

Inflation is when the cost of goods and services rises over time. Supply, demand, and money reserves all affect inflation. Generally, inflation is measured by the Consumer Price Index (CPI), a tool that tracks price changes of goods and services over time. The CPI measures goods and services such as gasoline, cars, clothing, and food. When the cost of goods and services increase, the quantity of the product may decrease due to lower demand. This decreases the purchasing power of the consumer. Basically, as more money is printed, it holds less and less value in purchasing power. For example, when inflation occurs, consumers in the United States are spending and saving less because the U.S. dollar is worth less, and therefore the consumer cannot buy or save as much money. However, if inflation occurs steadily over time, the people can better plan and prepare for future necessities.

Inflation can vary from year to year, usually never fluctuating more than 2 percent. Central banks try to prevent drastic increases or decreases of inflation to prohibit prices from rising or falling far from the minimum. Inflation can also vary based on different monetary currencies. Although rare, any country's economy may experience hyperinflation (when inflation rates increase to over 50 percent), while other economies may experience deflation (when the cost of goods and services decrease over time). Deflation occurs when the inflation rate drops below zero percent.

Business Cycle

A business cycle is when the Gross Domestic Product (GDP) moves downward and upward over a long-term growth trend. These cycles help determine where the economy currently stands, as well as where the economy could be heading. Business cycles usually occur almost every six years, and have four phases: expansion, peak, contraction, and trough. Here are some characteristics of each phase:

- Expansion:
- Increased employment rates and economic growth
- Production and sales increase
- On a graph, expansion is where the lines climb.
- Peak:
- Employment rates are at or above full employment and the economy is at maximum productivity.
- On a graph, the peak is the top of the hill, where expansion has reached its maximum.
- Contraction:
- When growth starts slowing
- Unemployment is on the rise.
- On a graph, contraction is where the graph begins to slide back down or contract.
- Trough:
- The cycle has hit bottom and is waiting for the next cycle to start again.
- On a graph, the trough is the bottom of the contraction prior to when it starts to climb back up.

When the economy is expanding or "booming," the business cycle is going from a trough to a peak. When the economy is headed down and toward a recession, the business cycle is going from a peak to a trough.

Four phases of a business cycle:

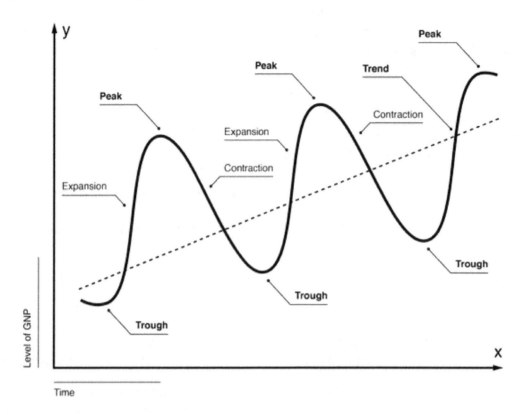

Economic Growth

Economic growth is measured by the increase in the Gross National Product (GNP) or Gross Domestic Product (GDP). The increase of goods and services over time indicates positive movement in economic growth. Keep in mind that the quantity of goods and services produced is not necessarily an indicator of economic growth. The value of the goods and services produced matters more than the quantity.

There are many causes of economic growth, which can be short- or long-term. In the short term, if aggregate demand (the total demand for goods and services produced at a given time) increases, then the overall Gross Domestic Product (GDP) increases as well. Not only will the GDP increase, interest rates may decrease. With reduced interest rates, spending and investing will increase. Consumer and government spending will also increase because there will be more disposable income. Real estate prices will rise, and there will be lower income taxes. All of these short-term factors can stimulate economic growth.

In the long term, if aggregate supply (the total supply of goods or services in a given time period) increases, then there is potential for an increase in capital as well. With more working capital, more infrastructure and jobs can be created. With more jobs, there is an increased employment rate, and education and training for jobs will improve. New technologies will be developed, and new raw materials may be discovered. All of these long-term factors can also stimulate economic growth.

Outside of the short- and long-term causes for economic growth, other factors include low inflation and stability. Lower inflation rates encourage more investing versus higher inflation rates that cause instability in the market. Stability encourages businesses to continue investing. If the market is unstable, investors may question the volatility of the market.

Potential Costs of Economic Growth:

- Inflation: When economic growth occurs, inflation tends to be high. If supply cannot keep up with demand, then the inflation rate may be unmanageable.

- Economic booms and recessions: The economy goes through cycles of booms and recessions. This causes inflation to increase and decrease over time, which puts the economy into a continuous cycle of rising and falling.
- Account inefficiencies: When the economy grows, consumers and businesses increase their import spending. The increase of import spending affects the current account and causes a shortage.
- Environmental costs: When the economy is growing, there is an abundance of output, which may result in more pollutants and a reduction in quality of life.
- Inequalities: Growth occurs differently among members of society. While the wealthy may be getting richer, those living in poverty may just be getting on their feet. So, while economic growth is happening, it may happen at two very different rates.

While these potential costs could affect economic growth, if the growth is consistent and stable, then growth can occur without severe inflation swings. Also, as technology improves, new ways of production can reduce negative environmental factors as well.

Government Involvement in the Economy

Governments have considerable influence over the flow of economies, which makes it important to understand the relationships between them. When a government has full control over the economic decisions of a nation, it is called a command system. This was the case in many absolute monarchies such as eighteenth-century France; King Louis XIV built his economy on the concept of mercantilism, which believed that the state should manage all resources, particularly by accumulating gold and silver. This system of economics discouraged exports and thereby limited trade.

In contrast, the market system is guided by the concept of capitalism, in which individuals and businesses have the freedom to manage their economic decisions. This allows for private property and increases the opportunities for entrepreneurship and trade. Early proponents of capitalism emphasized *laissez-faire* policies, which means "let it be," and argued that the government should not be involved with the economy at all. They believe the market is guided by the concept of self-interest and that individuals will optimally work for their personal success. However, individuals' interests do not necessarily correlate with the needs of the overall economy. For example, during a financial recession, consumers may decide to save up their money rather than make purchases; doing so helps them in the short run but further reduces demand in a slumping economy. Therefore, most capitalist governments still assert a degree of control over their economies while still allowing for private business.

Likewise, many command system economies, such as monarchical France, still relied heavily on private businesses maintained by wealthy businessmen. With the end of most absolute monarchies, communism has been the primary form of command system economies in the modern era. Communism is a form of socialism that emphasizes communal ownership of property and government control over production. The high degree of government control gives more stability to the economy, but it also creates considerable flaws. The monopolization of the economy by the government limits its ability to respond to local economic conditions because certain regions often have unique resources and needs.

With the collapse of the Soviet Union and other communist states, command systems have been largely replaced with market systems.

The U.S. government helps to manage the nation's economy through a market system in several ways. First and foremost, the federal government is responsible for the production of money for use within the economy; depending on how the government manages the monetary flow, it may lead to a stable economy, deflation, or inflation. Second, state and federal governments impose taxes on individuals, corporations, and goods. For example, a tariff might be imposed on imports in order to stimulate demand for local goods in the economy. Third, the government can pass laws that require additional regulation or inspections. In addition, the government has passed antitrust laws to inhibit the growth of private monopolies, which could limit free growth in the market system. Debates continue over whether the government should take further action to manage private industries or reduce its control over the private sector.

Just as governments can affect the direction of the economy, the state of the economy can have significant implications on government policies. Financial stability is critical in maintaining a prosperous state. A healthy economy will allow for new developments that contribute to the nation's growth and create jobs. On the other hand, an economic crisis, such as a recession or depression, can gravely damage a government's stability. Without a stable economy, business opportunities plummet, and people begin to lose income and employment. This, in turn, leads to frustration and discontent in the population, which can lead to criticism of the government. This could very well lead to demands for new leadership to resolve the economic crisis.

The dangers of a destabilized economy can be seen with the downfall of the French monarchy. The mercantilist approach to economics stifled French trade. Furthermore, regional aristocracies remained exempt from government taxes, which limited the government's revenues. This was compounded by expensive wars and poor harvests that led to criticism of King Louis XIV's government. The problems persisted for decades, and Louis XIV was forced to convene the Estates-General, a legislative body of representatives from across France, to address the crisis. The economic crises at the end of the eighteenth century were critical in the beginning of the French Revolution. Those financial issues, in turn, at least partially stemmed from both the government's control of the economy through mercantilism and its inability to impose economic authority over local regions.

Economic Systems
Economic systems determine what is being produced, who is producing it, who receives the product, and the money generated by the sale of the product. There are two basic types of economic systems: market economies (including free and competitive markets), and planned or command economies.

- Market Economies are characterized by:

- Privately owned businesses, groups, or individuals providing goods or services based on demand.

- The types of goods and services provided (supply) are based on that demand.

- Two types: competitive market and free market.

Competitive Market	Free Market
Due to the large number of both buyers and sellers, there is no way any one seller or buyer can control the market or price.	Voluntary private trades between buyers and sellers determine markets and prices without government intervention or monopolies.

- Planned or Command Economies:

- In planned or command economies, the government or central authority determines market prices of goods and services.

- The government or central authority determines what is being produced as well as the quantity of production.

- Some advantages to command economies include a large number of shared goods such as public services (transportation, schools, or hospitals).

- Disadvantages of command economies include wastefulness of resources.

Market Efficiency and the Role of Government (Taxes, Subsidies, and Price Controls)

Market efficiency is directly affected by supply and demand. The government can help the market stay efficient by either stepping in when the market is inefficient and/or providing the means necessary for markets to run properly. For example, society needs two types of infrastructure: physical (bridges, roads, etc.) and institutional (courts, laws, etc.). The government may impose taxes, subsidies, and price controls to increase revenue, lower prices of goods and services, ensure product availability for the government, and maintain fair prices for goods and services.

The Purpose of Taxes, Subsidies, and Price Controls

Taxes	Subsidies	Price Controls
-Generate government revenue -Discourage purchase or use of "bad" products such as alcohol or cigarettes	-Lower the price of goods and services -Reassure the supply of goods and services -Allow opportunities to compete with overseas vendors	-Act as emergency measures when government intervention is necessary -Set a minimum or maximum price for goods and services

Money and Banking

Money is the universal form of currency used throughout goods and services exchanges that holds its value over time. Money provides a convenient way for sellers and consumers to understand the value of their goods and services. As opposed to bartering (when sellers and consumers exchange goods or services as equal trades), money is quick and easy for both buyers and sellers.

There are three main forms of money: commodity, fiat, and bank. Here are characteristics of each form:

- Commodity money: Money as a valuable good, such as precious metals
- Fiat money: The value of the good set by supply and demand rather than the actual value it represents, such as paper money

- Bank money: Money that is credited by a bank to those who deposit it into bank accounts, such as checking and savings accounts or credit

While price levels within the economy set the demand for money, most countries have central banks that supply the actual money. Essentially, banks buy and sell money. Borrowers can take loans and pay back the bank, with interest, providing the bank with extra capital.

A central bank has control over the printing and distribution of money. Central banks serve three main purposes: manage monetary growth to help steer the direction of the economy, be a backup to commercial banks that are suffering, and provide options and alternatives to government taxation.

The Federal Reserve is the central bank of the United States. The Federal Reserve controls banking systems and determines the value of money in the United States. Basically, the Federal Reserve is the bank for banks.

All Western economies have to keep a minimum amount of protected cash called *required reserve*. Once banks meet those minimums, they can then lend or loan the excess to consumers. The required reserves are used within a fractional reserve banking system (fractional because a small portion is kept separate and safe). Not only do banks reserve, manage, and loan money, but they also help form monetary policies.

Monetary Policy

The central bank and other government committees control the amount of money that is made and distributed. The money supply determines monetary policy. Three main features sustain monetary policy:

- Assuring the minimum amount held within banks (bank reserves). When banks are required to hold more money in reserve funds, banks are less willing to lend money to help control inflation.

- Adjusting interest rates. For example, if the value of money is low, companies and consumers buy more (products, employees, stocks) because prices are cheap. Just like an investment, it is risky, but can pay off in the long term.

- The purchase and sales of bonds (otherwise known as open market operations). When buying bonds to increase money and selling bonds to reduce the supply, the central bank helps control the money supply.

In the United States, the Federal Reserve maintains monetary policy. There are two main types of monetary policy: expansionary monetary policy and contractionary monetary policy.

- Expansionary monetary policy:
- Increases the money supply
- Lowers unemployment
- Increases consumer spending
- Increases private sector borrowing
- Possibly decreases interest rates to very low levels, even near zero
- Decreases reserve requirements and federal funds
- Contractionary monetary policy:
- Decreases the money supply
- Helps control inflation

- Possibly increases unemployment due to slowdowns in economic growth
- Decreases consumer spending
- Decreases loans and/or borrowing

The Federal Reserve uses monetary policy to try to achieve maximum employment and secure inflation rates. Because the Federal Reserve is the "bank of banks," it truly strives to be the last-resort option for distressed banks. This is because once these kinds of institutions begin to rely on the Federal Reserve for help, all parts of the banking industry—such as those dealing with loans, bonds, interest rates, and mortgages—are affected.

International Trade and Exchange Rates

International trade is when countries import and export goods and services. Countries often want to deal in terms of their own currency. Therefore, when importing or exporting goods or services, consumers and businesses need to enter the market using the same form of currency. For example, if the United States would like to trade with China, the U.S. may have to trade in China's form of currency, the *Yuan*, versus the dollar, depending on the business.

The exchange rate is what one country's currency will exchange for another. The government and the market (supply and demand) determine the exchange rate. There are two forms of exchange rates: fixed and floating. Fixed exchange rates involve government interventions (like central banks) to help keep the exchange rates stable. Floating or "flexible" exchange rates constantly change because they rely on supply and demand needs. While each type of exchange rate has advantages and disadvantages, the rate truly depends on the current state of each country's economy. Therefore, each exchange rate may differ from country to country.

Advantages and Disadvantages of Fixed Versus Floating Exchange Rates			
Fixed Exchange Rate: government intervention to help keep exchange rates stable		Floating or "Flexible" Exchange Rate: Supply and demand determines the exchange rate	
Advantages	*Disadvantages*	*Advantages*	*Disadvantages*
-Stable prices	-Requires a large amount of reserve funds	-Central bank involvement is not needed.	-Currency speculation
-Stable foreign exchange rates	-Possibly mispricing currency values	-Facilitates free trade	-Exchange rate risks
-Exports are more competitive and in turn more profitable	-Inflation increases		-Inflation increases

While each country may have differing economic statuses and exchange rates, countries rely on one another for goods and services. Prices of imports and exports are affected by the strength of another country's currency. For example, if the United States dollar is at a higher value than another country's currency, imports will be less expensive because the dollar will have more value than that of the country selling its good or service. On the other hand, if the dollar is at a low value compared to the currency of another country, importers will tend to defer away from buying international items from that country. However, U.S. exporters to that country could benefit from the low value of the dollar.

Fiscal Policy

A fiscal policy is when the government is involved in adjusting spending and tax rates to assist the way in which an economy financially functions. Fiscal policies can either increase or decrease tax rates and spending. These policies represent a tricky balancing act, because if the government increases taxes too much, consumer spending and monetary value will decrease. Conversely, if the government lowers taxes, consumers will have more money in their pockets to buy more goods and services, which increases demand and the need for companies to supply those goods and services. Due to the higher demand, suppliers can add jobs to fulfill that demand. While the increase of supply, demand, and jobs are positive for the overall economy, they may result in a devaluation of the dollar and less purchasing power.

Consumer Economics

Economics are closely linked with the flow of resources, technology, and population in societies. The use of natural resources, such as water and fossil fuels, has always depended in part on the pressures of the economy. A supply of a specific good may be limited in the market, but with sufficient demand the sellers are incentivized to increase the available quantity. Unfortunately, the demand for certain objects can often be unlimited, and a high price or limited supply may prevent consumers from obtaining the product or service. If the sellers succumb to the consumers' demand and continue to exploit a scarce resource, supply could potentially be exhausted.

The resources for most products, both renewable and nonrenewable, are finite. This is a particularly difficult issue with nonrenewable resources, but even renewable resources often have limits: organic products such as trees and animals require stable populations and sufficient habitats to support those populations. Furthermore, the costs of certain decisions can have detrimental effects on other resources. For example, industrialization provides economic benefits in many countries but also has had the negative effect of polluting surrounding environments; the pollution, in turn, often eliminates or harms fish, plants, and other potential resources.

The control of resources within an economy is particularly important in determining how resources are used. While the demand may change with the choices of consumers, the range of supply depends on the objectives of the people producing the goods. They determine how much of their supply they allot for sale, and in the case of monopolies, they might have sole access to the resource. They might choose to limit their use of the resources or instead gather more to meet the demand. As they pay for the products, consumers can choose which sellers they rely on for the supply. In the case of a monopoly, though, consumers have little influence over the company's decision because there is no alternative supplier. Therefore, the function of supply within an economy can drastically influence how the resources are exploited.

The availability of resources, in turn, affects the human population. Humans require basic resources such as food and water for survival, as well as additional resources for healthy lifestyles. Therefore, access to these resources helps determine the survival rate of humans. For much of human existence, economies have had limited ability to extract resources from the natural world, which restricted the growth rate of populations. However, the development of new technologies, combined with increasing demand for certain products, has pushed resource use to a new level. On the one hand, this led to higher living standards that ensured that fewer people would die. However, this has also brought mass population growth. Admittedly, countries with higher standards of living often have lower birthrates. Even so, the increasing exploitation of resources has sharply increased the world's population as a whole to unsustainable levels. The rising population leads, in turn, to more demand for resources that cannot be met. This creates poverty, reduced living conditions, and higher death rates. As a result, economics can significantly influence local and world population levels.

Technology is also intricately related to population, resources, and economics. The role of demand within economies has incentivized people to innovate new technologies that enable societies to have a higher quality of life and greater access to resources. Entrepreneurs expand technologies by finding ways to create new products for the market. The Industrial Revolution, in particular, illustrates the relationship between economics and technology because the ambitions of businessmen led to new infrastructure that enabled more efficient and sophisticated use of resources. Many of these inventions reduced the amount of work necessary for individuals and allowed the development of leisure activities, which in turn created new economic markets. However, economic systems can also limit the growth of technology. In the case of monopolies, the lack of alternative suppliers reduces the incentive to meet and exceed consumer expectations. Moreover, as demonstrated by the effects of economics on resources, technology's increasing ability to extract resources can lead to their depletion and create significant issues that need to be addressed.

Distribution of Income

Distribution of income refers to how wages are distributed across a society or segments of a society. If everyone made the same amount of money, the distribution of income would be equal. That is not the case in most societies. The wealth of people and companies varies. Income inequality gaps are present in America and many other nations. Taxes provide an option to redistribute income or wealth because they provide revenue to build new infrastructure and provide cash benefits to some of the poorest members in society.

Choice and Opportunity Costs

When an individual decides between possibilities, that individual is making a choice. Choices allow people to compare opportunity costs. Opportunity costs are benefits that a person could have received, but gave up, in choosing another course of action. What is an individual willing to trade or give up for a different choice? For example, if an individual pays someone to mow the lawn because he or she would rather spend that time doing something else, then the opportunity cost of paying someone to mow the lawn is worth the time gained from not doing the job himself or herself.

On a larger scale, governments and communities have to assess different opportunity costs when it comes to using taxpayers' money. Should the government or community build a new school, repair roads, or allocate funds to local hospitals are all examples of choices taxpayers may have to review at some point in time. How do they decide which choice is the best, since each one has a trade off? By comparing the opportunity cost of each choice, they may decide what they are willing to live without for the sake of gaining something else.

Geography

Concepts and Terminology of Physical and Human Geography

Geographers utilize a variety of different maps in their study of the spatial world. Projections are maps that represent the entire world (which is spherical) on a flat surface. *Conformal projections* preserve angles locally, maintaining the shape of a small area in infinitesimal circles of varying sizes on a two-dimensional map. Conformal projections tend to possess inherent flaws due to their two-dimensional nature. For example, the most well-known projection, the *Mercator projection*, drastically distorts the size of land areas at the poles. In this particular map, Antarctica, one of the smallest continents, appears massive, almost rivaling the size of North America. In contrast to the poles, the areas closer to the central portion of the globe are more accurate. Other projections attempt to lessen the amount of distortion; the *Equal-area projection*, for example, attempts to equally represent the size of landforms on the globe. Nevertheless, equal-area projections like the *Lambert projection* also inherently alter the size of continents, islands, and other landforms, both close to Earth's center and near the poles. Other projections are a hybrid of the two primary models. For example, the *Robinson projection*, also referred to as the *Goode's homolosine projection*, tries to balance form and area in order to create a more visually accurate representation of the spatial world. Despite the efforts to maintain consistency with shapes, projections cannot provide accurate representations of the Earth's surface, due to their flat, two-dimensional nature. In this sense, projections are useful symbols of space, but they do not always provide the most accurate portrayal of spatial reality.

Unlike projections, *topographic maps* display contour lines, which represent the relative elevation of a particular place and are very useful for surveyors, engineers, and/or travelers. Hikers of the Appalachian Trail or Pacific Crest Trail, for example, may call upon topographic maps to calculate their daily climbs.

Thematic maps are also quite useful to geographers because they use two-dimensional surfaces to convey complex political, physical, social, cultural, economic, or historical themes.

Thematic maps can be broken down into different subgroups: *dot-density maps* and *flow-line maps*. A *dot-density map* is a type of thematic map that illustrates the volume and density in a particular area. Although most dots on these maps represent the number of people in an area, they don't always have to do that. Instead, these maps may represent the number of events, such as lightning strikes, that have taken place in an area. *Flow-line maps* are another type of thematic map, which utilize both thin and thick lines to illustrate the movement of goods, people, or even animals between two places. The thicker the line, the greater the number of moving elements; a thinner line would, of course, represent a smaller number.

Similar to topographic maps, an *isoline map* is also useful for calculating data and differentiating between the characteristics of two places. In an *isoline map*, symbols represent values, and lines can be drawn between two points in order to determine differences. For example, average temperature is commonly measured on isoline maps. Point A, which is high in the mountains, may have a value of 33 degrees, while point B, which is in the middle of the Mojave Desert, may have a value of 105 degrees. Using the different values, it is easy to determine that temperatures in the mountains are 72 degrees cooler than in the desert. Additionally, isoline maps help geographers study the world by creating questions. For example, is it only elevation that is responsible for the differences in temperature? If not, what other factors could cause such a wide disparity in the two values? Utilizing these, and other sorts of maps, is essential in the study of geography.

Using Geographic Concepts to Analyze Spatial Phenomena

Using Mental Maps to Organize Spatial Information

Mental maps are exactly what they sound like—maps that exist within someone's mind. The cognitive image of a particular place may differ from person to person, but the concept of remembering important places does not. For example, the commonalities usually emerge relative to the knowledge of one's workplace, school, home, or favorite restaurants. Furthermore, mental maps also embody the means of travelling from point A to point B. One may know the best route on public transit, the least hilly bike path, or the roadways that have the least amount of traffic. In places where someone has very little interaction, mental maps usually tend to be minimally informative, due to the absence of any personal experience in a particular place.

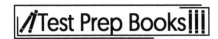

Maps are also organized through scale. Scale is simply the ratio of a distance on the ground to the corresponding distance on paper. Geographers and cartographers attempt to make the image on paper representative of the actual place. For example, the United States Geological Survey (USGS) utilizes the mathematical ratio of 1/24,000 in all of its topographical maps. This scale means that one inch on the map is equivalent to 24,000 inches—or nearly two-thirds of a mile—on the ground. The two primary types of maps, *large scale* and *small scale*, essentially serve the same purpose, but for two different types of places. Large-scale maps represent a much smaller area with greater detail, while small-scale maps are representative of much larger areas with less detail.

Recognizing and Interpreting Spatial Patterns

Two primary realms exist within the study of geography. The first realm, *physical geography*, essentially correlates with the land, water, and foliage of the Earth. The second realm, *human geography*, is the study of the Earth's people and how they interact with their environment. Like land and water on Earth, humans are also impacted by different forces such as culture, history, sociology, technological advancement and changes, and access to natural resources. For example, human populations tend to be higher around more reliable sources of fresh water. The metropolitan area of New York City, which has abundant freshwater resources, is home to nearly 20 million people, whereas Australia, both a continent and a country, has almost the same population. Although water isn't the only factor in this disparity, it certainly plays a role in a place's *population density*—the total number of people in a particular place divided by the total land area, usually square miles or square kilometers. Australia's population density stands at 8.13 people per square mile, while the most densely populated nation on Earth, Bangladesh, is home to 2,894 people per square mile.

Population density can have a devastating impact on both the physical environment/ecosystem and the humans who live within the environment/ecosystem of a particular place. For example, Delhi, one of India's most populated cities, is home to nearly five million gasoline-powered vehicles. Each day, those vehicles emit an enormous amount of carbon monoxide into the atmosphere, which directly affects the quality of life of Delhi's citizens. In fact, the problem of the smog and pollution has gotten so severe that many drivers are unable to see fifty feet in front of them. Additionally, densely populated areas within third-world nations, or developing nations, struggle significantly in their quest to balance the demands of the modern economy with their nation's lack of infrastructure. For example, nearly as many automobiles operate every day in major American cities like New York and Los Angeles as they do in Delhi, but they create significantly less pollution due to cleaner burning engines, better fuels, and governmental emission regulations.

Although it's a significant factor, population density is not the only source of strain on the resources of a particular place. Historical forces such as civil war, religious conflict, genocide, and government corruption can also alter the lives of a nation's citizens in a profound manner. For example, the war-torn nation on the Horn of Africa, Somalia, has not had a functioning government for nearly three decades. As a result, the nation's citizens have virtually no access to hospital care, vaccinations, or proper facilities for childbirth. Due to these and other factors, the nation's *infant mortality rate*, or the total number of child deaths per 1,000 live births, stands at a whopping 98.39/1000. When compared to Iceland's 1.82/1000, it's quite evident that Somalia struggles to provide basic services in the realm of childbirth and there is a dire need for humanitarian assistance.

Literacy rates, like the infant mortality rate, are also an excellent indicator of the relative level of development in a particular place. Like Somalia, other developing nations have both economic and social factors that hinder their ability to educate their own citizens. Due to radical religious factions within some nations like Afghanistan and Pakistan, girls are often denied the ability to attend school, which further reduces the nation's overall literacy rate. For example, girls in Afghanistan, which spent decades under Taliban control, have a 24.2 percent literacy rate, one of the lowest rates of any nation on Earth that keeps records (Somalia's government is so dysfunctional records don't exist).

Although literacy rates are useful in determining a nation's development level, high literacy rates do exist within developing nations. For example, Suriname, which has a significantly lower GDP than Afghanistan, enjoys a nearly 96 percent literacy rate among both sexes. Utilizing this and other data, geographers can create questions regarding how such phenomena occur. How is Suriname able to educate its population more effectively with fewer financial resources? Is it something inherent within their culture? Demographic data, such as population density, the infant mortality rate, and the literacy rate all provide insight into the characteristics of a particular place and help geographers better understand the spatial world.

Locating and Using Sources of Geographic Data

Geographic data is essential to fully understanding both the spatial and human realms of geography. In reference to the human population, different factors affect the quality of life one experiences during their lifetime. Geographers attempt to understand why those differences exist through data utilization and comparative analysis. For example, as has been previously mentioned, population density, infant mortality rates, and literacy rates are all useful tools in analyzing human characteristics of a place; however, those are not the only tools geographers utilize. In fact, organizations such as the *Population Reference Bureau* and the *Central Intelligence Agency* both provide an incredible amount of *demographic* data useful to researchers, students, or really anyone curious about the world in which they live.

The *CIA World Factbook* is an indispensable resource for anyone interested in the field of human or physical geography. Providing information such as land area, literacy rates, birth rate, and economic data, this resource is one of the most comprehensive on the Internet. In addition to the CIA World Factbook, the *Population Reference Bureau* (*PRB*) also provides students of geography with an abundant supply of information. In contrast to the CIA source, the *PRB* provides a treasure trove of analyses related to human populations including HIV rates, immigration rates, poverty rates, etc.

In addition to the aforementioned sources, the *United States Census Bureau* provides similar information about the dynamics of the American population. Not only does this source focus on the data geographers need to understand the world, but it also provides information about upcoming classes, online workshops, and even includes an online library of resources for both students and teachers.

Websites for each source can be found below:

- Population Reference Bureau: www.prb.org
- United States Census Bureau: www.census.gov
- CIA World Factbook: https://www.cia.gov/library/publications/the-world-factbook/

Spatial Concepts

Location is the central theme in understanding spatial concepts. In geography, there are two primary types of locations that people utilize on a daily basis. The first type, *relative location*, is used frequently and involves locating objects by noting their proximity to another, better known object. For example, directions from person to person may relate directly to massive shopping centers, major highways, or well-known intersections. Although relative location is important, in the modern world, it's common to

use digital satellite-based technologies, which rely on *GPS (Global Positioning System)*. To determine *Absolute Location*, or the exact latitudinal and longitudinal position on the globe, GPS uses sensors that interact with satellites orbiting the Earth. *Coordinates* correspond with the positions on a manmade grid system using imaginary lines known as *latitude* (also known as *parallels*) and *longitude* (also known as *meridians*).

In order to understand latitude and longitude, one should think of a simple X and Y-axis. The *equator* serves as the X-axis at zero degrees, and measures distance from north to south. The Y-axis is at zero degrees and is represented by the *Prime Meridian*.

In addition to anchoring the grid system to create the basis for absolute location, these major lines of latitude and longitude also divide the Earth into *hemispheres*. The Equator divides the Earth into the northern and southern hemispheres, while the Prime Meridian establishes the eastern and western hemispheres. Coordinates are always expressed in the following format:

Degrees north or south, degrees east or west, or 40°N, 50°E. Since lines of latitude and longitude are great distance from one another, absolute locations are often found in between two lines. In those cases, degrees are broken down into *minutes* and *seconds*, which are expressed in this manner: (40° 53' 44" N, 50° 22' 65" E).

In addition to the Equator and the Prime Meridian, other major lines of latitude and longitude exist to divide the world into regions relative to the direct rays of the sun. These lines correspond with the Earth's *tilt*, and are responsible for the seasons. For example, the northern hemisphere is tilted directly toward the sun from June 22 to September 23, which creates the summer season in that part of the world. Conversely, the southern hemisphere is tilted away from the direct rays of the sun and experiences winter during those same months.

The transitions from season to season involve two factors: the 23 ½ degree tilt of the Earth and the movement of the direct rays of the sun relative to the Earth's revolution. To clarify, the area between the *Tropic of Cancer* (23 ½ degrees north) and the *Tropic of Capricorn* (23 ½ degrees south) can be envisioned as the playing field for the direct rays of the sun. These rays never leave the playing field, and, as a result, the area between those two lines of latitude—the *tropics*—tends to be warmer and experience fewer variations in seasonal temperatures.

In contrast, the area between the Tropic of Cancer and the *Arctic Circle* (66 ½ degrees north) is in the *middle latitudes*—the region where most of the Earth's population resides. In the Southern Hemisphere, the middle latitudes exist between the Tropic of Capricorn and the *Antarctic Circle* (66 ½ degrees south). In both of these places, indirect rays of the sun strike the Earth, so seasons are more pronounced and milder temperatures generally prevail. The final region, known as the *high latitudes*, is found north of the Arctic Circle and south of the Antarctic Circle. These regions generally tend to be cold all year, and experience nearly twenty-four hours of sunlight during their respective *summer solstice* and twenty-four hours of darkness during the *winter solstice*.

Regarding the seasons, it is important to understand that those in the Southern Hemispheres are opposite of those in the Northern Hemisphere, due to the position of the direct rays of the sun. When the sun's direct rays are over the Equator, it is known as an *equinox*, and day and night are almost of equal length throughout the world. Equinoxes occur twice a year; the fall, or autumnal equinox, occurs on September 22nd, while the spring equinox occurs on March 20th. Obviously, if seasons are opposite of one another depending on the hemisphere, the corresponding names flip-flop depending on one's location (i.e. when the Northern Hemisphere is experiencing summer, it is winter in the Southern Hemisphere).

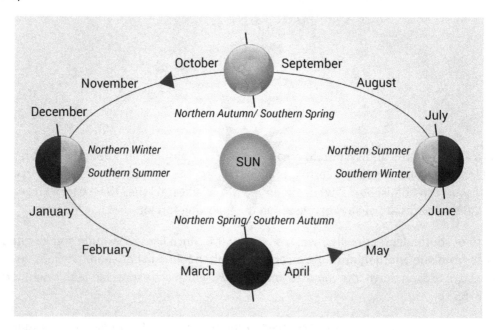

Place

Both absolute and relative location help humans understand their sense of place. Place is a simple concept that helps to define the characteristics of the world around us. For example, people may create *toponyms* to further define and orient themselves with their sense of place. Toponyms are simply names given to locations to help develop familiarity within a certain location. Although not always the case, toponyms generally utilize geographical features, important people in an area, or even wildlife commonly found in a general location. For example, many cities in the state of Texas are named in honor of military leaders who fought in the Texas Revolution of 1836 (such as Houston and Austin), while other places, such as Mississippi and Alabama, utilize Native American toponyms to define their sense of place.

Regions

In addition to location and place, geographers also divide the world into regions in order to more fully understand differences inherent with the world, its people, and its environment. As mentioned previously, lines of latitude such as the Equator, the Tropics, and the Arctic and Antarctic Circles already divide the Earth into solar regions relative to the amount of either direct or indirect sunlight that they receive. Although not the same throughout, the middle latitudes generally have a milder climate than

areas found within the tropics. Furthermore, tropical locations are usually warmer than places in the middle latitudes, but that is not always the case. For example, the lowest place in the United States—Death Valley, California—is also home to the nation's highest-ever recorded temperature. Likewise, the Andes Mountains in Peru and Ecuador, although found near the Equator, are also home to heavy snow, low temperatures, and dry conditions, due to their elevation.

Formal regions are spatially defined areas that have overarching similarities or some level of *homogeneity* or *uniformity*. Although not exactly alike, a formal region generally has at least one characteristic that is consistent throughout the entire area. For example, the United States could be broken down into one massive formal region due to the fact that in all fifty states, English is the primary language. Of course, English isn't the only language spoken in the United States, but throughout that nation, English is heavily used. As a result, geographers are able to classify the United States as a formal region; but, more specifically, the United States is a *linguistic region*—a place where everyone generally speaks the same language.

Functional regions are similar to formal regions in that they have similar characteristics, but they do not have clear boundaries. The best way to understand these sorts of regions is to consider large cities. Each large city encompasses a large *market area*, whereby people in its vicinity generally travel there to conduct business, go out to eat, or watch a professional sporting event. However, once anyone travels farther away from that *primate city*, they transition to a different, more accessible city for their needs. The functional region, or *area of influence*, for that city, town, or sports team transitions, depending upon the availability of other primate cities. For example, New York City has two primary professional baseball, basketball, and football teams. As a result, its citizens may have affinities for different teams even though they live in the same city. Conversely, a citizen in rural Idaho may cheer for the Seattle Seahawks, even though they live over 500 miles from Seattle, due to the lack of a closer primate city.

Discussing Economic, Political, and Social Factors

Effects of Physical Processes, Climate Patterns, and Natural Hazards on Human Societies

The Earth's surface, like many other things in the broader universe, does not remain the same for long; in fact, it changes from day to day. The Earth's surface is subject to a variety of physical processes that continue to shape its appearance. In each process, water, wind, temperature, or sunlight play a role in continually altering the Earth's surface.

Erosion can be caused by a variety of different stimuli including ice, snow, water, wind, and ocean waves. *Wind erosion* is a specific phenomenon that occurs in generally flat, dry areas with loose topsoil. Over time, the persistent winds can dislodge significant amounts of soil into the air, reshaping the land and wreaking havoc on those who depend on agriculture for their livelihoods. Erosion can also be caused by water and is responsible for changing landscapes as well. For example, the Grand Canyon was carved over thousands of years by the constant movement of the Colorado River. Over time, the river moved millions of tons of soil, cutting a huge gorge in the Earth along the way. In all cases, erosion involves the movement of soil from one place to another.

In water erosion, material carried by the water is referred to as *sediment*. With time, some sediment can collect at the mouths of rivers, forming *deltas*, which become small islands of fertile soil. This process of detaching loose soils and transporting them to a different location where they remain for an extended period of time is referred to as *deposition*, and is the end result of the erosion process.

In contrast to erosion, *weathering* does not involve the movement of any outside stimuli. In this physical process, the surface of the Earth is either broken down physically or chemically. *Physical weathering* involves the effects of atmospheric conditions such as water, ice, heat, or pressure. Through the process of weathering over the course of centuries, large rocks can be broken down with the effects of icy conditions. *Chemical weathering* generally occurs in warmer climates and involves organic material that breaks down rocks, minerals, or soil. This process is what scientists believe led to the creation of fossil fuels such as oil, coal, and natural gas.

Climate Patterns

Weather is defined as the condition of the Earth's atmosphere at a particular time. *Climate* is different; instead of focusing on one particular day, climate is the relative pattern of weather in a place for an extended period of time. For example, the city of Atlanta is in the American South and generally has a humid subtropical climate; however, Atlanta also occasionally experiences snowstorms in the winter months. Despite the occasional snow and sleet storm, over time, geographers, meteorologists, and other Earth scientists have determined the patterns that are indicative to north Georgia, where Atlanta is located. Almost all parts of the world have predictable climate patterns, which are influenced by the surrounding geography.

The Central Coast of California is an example of a place with a predictable climate pattern. Santa Barbara, California, one of the region's larger cities, has almost the same temperature for most of the summer, spring, and fall, with only minimal fluctuation during the winter months. The temperatures there, which average between 75 and 65 degrees Fahrenheit daily regardless of the time of year, are influenced by a variety of different climatological factors including elevation, location relative to the mountains and ocean, and ocean currents. In the case of Santa Barbara, the city's location on the Pacific Coast and its position near mountains heavily influences its climate. The cold California current, which sweeps down the west coast of the United States, causes the air near the city to be temperate, while the mountains trap cool air over the city and the surrounding area. This pattern, known as the *orographic effect*, or *rain shadow*, also affects temperatures on the leeward side of the mountains by blocking most of the cool air and causing dry conditions to dominate. Temperatures can fluctuate by more than 20 degrees Fahrenheit on opposite sides of the mountain.

Other factors affecting climate include elevation, prevailing winds, vegetation, and latitudinal position on the globe.

Like climate, *natural hazards* also affect human societies. In tropical and subtropical climates, hurricanes and typhoons form over warm water and can have devastating effects. Additionally, tornadoes, which are powerful cyclonic windstorms, also are responsible for widespread destruction in many parts of the United States and in other parts of the world. Like storms, earthquakes, usually caused by shifting plates along faults deep below the Earth's surface, also cause widespread devastation, particularly in nations with a poor or crumbling infrastructure. For example, San Francisco, which experiences earthquakes regularly due to its position near the San Andreas Fault, saw relatively little destruction and deaths (67 total) as a result of the last major earthquake to strike there. However, in 2010, an earthquake of similar magnitude reportedly killed over 200,000 people in the western hemisphere's poorest nation, Haiti.

Although a variety of factors may be responsible for the disparity, modern engineering methods and better building materials most likely helped to minimize destruction in San Francisco. Other natural hazards, such as tsunamis, mudslides, avalanches, forest fires, dust storms, flooding, volcanic eruptions, and blizzards, also affect human societies throughout the world.

Characteristics and Spatial Distribution of Earth's Ecosystems

Earth is an incredibly large place filled with a variety of different land and water *ecosystems*. *Marine ecosystems* cover over 75 percent of the Earth's surface and contain over 95 percent of the Earth's water. Marine ecosystems can be broken down into two primary subgroups: *freshwater ecosystems*, which only encompass around 2 percent of the earth's surface; and *ocean ecosystems*, which make up over 70 percent. On land, *terrestrial ecosystems* vary depending on a variety of factors, including latitudinal distance from the equator, elevation, and proximity to mountains or bodies of water. For example, in the high latitudinal regions north of the Arctic Circle and south of the Antarctic Circle, frozen *tundra* dominates. Tundra, which is characterized by low temperatures, short growing seasons, and minimal vegetation, is only found in regions that are far away from the direct rays of the sun.

In contrast, *deserts* can be found throughout the globe and are created by different ecological factors. For example, the world's largest desert, the Sahara, is almost entirely within the tropics; however, other deserts like the Gobi in China, the Mojave in the United States, and the Atacama in Chile, are the result of the orographic effect and their close proximity to high mountain ranges such as the Himalayas, the Sierra Nevada, and the Andes, respectively. In the Middle Latitudes, greater varieties of climatological zones are more common due to fluctuations in temperatures relative to the sun's rays, coupled with the particular local topography.

In the Continental United States, *temperate deciduous forest* dominates the southeastern portion of the country. However, the Midwestern states such as Nebraska, Kansas, and the Dakotas, are primarily *grasslands*. Additionally, the states of the Rocky Mountains can have decidedly different climates relative to elevation. In Colorado, Denver, also known as the "Mile High City," will often see snowfalls well into late April or early May due to colder temperatures, whereas towns and cities in the eastern part of the state, with much lower elevations, may see their last significant snowfall in March.

In the tropics, which are situated between the Tropics of Cancer and Capricorn, temperatures are generally warmer, due to the direct rays of the sun's persistence. However, like most of the world, the tropics also experience a variety of climatological regions. In Brazil, Southeast Asia, Central America, and even Northern Australia, tropical rainforests are common. These forests, which are known for abundant vegetation, daily rainfall, and a wide variety of animal life, are absolutely essential to the health of the world's ecosystems. For example, the *Amazon Rain Forest* is also referred to as "the lungs of the world," as its billions of trees produce substantial amounts of oxygen and absorb an equivalent amount of carbon dioxide—the substance that many climatologists assert is causing climate change or *global warming*.

Unlike temperate deciduous forests whose trees lose their leaves during the fall and winter months, *tropical rain forests* are always lush, green, and warm. In fact, some rainforests are so dense with vegetation that a few indigenous tribes have managed to exist within them without being influenced by any sort of modern technology, virtually maintaining their ancient way of life in the modern era.

The world's largest ecosystem, the *taiga*, is found primarily in high latitudinal areas, which receive very little of the sun's indirect rays. These forests are generally made up of *coniferous* trees, which do not lose their leaves at any point during the year as *deciduous* trees do. Taigas are cold-climate regions that make up almost 30 percent of the world's land area. These forests dominate the northern regions of Canada, Scandinavia, and Russia, and provide the vast majority of the world's lumber.

Overall, it is important to remember that climates are influenced by five major factors: elevation, latitude, proximity to mountains, ocean currents, and wind patterns. For example, the cold currents off

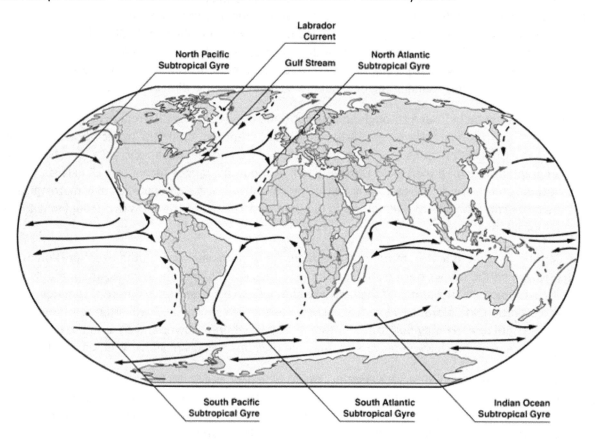

the coast of California provide the West Coast of the United States with pleasant year-round temperatures. Conversely, Western Europe, which is at the nearly the same latitude as most of Canada, is influenced by the warm waters of the *Gulf Stream*, an ocean current that acts as a conveyor belt, moving warm tropical waters to the icy north. In fact, the Gulf Stream's influence is so profound that it even keeps Iceland—an island nation in the far North Atlantic—relatively warm.

Interrelationships Between Humans and Their Environment

Like any other animal, humans adapt to their environment; but, unlike other animals, humans also adapt their environment to suit their needs. For example, human social systems are created around the goal of providing people with access to what they need to live more productive, fulfilling, and meaningful lives. Sometimes, humans create systems that are destructive, but generally speaking, humans tend to use their environment to make their lives easier. For example, in warmer climates, people tend to wear more comfortable clothing such as shorts, linen shirts, and hats. Additionally, in the excessively sun-drenched nations of the Middle East, both men and women wear flowing white clothing complete with both a head and neck covering, in order to prevent the blistering effects of sun exposure to the skin. Likewise, the native Inuit peoples of northern Canada and Alaska use the thick furs from the animals they kill to insulate their bodies against the bitter cold.

Humans also adapt to their environment to ensure that they have access to enough food and water for survival. Irrigation, or the process of moving water from its natural location to where it's needed, is an example of how humans change their environment in order to survive. For example, the city of Los Angeles, America's second most populous city, did not have adequate freshwater resources to sustain its population. However, city and state officials realized that abundant water resources existed approximately three hundred miles to the east. Rather than relocating some of its population to areas

with more abundant water resources, the State of California undertook one of the largest construction projects in the history of the world, the Los Angeles Aqueduct, which is a massive concrete irrigation ditch connecting water-rich areas with the thirsty citizens of Los Angeles.

The Los Angeles Aqueduct is just one example of a human-environment interaction. In other cases, humans utilize what nature provides in close proximity. For example, the very first permanent British Colony in North America, Jamestown, VA, was heavily influenced by its environment. In contrast to the Pilgrims who settled in Plymouth, Massachusetts, Jamestown settlers found themselves in a hot, humid climate with fertile soil. Consequently, its inhabitants engaged in agriculture for both food and profit. Twelve years after Jamestown's foundation in 1607, it was producing millions of dollars of tobacco each year. In order to sustain this booming industry, over time, millions of African slaves and indentured servants from Europe were imported to provide labor.

Conversely, the poor soils around Plymouth did not allow for widespread cash crop production, and the settlers in New England generally only grew enough food for themselves on small subsistence farms. Furthermore, slavery failed to take a strong foothold in the New England states, thus creating significantly different cultures within the same country, all due in part to human interaction with the environment.

Renewable and Nonrenewable Resources

When gas prices are high, prices on virtually everything increase. After all, there are very few products that humans can buy that are not transported by either a gasoline- or diesel-powered engine. As a result, an increase in fuel prices leads to an increase in the price of food, goods, or other cargo. Recently, there has been considerable debate regarding the reliance on *nonrenewable resources* like oil, natural gas, and coal. These resources, which are also known as *fossil fuels*, are quite common throughout the world and are generally abundant, and cheaper to use than *renewable resources* like solar, wind, and geothermal energy.

While solar energy is everywhere, the actual means to convert the sun's rays into energy is not. Conversely, coal-fired power plants and gasoline-powered engines, which are older technologies in use during the industrial revolution, remain quite common throughout the world. In fact, reliance on non-renewable resources continues to grow, due to the availability coupled with the existing infrastructure. However, use of renewable energy is increasing, as it becomes more economically competitive with nonrenewable resources.

In addition to sources of energy, nonrenewable resources also include anything that can be exhausted. These can include precious metals like gold, silver, and platinum, freshwater underground aquifers, and precious stones such as diamonds, emeralds, and opals. Although abundant, most nonrenewable sources of energy are not sustainable because their creation takes so long that they cannot be reproduced. Renewable resources are sustainable, but must be properly overseen so that they remain renewable. For example, the beautiful African island of Madagascar is home to some of the most amazing rainforest trees in the world. As a result, logging companies cut, milled, and sold thousands of them in order to make quick profits without planning how to ensure the continued health of the forests.

As a result of severe deforestation on the island, mudslides became more and more common as the forests gradually shrank from widespread logging. In this case, renewable resources were mismanaged, and thus essentially became nonrenewable, due to the length of time for growth for the replacement of rainforest trees. In the United States, paper companies harvest pine trees to create paper; and because it can take almost twenty years for a pine tree to reach maturity, most of the companies utilize planning techniques to ensure that mature pine trees will always be available. In this manner, these resources remain renewable for human use in a sustainable fashion.

Renewable sources of energy are relatively new in the modern economy. Even though electric cars, wind turbines, and solar panels are becoming more common, they still do not provide enough energy to power the world's economy. As a result, reliance on older, reliable forms of energy continues, which has a devastating effect on the environment. Beijing, China, which has seen a massive boom in industrial jobs, is also one of the most polluted places on Earth. Furthermore, developing nations with very little modern infrastructure also rely heavily on fossil fuels, due to the ease in which they are converted into usable energy. Even the United States, which has one of the most developed infrastructures in the world, still relies almost exclusively on fossil fuels, with only ten percent of the required energy coming from renewable sources.

Spatial Patterns of Cultural and Economic Activities

Spatial patterns refer to where things are in the world. Biomes, regions, and landforms all have spatial patterns regarding where they exist. Additionally, elements of *human geography*—the study of human culture and its effect on the world—also have certain patterns regarding where they appear on Earth.

Ethnicity

An ethnic group, or ethnicity, is essentially a group of people with a common language, society, culture, or ancestral heritage. Different ethnicities developed over centuries through historical forces, the impact of religious traditions, and other factors. Thousands of years ago, it was more common for ethnic groups to remain in one area with only the occasional interaction with outside groups.

In the modern world, different ethnicities interact on a daily, if not hourly, basis, due to better transportation resources and the processes of globalization. For example, in the United States, it is not uncommon for a high school classroom to encompass people of Asian, African, Indian, European, or Native descent. That's not to suggest that all American classrooms have ethnic diversity, but, in general, due to a variety of pull-factors, the United States continues to attract people from all over the world.

In less developed parts of the world, travel is limited due to the lack of infrastructure. Consequently, ethnic groups develop in small areas that can differ greatly from other people just a few miles away. For example, on the Balkan Peninsula in southeastern Europe, a variety of different ethnic groups live in close proximity to one another. Croats, Albanians, Serbs, Bosnians, and others all share the same land, but have very different worldviews, traditions, and religious influences. In the case of the Balkan Peninsula, such diversity has not always been a positive characteristic. For example, the First World War began there in 1914 related to a dispute regarding Serbia's national independence. Additionally, Bosnia was the scene of a horrible genocide against Albanians in an "ethnic cleansing" effort that continued throughout the late 20th century.

Linguistics

Linguistics, or the study of language, groups certain languages together according to their commonalities. For example, the Romance Languages—French, Spanish, Italian, Romansh, and Portuguese—all share language traits from Latin, the language of the former Roman Empire. These languages, also known as *vernaculars*, or more commonly spoken *dialects*, evolved over centuries of physical isolation on the European continent. The Spanish form of Latin emerged into today's Spanish language.

In other parts of the world, the same pattern is true. The Bantu people of Africa travelled extensively and spread their language, now called *Swahili*, which is the first Pan-African language. When thinking of the world as a whole, it is important to understand that thousands of languages exist; however, to interconnect the world, it is important to have a means of communication with which everyone is at least somewhat familiar. A *lingua franca* is essentially the language of business. In other words, when executives from multinational corporations need to communicate regarding business, they often communicate in English, which is considered to be the world's lingua franca, due to the economic dominance of the United States.

Religion
Religion has played a tremendous role in creating the world's cultures. Devout Christians crossed the Atlantic in hopes of finding religious freedom in New England, Muslim missionaries and traders travelled to the Spice Islands of the East Indies to teach about the Koran, and Buddhist monks traversed the Himalayan Mountains into Tibet to spread their faith.

In some countries, religion helps to shape legal systems. These nations, termed *theocracies*, have no separation of church and state and are more common in Islamic nations such as Saudi Arabia, Iran, and Qatar. In contrast, even though religion has played a tremendous role in the history of the United States, its government remains *secular*, or nonreligious, due to the influence of European Enlightenment philosophy at the time of its inception.

Like ethnicity and language, religion is also a primary way that people self-identify. As a result, religious influences can shape a region's laws, architecture, literature, and music. For example, when the Ottoman Turks, who are Muslim, conquered Constantinople, which was once the home of the Eastern Orthodox Christian Church, they replaced Christian places of worship with mosques. Additionally, different forms of Roman architecture were replaced with those influenced by Arabic traditions.

Economics
Economic activity also has a spatial component. For example, nations with few natural resources generally tend to import what they need from nations willing to export raw materials to them. Furthermore, areas that are home to certain raw materials generally tend to alter their environment in order to maintain production of those materials. In the San Joaquin Valley of California, an area known for extreme heat and desert-like conditions, local residents have engineered elaborate drip irrigation systems to adequately water lemon, lime, olive, and orange trees, utilizing the warm temperatures to constantly produce citrus fruits. Additionally, other nations with abundant petroleum reserves build elaborate infrastructures in order to pump, house, refine, and transport their materials to nations who require gasoline, diesel, or natural gas.

Essentially, different spatial regions on Earth create jobs, infrastructure, and transportation systems that seek to ensure the continued flow of goods, raw materials, and resources out of their location, so long as financial resources keep flowing into the area.

Patterns of Migration and Settlement
Migration is governed by two primary causes: *push factors*, which are reasons causing someone to leave an area, and *pull factors*, which are factors luring someone to a particular place. These two factors often work in concert with one another. For example, the United States of America has experienced significant *internal migration* from the industrial states in the Northeast (such as New York, New Jersey, Connecticut) to the Southern and Western states. This massive migration, which continues into the present-day, is due to high rents in the northeast, dreadfully cold winters, and lack of adequate

retirement housing, all of which are push factors. These push factors lead to migration to the *Sunbelt*, a term geographers use to describe states with warm climates and less intense winters.

In addition to internal migrations within nations or regions, international migration also takes place between countries, continents, and other regions. The United States has long been the world's leading nation in regard to *immigration*, the process of having people come into a nation's boundaries. Conversely, developing nations that suffer from high levels of poverty, pollution, warfare, and other violence all have significant push factors, which cause people to leave and move elsewhere. This process, known as *emigration*, is when people in a particular area leave in order to seek a better life in a different—usually better—location.

The Development and Changing Nature of Agriculture

Agriculture is essential to human existence. The *Neolithic Revolution*, or the use of farming to produce food, had a profound effect on human societies. Rather than foraging and hunting for food, human societies became more stable and were able to grow due to more consistent food supplies. In modern times, farming has changed drastically in order to keep up with the increasing world population.

Until the twentieth century, the vast majority of people on Earth engaged in *subsistence farming*, or the practice of growing only enough food to feed one's self, or one's family. Over time, due to inventions such as the steel plow, the mechanical reaper, and the seed drill, farmers were able to produce more crops on the same amount of land. As food became cheaper and easier to obtain, populations grew, but rather than leading to an increase in farmers, fewer people actually farmed. After the advent of mechanized farming in developed nations, small farms became less common, and many were either abandoned or absorbed by massive commercial farms producing both foodstuffs, staple crops, and cash crops.

In recent years, agricultural practices have undergone further changes in order to keep up with the rapidly growing population. Due in part to the *Green Revolution*, which introduced the widespread use of fertilizers to produce massive amounts of crops, farming techniques and practices continue to evolve. For example, *genetically modified organisms*, or *GMOs*, are plants or animals whose genetic makeup has been modified using different strands of DNA in hopes of producing more resilient strains of staple crops, livestock, and other foodstuffs. This process, which is a form of *biotechnology*, attempts to solve the world's food production problems through the use of genetic engineering. Although these crops are abundant and resistant to pests, drought, or frost, they are also the subject of intense scrutiny.

For example, the international food company, Monsanto, has faced an incredible amount of criticism regarding its use of GMOs in its products. Many activists assert that "rewiring" mother nature is inherently problematic and that foods produced through such methods are dangerous to human health. Despite the controversy, GMOs and biotechnologies continue to change the agricultural landscape by changing the world's food supply.

Like Monsanto, other agribusinesses exist throughout the world. Not only do these companies produce food for human consumption, but they also provide farming equipment, fertilizers, agrichemicals, and breeding and slaughtering services for livestock. While these companies are found all over the world, they are generally headquartered near the product they produce. For example, General Mills, a cereal manufacturer, is headquartered in the Midwestern United States, near its supply of wheat and corn—the primary ingredients in its cereals.

Contemporary Patterns and Impacts of Development, Industrialization, and Globalization

As mentioned previously, *developing nations* are nations that are struggling to modernize their economy, infrastructure, and government systems. Many of these nations may struggle to provide basic services to their citizens like clean water, adequate roads, or even police protection. Furthermore, government corruption makes life even more difficult for these countries' citizens.

In contrast, *developed nations* are those who have relatively high *Gross Domestic Products (GDP)*, or the total value of all goods and services produced in the nation in a given year. To elucidate, the United States, which is one of the wealthiest nations on Earth when ranked by overall GDP, has nearly a 19 trillion dollar GDP; while Haiti, one of the poorest nations in the Western Hemisphere, has nearly a nine billion dollar GDP. This is a difference of almost seventeen trillion dollars. This is not to disparage Haiti or other developing nations; the comparison is simply used to show that extreme inequities exist in very close proximity to one another, and it may be difficult for developing nations to meet the needs of their citizens and move their economic infrastructure forward toward modernization.

In the modern world, industrialization is the initial key to modernization and development. For developed nations, the process of industrialization took place centuries ago. England, where the *Industrial Revolution* began, actually began to produce products in factories in the early 1700s. Later, the United States and some nations of Western Europe followed suit, using raw materials brought in from their colonies abroad to make finished products. For example, cotton was spun into fabric on elaborate weaving machines that mass-produced textiles. As a result, nations that perfected the textile process were able to sell their products around the world, which produced enormous profits. Over time, those nations were able to accumulate wealth, improve their nation's infrastructure, and provide more services for their citizens.

Similar to the events of the eighteenth and nineteenth centuries, nations throughout the world are undergoing the same process in today's world. China exemplifies this concept. In China, agriculture was once the predominant occupation, and although it is true that agriculture is still a dominant sector of the Chinese economy, millions of Chinese citizens are flocking to major cities like Beijing, Shanghai, and Hangzhou, due to the availability of factory jobs that allow its workers a certain element of *social mobility*, or the ability to rise up out of one's socioeconomic situation.

Due to improvements in transportation and communication, the world has become figuratively smaller. For example, university students on the Indian Subcontinent now compete directly with students all over the world to obtain the skills employers desire to move their companies forward. Additionally, many corporations in developed nations have begun to *outsource* labor to nations with high levels of educational achievement but lower wage expectations. The process of opening the marketplace to all nations throughout the world, or *globalization*, has only just started to take hold in the modern economy. As industrial sites shift to the developing world, so does the relative level of opportunity for those nation's citizens.

However, due to the massive amounts of pollution produced by factories, the process of globalization also has had significant ecological impacts. The most widely known impact, *climate change*, which most climatologists assert is caused by an increase of carbon dioxide in the atmosphere, remains a serious problem that has posed challenges for developing nations, who need industries in order to raise their standard of living, and developed nations, whose citizens use a tremendous amount of fossil fuels to run their cars, heat their homes, and maintain their ways of life.

Demographic Patterns and Demographic Change

Demography, or the study of human populations, involves a variety of closely related stimuli. First, as has been previously addressed, economic factors play a significant role in the movement of people, as do climate, natural disasters, or internal unrest. For example, in recent years, millions of immigrants from the war-torn country of Syria have moved as far as possible from danger.

Although people are constantly moving, some consistencies remain throughout the world. First, people tend to live near reliable sources of food and water, which is why the first human civilizations sprung up in river valleys like the Indus River Valley in India, the Nile River Valley in Egypt, and the Yellow River Valley in Asia. Second, extreme temperatures tend to push people away, which is why the high latitudinal regions near the North and South Poles have such few inhabitants. Third, the vast majority of people tend to live in the Northern Hemisphere, due to the simple fact that more land lies in that part of the Earth.

In keeping with these factors, human populations tend to be greater where human necessities are easily accessible, or at least more readily available. In other words, such areas have a greater chance of having a higher population density than places without such characteristics.

Demographic patterns on Earth are not always stagnant. In contrast, people move and will continue to move as both push and pull factors fluctuate along with the flow of time. For example, in the 1940s, thousands of Europeans fled their homelands due to the impact of the Second World War.

Today, thousands of migrants arrive on European shores each month due to conflicts in the Levant and difficult economic conditions in Northern Africa. Furthermore, as previously discussed, people tend to migrate to places with a greater economic benefit for themselves and their families. As a result, developed nations such as the United States, Germany, Canada, and Australia have a net gain of migrants, while developing nations such as Somalia, Zambia, and Cambodia generally tend to see thousands of their citizens seek better lives elsewhere.

It is important to understand the key variables in changes regarding human population and its composition worldwide. Religion and religious conflict play a role in where people choose to live. For example, the Nation of Israel won its independence in 1948 and has since attracted thousands of people of Jewish descent from all over the world. Additionally, the United States has long attracted people from all over the world, due to its promise of religious freedom inherent within its own Constitution. In contrast, nations like Saudi Arabia and Iran do not typically tolerate different religions, resulting in a decidedly uniform religious—and oftentimes ethnic—composition. Other factors such as economic opportunity, social unrest, and cost of living also play a vital role in demographic composition.

Basic Concepts of Political Geography

Nations, states, and nation-states are all terms with very similar meanings, but knowing the differences aids in a better understanding of geography. A nation is a people group with similar cultural, linguistic, and historical experiences. A state is a political unit with sovereignty, or the ability to make its own decisions within defined borders; and a nation-state is an entity that combines states into one, singular government system. For example, in the United States, the state of Texas is not an independent state. Instead, it is part of the United States and thus, is subject to its laws.

In a similar fashion, the United Kingdom encompasses four member states: England, Wales, Northern Ireland, and Scotland. Although people in those states may consider themselves to be *sovereign*, or self-governing, the reality is that those states cannot make decisions regarding international trade,

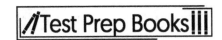

declarations of war, or other important decisions regarding the rest of the world. Instead, they are *semi-autonomous*, meaning that they can make some decisions regarding how their own state is run, but must yield more major powers to a centralized authority. In the United States, this sort of system is called *Federalism*, or the sharing of power among Local, State, and Federal entities, each of whom is assigned different roles in the overall system of government.

Nation-states, and the boundaries that define where they are, are not always permanent. For example, after the fall of the Soviet Union in 1991, new nations emerged that had once been a part of the larger entity called the Union of Soviet Socialists Republics. These formerly sovereign nations were no longer forced to be a part of a unifying communist government, and as a result, they regained their autonomy and became newly independent nations that were no longer *satellite nations* of the Soviet Union.

In a historical sense, the United States can be seen as a prime example of how national boundaries change. After the conclusion of the American Revolution in 1781, the Treaty of Paris defined the United States' western boundary as the Mississippi River; today, after a series of conflicts with Native American groups, the Mexican government, Hawaiian leadership, the Spanish, and the purchase of Alaska from the Russians, the boundaries of the United States have changed drastically. In a similar fashion, nations in Europe, Africa, and Asia have all shifted their boundaries due to warfare, cultural movements, and language barriers.

In the modern world, boundaries continue to change. For example, the Kurds, an ethnic minority and an excellent example of a nation, are still fighting for the right to control their people's' right to *self-determination*, but have not yet been successful in establishing a state for themselves. In contrast, the oil-rich region of South Sudan, which has significant cultural, ethnic, and religious differences from Northern Sudan, successfully won its independence in a bloody civil war, which established the nation's newest independent state.

In recent years, Russia has made the world nervous by aggressively annexing the Crimean Peninsula, an area that has been part of the Ukraine since the end of the Cold War. Even the United Kingdom and Canada have seen their own people nearly vote for their own rights to self-determination. In 1995, the French-speaking Canadian province of Quebec narrowly avoided becoming a sovereign nation through a tightly contested referendum. In a similar fashion, Scotland, which is part of the UK, also voted to remain a part of the Crown, even though many people in that state see themselves as inherently different from those in other regions within the nation.

Political geography is constantly changing. Boundaries on maps from ten years ago are not consistent with those of 2016 and beyond. *Decolonization*, or the removal of dependency on colonizers, has altered the political landscape of Africa, allowed more autonomy for the African people, and has forever redefined the boundaries of the entire continent.

Interpreting Maps

Geography is visually conveyed using maps, and a collection of maps is called an atlas. To illustrate some key points about geography, please refer to the map below.

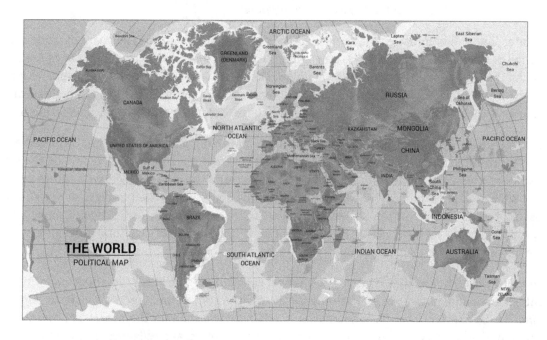

This is a traditional map of the world that displays all of the countries and six of the seven continents. Countries, the most common approach to political regions, can be identified by their labels. The continents are not identified on this map, with the exception of Australia, but they are larger landmasses that encompass most of the countries in their respective areas; the other five visible continents are North America, South America, Europe, Africa, and Asia. The seventh continent, Antarctica, is found at the South Pole and has been omitted from the map.

The absence of Antarctica leads into the issues of distortion, in which geographical features are altered on a map. Some degree of distortion is to be expected with a two-dimensional flat map of the world because the earth is a sphere. A map projection transforms a spherical map of the world into a flattened perspective, but the process generally alters the spatial appearance of landmasses. For instance, Greenland often appears, such as in the map above, larger than it really is.

Furthermore, Antarctica's exclusion from the map is, in fact, a different sort of distortion—that of the mapmakers' biases. Mapmakers determine which features are included on the map and which ones are not. Antarctica, for example, is often missing from maps because, unlike the other continents, it has a limited human population. Moreover, a study of the world reveals that many of the distinctions on maps are human constructions.

Even so, maps can still reveal key features about the world. For instance, the map above has areas that seem almost three-dimensional and jut out. They represent mountains and are an example of

topography, which is a method used to display the differing elevations of the terrain. A more detailed topographical map can be viewed below.

On some colored maps, the oceans, represented in blue between the continents, vary in coloration depending on depth. The differences demonstrate *bathymetry*, which is the study of the ocean floor's depth. Paler areas represent less depth, while darker spots reflect greater depth.

Please also note the many lines running horizontally and vertically along the map. The horizontal lines, known as *parallels,* mark the calculated latitude of those locations and reveal how far north or south these areas are from the equator, which bisects the map horizontally. Generally, with exceptions depending on specific environments, climates closer to the equator are warmer because this region receives the most direct sunlight. The equator also serves to split the globe between the Northern and Southern hemispheres.

Longitude, as signified by the vertical lines, determines how far east or west different regions are from each other. The lines of longitude, known as *meridians,* are also the basis for time zones, which allocate different times to regions depending on their position eastward and westward of the prime meridian. As one travels west between time zones, the given time moves backward accordingly. Conversely, if one travels east, the time moves forward.

There are two particularly significant longitude-associated dividers in this regard. The prime [Greenwich] meridian, as displayed on the next page, is defined as zero degrees in longitude, and thus determines the other lines. The line, in fact, circles the globe north and south, and it therefore divides the world into the Eastern and Western hemispheres. It is important to not confuse the Greenwich meridian with the International Date Line, which is an invisible line in the Pacific Ocean that was created to represent the change between calendar days. By traveling westward across the International Date Line, a traveler would essentially leap forward a day. For example, a person departing from the United States on Sunday

would arrive in Japan on Monday. By traveling eastward across the line, a traveler would go backward a day. For example, a person departing from China on Monday would arrive in Canada on Sunday.

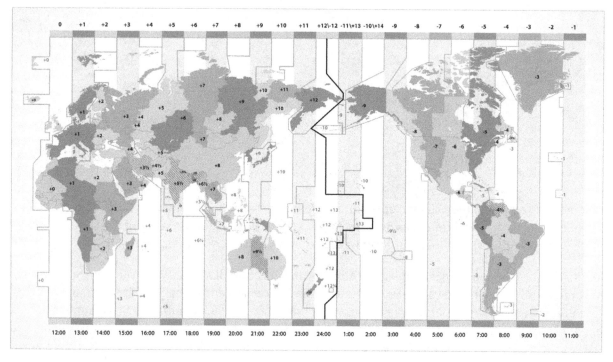

Although world maps are useful in showing the overall arrangement of continents and nations, it is also important at times to look more closely at individual countries because they have unique features that are only visible on more detailed maps.

For example, take the following map of the United States of America. It should be noted that the country is split into multiple states that have their own culture and localized governments. Other countries are often split into various divisions, such as provinces, and while these features are ignored for the sake of clarity on larger maps, they are important when studying specific nations. Individual states can be further subdivided into counties and townships, and they may have their own maps that can be examined for closer analysis.

Finally, one of the first steps in examining any map should be to locate the map's key or legend, which will explain what features different symbols represent on the map. As these symbols can be arbitrary depending on the maker, a key will help to clarify the different meanings.

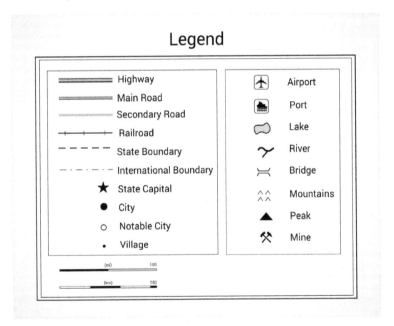

Practice Questions

1. Which of the following is the primary problem with map projections?
   a. They are not detailed.
   b. They do not include physical features.
   c. They distort areas near the poles.
   d. They only focus on the Northern Hemisphere.

2. Which type of map illustrates the world's climatological regions?
   a. Topographic Map
   b. Conformal Projection
   c. Isoline Map
   d. Thematic Map

3. Latitudinal lines are used to measure distance in which direction?
   a. East to west
   b. North to south
   c. Between two sets of coordinates
   d. In an inexact manner

4. Literacy rates are more likely to be higher in which area?
   a. Developing nations
   b. Northern Hemispherical Nations
   c. Developed Nations
   d. Near centers of trade

5. All of the following are negative demographic indicators EXCEPT which of the following?
   a. High Infant Mortality Rates
   b. Low Literacy Rates
   c. High Population Density
   d. Low Life Expectancy

6. Which of the following is NOT a factor in a location's climate?
   a. Latitudinal position
   b. Elevation
   c. Longitudinal position
   d. Proximity to mountains

7. All but which of the following are true of the Tropics?
   a. They are consistently hit with direct rays of the sun.
   b. They fall between the Tropics of Cancer and Capricorn.
   c. They are nearer the Equator than the Middle Latitudes.
   d. They are always warmer than other parts of the Globe.

8. A developing nation is more likely to have which of the following?
   a. Complex highway networks
   b. Higher rates of subsistence farmers
   c. Stable government systems
   d. Little economic instability

9. Which best describes ethnic groups?
    a. Subgroups within a population who share a common history, language, or religion
    b. Divisive groups within a nation's boundaries seeking independence
    c. People who choose to leave a location
    d. Any minority group within a nation's boundaries

10. Which of the following could be considered a pull factor for a particular area?
    a. High rates of unemployment
    b. Low GDP
    c. Educational opportunity
    d. High population density

11. In recent years, agricultural production has been affected by which of the following?
    a. The prevalence of biotechnology and GMOs
    b. Weaker crop yields due to poor soil
    c. Plagues of pests, which have limited food production
    d. Revolutions in irrigation, which utilize salinated water

12. Which of the following is true of political boundaries?
    a. They have remained static for centuries.
    b. They are generally visible on Earth.
    c. They are constantly changing.
    d. They are never disputed among nations.

13. Which of the following is the subgroup of economics that studies large-scale economic issues such as unemployment, interest rates, price levels, and national income?
    a. Microeconomics
    b. Macroeconomics
    c. Scarcity
    d. Supply and demand

14. A homeowner hires a landscape company to mow the grass because he or she would like to use that time to do something else. The trade-off of paying someone to do a job to make more valuable use of time is an example of what?
    a. Economic systems
    b. Supply and demand
    c. Opportunity cost
    d. Inflation

15. Which kind of market does not involve government interventions or monopolies while trades are made between suppliers and buyers?
    a. Free
    b. Command
    c. Gross
    d. Exchange

16. Which is NOT an indicator of economic growth?
    a. GDP (Gross Domestic Product)
    b. Unemployment
    c. Inflation
    d. Theory of the Firm

17. In a business cycle, a recession occurs between which cycles?
    a. Expansion, peak
    b. Peak, contraction
    c. Contraction, trough
    d. Trough, expansion

18. What is the name of the central bank that controls the value of money in the United States?
    a. Commodity Reserve
    b. Central Reserve
    c. Federal Reserve
    d. Bank Reserve

19. Which option does NOT sustain monetary policies?
    a. Closed market operations
    b. Open market operations
    c. Assuring bank reserves
    d. Adjusting interest rates

20. What determines the exchange rate in a "floating" or "flexible" exchange?
    a. The government
    b. Taxes
    c. The Federal Reserve
    d. The market

21. Which statement is true about inflation and purchasing power?
    a. As inflation decreases, purchasing power increases.
    b. As inflation increases, purchasing power decreases.
    c. As inflation increases, purchasing power increases.
    d. As inflation decreases, purchasing power decreases.

22. Which statement is true about goods and services?
    a. The quantity of goods and services matters more than their value.
    b. The value of goods and services matters more than their quantity.
    c. The quality of goods and services matters more than their production.
    d. The production of goods and services matters more than their quality.

23. What caused the end of the Western Roman Empire in 476 CE?
    a. Invasions by Germanic tribes
    b. The Mongol invasion
    c. The assassination of Julius Caesar
    d. Introduction of Taoism in Rome

24. Which of the following statements best describes King Louis XIV of France?
   a. He abdicated his throne during the French Revolution.
   b. He supported the American Revolution.
   c. He was the ultimate example of an absolute monarch.
   d. He created the concept of the Mandate of Heaven.

25. Which of the following consequences did NOT result from the discovery of the New World in 1492 CE?
   a. Proof that the world was round instead of flat
   b. The deaths of millions of Native Americans
   c. Biological exchange between Europe and the New World
   d. The creation of new syncretic religions

26. Which of the following statements best describes the relationship, if any, between the revolutions in America and France?
   a. The French Revolution inspired the American Revolution.
   b. The American Revolution inspired the French Revolution.
   c. They both occurred simultaneously.
   d. There was no connection between the French and American revolutions.

27. Which of the following was a consequence of industrialization in Europe during the 1800s?
   a. The birth of the working class
   b. The expansion of European empires in Africa and Asia
   c. Improved transportation and economic efficiency
   d. All of the above

28. Which of the following military technologies did NOT play a role in World War I from 1914 to 1918?
   a. The atomic bomb
   b. Poison gas
   c. Armored tanks
   d. Aircraft

29. Which of the following statements best describes international affairs between World War I and World War II?
   a. A lenient World War I peace treaty for Germany delayed the start of World War II.
   b. The policy of appeasement only encouraged further aggression by Hitler.
   c. A powerful League of Nations fostered increased cooperation and negotiation.
   d. Tensions grew between Germany and Japan.

30. Which of the following trends did NOT occur after the end of the Cold War in 1991?
   a. A decrease in nationalistic tension
   b. An increase in cultural and economic globalization
   c. An increase in religious fundamentalism
   d. An increase in environmentalism

31. Which of the following correctly lists the Thirteen Colonies?
    a. Connecticut, Delaware, Georgia, Maryland, Massachusetts, New Hampshire, New Jersey, New York, North Carolina, Pennsylvania, Rhode Island, South Carolina, Virginia
    b. Carolina, Connecticut, Delaware, Maryland, Massachusetts, New Hampshire, New Jersey, New York, Ohio, Pennsylvania, Rhode Island, Virginia, West Virginia
    c. Connecticut, Delaware, Georgia, Maine, Massachusetts, New Hampshire, New Jersey, New York, North Carolina, South Carolina, Pennsylvania, Vermont, Virginia
    d. Canada, Connecticut, Delaware, Georgia, Florida, Maryland, Massachusetts, New Hampshire, New York, North Carolina, Rhode Island, South Carolina, Virginia

32. Which of the following was NOT an issue contributing to the American Revolution?
    a. Increased taxes on the colonies
    b. Britain's defeat in the French and Indian War
    c. The stationing of British soldiers in colonists' homes
    d. Changes in class relations

33. The election of a presidential candidate from which party led to the Civil War?
    a. Democrat
    b. Whig
    c. Republican
    d. Federalist

34. Which of the following sets comprises a primary cause and effect of the American Revolution?
    a. A cause was the taxation of the colonies, and an effect was the civil rights movement.
    b. A cause was the Declaration of Independence, and an effect was the Constitution.
    c. A cause was the French and Indian War, and an effect was the Bill of Rights.
    d. A cause was the debate over slavery, and an effect was the Seven Years' War.

35. What are the two main parts of the federal legislative branch?
    a. President and vice president
    b. Federal and state
    c. District court and court of appeals
    d. Senate and House of Representatives

36. What is NOT a responsibility for citizens of democracy?
    a. To stay aware of current issues and history
    b. To avoid political action
    c. To actively vote in elections
    d. To understand and obey laws

37. Which of the following advancements was NOT invented by Greek culture?
    a. The alphabet
    b. The Hippocratic Oath
    c. Democratic government
    d. Theater

38. Which of the following was an important development in the twentieth century?
    a. The United States and the Soviet Union officially declared war on each other in the Cold War.
    b. The League of Nations signed the Kyoto Protocol.
    c. World War I ended when the United States defeated Japan.
    d. India violently partitioned into India and Pakistan after the end of colonialism.

39. What is NOT an effect of monopolies?
    a. Promote a diverse variety of independent businesses
    b. Inhibit developments that would be problematic for business
    c. Control the supply of resources
    d. Limit the degree of choice for consumers

40. Which method is NOT a way that governments manage economies in a market system?
    a. Laissez-faire
    b. Absolute Monarchy
    c. Capitalism
    d. Self-interest

41. Which of the following nations did NOT establish colonies in what would become the United States?
    a. Italy
    b. England
    c. France
    d. Spain

42. Which of the following statements about the U.S. Constitution is true?
    a. It was signed on July 4, 1776.
    b. It was enacted at the end of the Revolutionary War.
    c. New York failed to ratify it, but it still passed by majority.
    d. It replaced the Articles of Confederation.

43. What is recognized as contributing to democratization after an authoritarian regime such as post-Communism?
    a. Independent media
    b. Corporatism
    c. Socialism
    d. Devolution effects

44. Which term is best defined as a group of people joined by a common culture, language, heritage, history, and religion?
    a. State
    b. Nation
    c. Regime
    d. Government

45. What has been described as the exercise of power to achieve a predetermined end?
    a. Sovereignty
    b. Authority
    c. Politics
    d. Regime Change

46. The way or ways citizens are taught and encouraged to participate in the structures and processes of government is known as what?
    a. State Craft
    b. Nation Building
    c. Globalization
    d. Political Culture

47. Regime types fall along a continuum between which two extremes?
    a. Constitutional and non-constitutional
    b. Military and judicial
    c. Federal and communist
    d. Authoritarian and democratic

48. The European Union and United Nations are examples of what type of government?
    a. Regional
    b. Federal
    c. Supranational
    d. National

49. Which check does the legislative branch possess over the judicial branch?
    a. Appoint judges
    b. Call special sessions of Congress
    c. Rule legislation unconstitutional
    d. Determine the number of Supreme Court judges

50. Which of the following is NOT included in the Bill of Rights?
    a. Freedom to assemble
    b. Freedom against unlawful search
    c. Freedom to vote
    d. Reservation of non-enumerated powers to the states or the people

51. Which form of government divides power between a regional and central government?
    a. Democracy
    b. Constitutional monarchy
    c. Federalism
    d. Feudalism

52. Under Federalism, which is considered a concurrent power held by both the states and the federal government?
    a. Hold elections
    b. Regulate immigration
    c. Expand the territories of a state
    d. Pass and enforce laws

53. The United States elects the president by which of the following ways?
    a. Popular majority vote
    b. Plurality vote
    c. Electoral College
    d. Party list system

54. In the American election system, where do the candidates ultimately receive the nomination from their party?
    a. At the primary
    b. At the caucus
    c. At the debates
    d. At the party convention

55. Which of the following are reasons that geography is important to the examination of history?
    I. Historians make use of maps in their studies to get a clear picture of how history unfolded.
    II. Knowing the borders of different lands helps historians learn different cultures' interactions.
    III. Geography is closely linked with the flow of resources, technology, and population in societies.
    IV. Environmental factors, such as access to water and proximity of mountains, help shape the course of civilization.

    a. I, II, and III only
    b. II, III, and IV only
    c. I, II, and IV only
    d. I, III, and IV only

56. Which of the following was NOT an important invention in the twentieth century?
    a. Airplanes
    b. Telegraph
    c. Television
    d. Computers

57. What was a concern that George Washington warned of in his Farewell Address?
    a. The danger of political parties
    b. To be prepared to intervene in Europe's affairs
    c. The abolition of slavery
    d. To protect states' rights through sectionalism

58. The presidential and parliamentary systems differ in which of the following ways?
    a. The presidential system establishes a separation of powers.
    b. The legislature elects the chief executive in a presidential system.
    c. Voters directly elect the prime minister in a parliamentary system.
    d. The parliamentary system never includes a president.

59. Which political concept describes a ruling body's ability to influence the actions, behaviors, or attitudes of a person or community?
    a. Authority
    b. Sovereignty
    c. Power
    d. Legitimacy

60. Which part of the legislative process differs in the House and the Senate?
    a. Who may introduce the bill
    b. How debates about a bill are conducted
    c. Who may veto the bill
    d. What wording the bill contains

Answer Explanations

1. C: Map projections, such as the Mercator Projection, are useful for finding positions on the globe, but they attempt to represent a spherical object on a flat surface. As a result, they distort areas nearest the poles, which misrepresent the size of Antarctica, Greenland, and other high latitudinal locations. Map projects can include great detail; some illustrate the physical features in an area, and most include both the northern and southern hemispheres.

2. D: Thematic maps create certain themes in which they attempt to illustrate a certain phenomenon or pattern. The obvious theme of a climate map is the climates in the represented areas. Thematic maps are very extensive and can include thousands of different themes, which makes them quite useful for students of geography. Topographic maps are utilized to show physical features, conformal projections attempt to illustrate the globe in an undistorted fashion, and isoline maps illustrate differences in variables between two points on a map.

3. B: Lines of latitude measure distance North and South. The Equator is zero degrees, and the Tropic of Cancer is 23 ½ degrees north of the Equator. The distance between those two lines measures degrees North to South, as with any other two lines of latitude. Longitudinal lines, or meridians, measure distance East and West, even though they run north and south down the Globe. Latitude is not inexact, in that there are set distances between the lines. Furthermore, coordinates can only exist with the use of longitude and latitude.

4. C: Developed Nations have better infrastructural systems, which can include government, transportation, financial, and educational institutions. Consequently, its citizens tend to have higher rates of literacy, due to the sheer availability of educational resources and government sanctioned educational systems. In contrast, developing nations struggle to provide educational resources to their citizens. Nations in the Northern Hemisphere have no greater availability to educational resources than those in the Southern Hemisphere, and centers of trade don't necessarily equate to higher levels of education as many may exist in poorer nations with fewer resources.

5. C: Although it can place a strain on some resources, population density is not a negative demographic indicator. For example, New York City, one of the most densely populated places on Earth, enjoys one of the highest standards of living in the world. Other world cities such as Tokyo, Los Angeles, and Sydney also have tremendously high population densities and high standards of living. High infant mortality rates, low literacy rates, and low life expectancies are all poor demographic indicators that suggest a low quality of life for the citizens living in those areas.

6. C: Longitudinal position, or a place's location either east or west, has no bearing on the place's climate. In contrast, a place's latitudinal position, or its distance away from the direct rays of the sun in the Tropics, greatly affects its climate. Additionally, proximity to mountains, which can block wind patterns, and elevation, which generally lowers temperature by three degrees for every one thousand feet gained, also impacts climate.

7. D: Although nearest the direct rays of the sun, the Tropics are not always warm. In fact, the nations of Ecuador and Peru, which are entirely within the Tropics, are home to the Andes Mountains, which remain snowcapped the entire year. This climatological anomaly is also due to cooler ocean currents and the orographic effect. Choices *A, B,* and *C* are all true of the tropics.

8. B: Developing nations tend to have higher levels of impoverished citizens. As a result, many of their citizens must rely on subsistence farming, or producing enough food to feed their families, in order to survive. In contrast, developed nations tend to produce surpluses of food and very few, if any, of its citizens engage in subsistence farming. Developing nations are less likely to have complex highway systems, stable governments, and economic stability due to financial pressures.

9. A: Although some ethnic groups throughout the world do engage in armed conflicts, the vast majority do not. Most ethnic groups tend to live in relative harmony with others with whom they share differences. Ethnic groups are simply a group of people with a religious, cultural, economic, or linguistic commonality. Additionally, ethnic groups don't always choose to leave places. Many have called certain locations home for centuries. Also, some ethnic groups actually make up the majority in some countries and are not always minority groups.

10. C: Pull factors are reasons people immigrate to a particular area. Obviously, educational opportunities attract thousands of people on a global level and on a local level. For example, generally areas with strong schools have higher property values, due to the relative demand for housing in those districts. The same is true for nations with better educational opportunities. Unemployment, low GDP, and incredibly high population densities may serve to deter people from moving to a certain place and can be considered push factors.

11. A: The use of biotechnology and GMOs has increased the total amount of food on Earth. Additionally, it has helped to sustain the Earth's growing population; however, many activists assert that scientists are creating crops that, in the long run, will be destructive to human health, even though not enough evidence exists to prove such an allegation. Agricultural production has not been affected by poorer soil, plagues of pests, or the use of saline for irrigation purposes.

12. C: Like the boundaries of the United States, political boundaries are constantly changing due to war (South Sudan), religious conflict (India and Pakistan, Israel, East Timor), and differing political ideologies (North and South Korea, Reunification of Germany after the Cold War). The only constant with political boundaries is change. It is not possible to see manmade lines separating countries on Earth, unless they are natural boundaries. Additionally, boundaries are always under dispute, and they have not remained static for centuries.

13. B: Macroeconomics. Macroeconomics studies the economy on a large scale and focuses on issues such as unemployment, interest rates, price levels, and national income. Microeconomics studies more individual or small group behaviors such as scarcity or supply and demand. Scarcity is incorrect because it refers to the availability of goods and services. Supply and demand is also incorrect because it refers to the quantity of goods and services that is produced and/or needed.

14. C: Opportunity cost. Opportunity cost can trade time, power, or anything else of value in exchange for something else. Economic systems determine what is being produced and by whom. Supply and demand refers to the quantity of goods and services that is produced or needed. Finally, inflation refers to how the cost of goods and services increases over time.

15. A: Free. A free market does not involve government interventions or monopolies while trading between buyers and suppliers. However, in a command market, the government determines the price of goods and services. Gross and exchange markets refer to situations where brokers and traders make exchanges in the financial realm.

16. D: Theory of The Firm. Behaviors of firms is not an indicator of economic growth because it refers to the behavior that firms follow to reach their desired outcome. GDP, unemployment, and inflation are all indicators that help determine economic growth.

17. C: Contraction and trough. A recession occurs between the contraction and trough phases of the business cycle. Between expansion and peak phases, employment and productivity are on the rise, causing a "boom." Between the peak and contraction, unemployment rates are starting to fall, but have not yet hit an all-time low. Between trough and expansion phases, the economy is getting back on its feet and starting to increase employment again.

18. C: Federal Reserve. The Federal Reserve is the bank of banks. It is the central bank of the United States and controls the value of money. A commodity is the value of goods such as precious metals. While the Central Reserve and Bank Reserve may sound like good options, the term "bank reserve" refers to the amount of money a bank deposits into a central bank, and the Central Reserve is simply a fictitious name.

19. A: Closed market operations. Monetary policies are sustained by assuring bank reserves, adjusting interest rates, and open market operations. Closed market operations do NOT uphold monetary policies.

20. D: The market. The market, through supply and demand, determines the exchange rate with a "flexible" or "floating" exchange rate. The government is not a correct answer because it is involved in "fixed" exchange rates to help keep exchange rates stable. Taxes is also incorrect because they create government revenue. The Federal Reserve is the bank of banks.

21. B: As inflation increases, purchasing power decreases. As more money is printed, the monetary value of the dollar drops and, in turn, decreases the purchasing power of goods and services. So, as inflation increases, consumers are not spending as much and the value of the dollar is low.

22. B: The value of goods and services matters more than their quantity. For example, in the real estate industry, if a realtor sells ten houses valued at $200,000, his or her commission would be the same as a realtor who sells one house valued at $2,000,000. Even though one realtor sold more homes, the value of 10 houses adds up to the same amount as the single home that the other realtor sold. Therefore, the number of goods and services produced does not determine economic growth—the value of the goods and services does.

23. A: Invasions by Germanic tribes. Large numbers of Franks, Goths, Vandals, and other Germanic peoples began moving south in the fifth century CE. They conquered Rome twice, and the Western Roman Empire finally disintegrated. The Mongol invasion, Choice *B*, pushed westward in the thirteenth century, long after the western Roman Empire was gone. The assassination of Julius Caesar, Choice *C*, led to the end of the Roman Republic and the birth of the Roman Empire. Taoism never spread to Rome, making Choice *D* incorrect.

24. C: Louis the XIV was an absolute monarch who ruled during the sixteenth century. He concentrated power on the throne by forcing nobles to spend most of their time at the royal court. The French Revolution occurred about two hundred years after he died. Absolute monarchs like Louis the XIV bolstered their prestige by claiming they were appointed by God. The Mandate of Heaven was a similar concept, but it was developed by the Zhou Dynasty in China about two thousand years before Louis XIV was born.

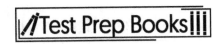

25. A: Most scholars already knew the world was round by 1492. On the other hand, the arrival of Europeans in North and South America introduced deadly diseases that killed millions of native peoples. Europeans had developed immunity to diseases such as smallpox, while Native Americans had not. In addition, Europeans introduced a number of new plants and animals to the New World, but they also adopted many new foods as well, including potatoes, tomatoes, chocolate, and tobacco. Finally, Europeans tried to convert Native Americans to Christianity, but Indians did not completely give up their traditional beliefs. Instead, they blended Christianity with indigenous and African beliefs to create new syncretic religions.

26. B: The American Revolution occurred first in 1775, and a number of European soldiers fought for the patriots. The American Revolution, in part, inspired the French Revolution. The Marquis de Lafayette came to America in 1777 and was wounded during the Battle of Brandywine. He returned to France after the American Revolution and became a leader in the French Revolution in 1789.

27. D: The Industrial Revolution is probably one of the most important turning points in world history. The United States and Western Europe, especially Britain, were the first areas to industrialize. Steam engines were used to improve economic and transportation efficiency. They also gave western empires a military advantage over less developed countries in Asia and Africa. Finally, industrialization required large amounts of unskilled labor, which created the working class.

28. A: The atomic bomb was created during World War II (1939 – 1945). Scientists and engineers did develop a number of other weapons in order to break through the heavily entrenched front lines during World War I. Poison gas killed or injured millions of men between 1914 and 1918. Aircraft were used to observe enemy positions and bombard enemy troops. Armored tanks were able to crush barbed wire fences and deflected machine gun bullets.

29. B: Eager to avoid another global conflict, European leaders tried to appease Hitler by letting him occupy Austria and Czechoslovakia. This policy failed because it only emboldened Hitler, and he invaded Poland in 1939. Rather than receiving leniency after World War I, Germany was forced to sign a humiliating peace treaty. Furthermore, the League of Nations failed to prevent conflict because it lacked any real power. This encouraged continued aggression from Italy, Germany, and Japan, which culminated in World War II.

30. A: Nationalism remains a powerful force to this day. Nationalism drove conflict in Ireland, Spain, Yugoslavia, and elsewhere. However, the end of the Cold War removed many of the political barriers that had prevented interaction between the western and Communist blocs. In addition, religious fundamentalism became an increasingly common response to the rapid changes that occurred during the late twentieth and early twenty-first centuries. There was also a rise in cultural and economic globalization, as well as in environmentalism.

31. A: Carolina is divided into two separate states—North and South. Maine was part of Nova Scotia and did not become an American territory until the War of 1812. Likewise, Vermont was not one of the original Thirteen Colonies. Canada remained a separate British colony. Finally, Florida was a Spanish territory. Therefore, by process of elimination, *A* is the correct list.

32. B: Britain was not defeated in the French and Indian War, and, in fact, disputes with the colonies over the new territories it won contributed to the growing tensions. All other options were key motivations behind the Revolutionary War.

33. C: Abraham Lincoln was elected president as part of the new Republican Party, and his plans to limit and potentially abolish slavery led the southern states to secede from the Union.

34. C: The Declaration of Independence occurred during the American Revolution, so it should therefore be considered an effect, not a cause. Similarly, slavery was a cause for the later Civil War, but it was not a primary instigator for the Revolutionary War. Although a single event can have many effects long into the future, it is also important to not overstate the influence of these individual causes; the civil rights movement was only tangentially connected to the War of Independence among many other factors, and therefore it should not be considered a primary effect of it. The French and Indian War (which was part of the Seven Years' War) and the Bill of Rights, on the other hand, were respectively a cause and effect from the American Revolution, making Choice *C* the correct answer.

35. D: The president and vice president are part of the executive branch, not the legislative branch. The question focuses specifically on the federal level, so state government should be excluded from consideration. As for the district court and the court of appeals, they are part of the judicial branch. The legislative branch is made up of Congress, which consists of the House of Representatives and the Senate.

36. B: To avoid involvement in political processes such as voting is antithetical to the principles of a democracy. Therefore, the principal responsibility of citizens is the opposite, and they should be steadily engaged in the political processes that determine the course of government.

37. A: Although Greeks used the alphabet as the basis for their written language, leading to a diverse array of literature, they learned about the alphabet from Phoenician traders. All the other options, in contrast, were invented in Greece.

38. D: It is important to realize that the Cold War was never an official war and that the United States and the Soviet Union instead funded proxy conflicts. The Kyoto Protocol was signed by members of the United Nations, as the League of Nations was long since defunct. While Japan was a minor participant in World War I, it was not defeated by America until World War II. The correct answer is *D*: India's partition between Hindu India and Islamic Pakistan led to large outbreaks of religious violence.

39. A: Rather than competition, a monopoly prevents other businesses from offering a certain product or service to consumers.

40. B: Absolute monarchy, which is built on the vision of full government control over the economy, is a hallmark of command system economies. Laissez-faire, capitalism, and self-interest, in contrast, are all fundamental concepts behind the market system.

41. A: England, France, and Spain all established North American colonies that would later be absorbed into the United States, but Italy, despite Christopher Columbus' role as an explorer, never established a colony in America.

42. D: The Constitution was signed in 1787; the Declaration of Independence was signed in 1776. It was successfully ratified by all the current states, including New York. Finally, the Articles of Confederation was established at the end of the American Revolution; the Constitution would replace the articles years later due to issues with the government's structure.

43. A: The media is tightly controlled and monitored in authoritarian regimes. Independent media is looked at as destabilizing to authoritarian regimes when there are reports on government abuse, corruption, and policy critiques. New information and views from other political parties can significantly impact the perceptions of local, national, and international publics. Independent media and media supported by the West helped to undermine the former USSR.

44. B: A Nation is defined as a group of people who have common traits, such as heritage, history, language, culture, and religion. It has nothing to do with borders, sovereignty, power, people in office, or the rules by which a government operates (all of which are found in the other answer terms of state, government, constitution, and regime).

45. C: Politics has been described by many as the exercise of power or use of force to achieve a particular end. Sovereignty is the ability of a particular government to gain and maintain the acquiescence of its populace. It exercises authority in this pursuit. If it fails, it may result in Regime change. However, it is politics that is used by governments in order to properly exercise Sovereignty and authority in meeting the needs of the people.

46. D: Political culture is the term used to describe the various ways a government seeks to encourage its people to participate in the political process. In liberal democracies, political culture encourages the free exchange of ideas, and high levels of education and input, including running for office. In authoritarian regimes, participation is discouraged apart from speaking well of the leader, the government, and its ideals. Some governments encourage direct participation, others discourage it based on the goals they have for their country and citizens.

47. D: Governmental regimes fall along a continuum between total authoritarianism and complete direct democracy. Most countries are neither totally authoritarian, nor a complete direct democracy, but they do all fall along this continuum. China and Iran would be towards the authoritarian end of the spectrum and the United States, United Kingdom, and Mexico would be towards the democratic end.

48. C: Supranational institutions are generally united over shared interests. As such, these institutions develop international laws, set international trade norms, and advocate for human rights and development.

49. D: The Constitution granted Congress the power to decide how many justices should be on the court, and Congress first decided on six judges in the Judiciary Act of 1789. The Constitution granted the power to appoint judges and to call special sessions of Congress to the president. Only the Supreme Court may interpret the laws enacted by Congress and rule a law unconstitutional and subsequently overturn the law.

50. C: The first ten amendments to the Constitution are collectively referred to as the Bill of Rights. The Founding Fathers did not support universal suffrage, and as such, the Bill of Rights did not encompass the freedom to vote. The Fifteenth Amendment provided that the right to vote shall not be denied on the basis of race, color, or previous condition of servitude, and women did not receive the right to vote until passage of the Nineteenth Amendment. The other three answer choices are included in the Bill of Rights—the freedom to assembly is established in the First Amendment; the freedom against unlawful search is established in the Fourth Amendment; and the reservation of non-enumerated powers to the states or the people is established in the Tenth Amendment.

51. C: Federalism divides power between regional and federal governments, and it is the form of government upon which the United States is structured, according to the Tenth Amendment. While a constitutional monarchy, Choice *B*, is typically divided between a monarch, the head of state, and a legislative body, usually a parliament, power is not reserved to the regional government. A democratic government, Choice *A*, is a government ruled by the people and does not specify division of powers. Feudalism, Choice *D*, is an economic system popular in medieval Europe where the monarchy granted the nobility land in exchange for military service, and the nobility allowed serfs to live on their land in exchange for labor or percentage of crops.

52. D: Both the states and the federal government may propose, enact, and enforce laws. States pass legislation that concerns the states in their state legislative houses, while the federal government passes federal laws in Congress. Only states may hold elections and determine voting procedures, even for federal offices such as the president of the United States, and only the federal government may expand any state territory, change state lines, admit new states into the nation, or regulate immigration and pass laws regarding naturalization of citizens.

53. C: The president of the United States is elected by the Electoral College. The number of electors for each state depends on the state's total number of senators and representatives. The president must receive a majority (270) of the electoral votes (538), and if this doesn't occur, the Twelfth Amendment empowers the House of Representatives to elect the president. Choices *A, B,* and *C* are different methods for electing candidates.

54. D: The two major political parties hold conventions to nominate their presidential candidate. The delegates are awarded based on candidates' performance in the primary elections or caucuses vote at the party convention to select the nominee. Primaries and caucuses are the democratic contests held by each state to award their delegates. The candidates participate in debates on the campaign issues, but they do not receive the nomination at debates.

55. C: I, II, and IV only. Historians make use of maps in their studies to get a clear picture of how history unfolded, knowing the borders of different lands helps historians learn different cultures' interactions, and environmental factors, such as access to water and the proximity of mountains, help determine the course of civilization. The phrase "Geography is closely linked with the flow of resources, technology, and population in societies" is a characteristic of economics.

56. B: Out of the four inventions mentioned, the first telegraphs were invented in the 1830s, not in the twentieth century. In contrast, the other inventions had considerable influence over the course of the twentieth century.

57. A: George Washington was a slave owner himself in life, so he did not make abolition a theme in his Farewell Address. On the other hand, he was concerned that sectionalism could potentially destroy the United States, and he warned against it. Furthermore, he believed that Americans should avoid getting involved in European affairs. However, one issue that he felt was especially problematic was the formation of political parties, and he urged against it in his farewell.

58. A: The presidential system establishes a separation of powers. In the presidential system, voters directly elect the chief executive, and the presidential system establishes a separation of powers between different branches of government. In contrast, the parliament elects the chief executive, and the increased collaboration and dependency creates a more responsive government. Choices *B* and *C* confuse how the executive is elected in each system. Choice *D* is incorrect because many parliamentary systems include a president, though the status of head of state is often purely ceremonial.

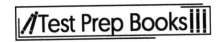

59. C: Power is the ability of a ruling body or political entity to influence the actions, behavior, and attitude of a person or group of people. Authority, Choice *A*, is the right and justification of the government to exercise power as recognized by the citizens or influential elites. Similarly, legitimacy, Choice *D*, is another way of expressing the concept of authority. Sovereignty, Choice *B*, refers to the ability of a state to determine and control their territory without foreign interference.

60. B: The process by which the House and Senate may debate a bill differs. In the House, how long a speaker may debate a bill is limited, while in the Senate, speakers may debate the bill indefinitely and delay voting on the bill by filibuster—a practice in which a speaker refuses to stop speaking until a majority vote stops the filibuster or the time for the vote passes. In both the House and the Senate, anyone may introduce a bill. Only the president of the United States may veto the bill, so neither the House nor Senate holds that power. Before the bill may be presented to the president to be signed, the wording of the bill must be identical in both houses. Another procedural difference is that the number of amendments is limited in the House but not the Senate; however, this does not appear as an answer choice.

Dear HiSET Test Taker,

We would like to start by thanking you for purchasing this study guide for your HiSET exam. We hope that we exceeded your expectations.

Our goal in creating this study guide was to cover all of the topics that you will see on the test. We also strove to make our practice questions as similar as possible to what you will encounter on test day. With that being said, if you found something that you feel was not up to your standards, please send us an email and let us know.

We would also like to let you know about other books in our catalog that may interest you.

Test Name	Amazon Link
GED	amazon.com/dp/1628458992
SAT	amazon.com/dp/1628457376
ACT	amazon.com/dp/1628459468
ACCUPLACER	amazon.com/dp/1628459344

We have study guides in a wide variety of fields. If the one you are looking for isn't listed above, then try searching for it on Amazon or send us an email.

Thanks Again and Happy Testing!
Product Development Team
info@studyguideteam.com

FREE Test Taking Tips DVD Offer

To help us better serve you, we have developed a Test Taking Tips DVD that we would like to give you for FREE. **This DVD covers world-class test taking tips that you can use to be even more successful when you are taking your test.**

All that we ask is that you email us your feedback about your study guide. Please let us know what you thought about it – whether that is good, bad or indifferent.

To get your **FREE Test Taking Tips DVD**, email freedvd@studyguideteam.com with "FREE DVD" in the subject line and the following information in the body of the email:

   a. The title of your study guide.

   b. Your product rating on a scale of 1-5, with 5 being the highest rating.

   c. Your feedback about the study guide. What did you think of it?

   d. Your full name and shipping address to send your free DVD.

If you have any questions or concerns, please don't hesitate to contact us at freedvd@studyguideteam.com.

Thanks again!

CPSIA information can be obtained
at www.ICGtesting.com
Printed in the USA
LVHW101637120621
690049LV00001B/6